Mercator

Also by Nicholas Crane

Clear Waters Rising
Two Degrees West

Mercator

THE MAN WHO MAPPED THE PLANET

NICHOLAS CRANE

Weidenfeld & Nicolson
LONDON

First published in Great Britain in 2002
by Weidenfeld & Nicolson

Third impression September 2002

A CIP catalogue record for this book
is available from the British Library.

ISBN 0 297 64665 6

Typeset by Selwood Systems, Midsomer Norton

Printed in Great Britain by Butler & Tanner Ltd,
Frome and London

Weidenfeld & Nicolson

The Orion Publishing Group Ltd
Orion House
5 Upper Saint Martin's Lane
London WC2H 9EA

Contents

Illustrations

To Annabel, and to
Imogen, Kit and Connie,
with love.

A Personal Note to the Reader

Maps codify the miracle of existence. And the man who wrote the codes for the maps we use today was Gerard Mercator, a cobbler's boy born five hundred years ago on a muddy floodplain in northern Europe. In his own time, Mercator was 'the prince of modern geographers', his depictions of the planet and its regions unsurpassed in accuracy, clarity and consistency. More recently, he was crowned by the American scholar Robert W. Karrow as 'the first modern, scientific cartographer'. Mercator was a humble man with a universal vision. Where his contemporaries had adopted a piecemeal approach to cartography, Mercator sought to wrap the world in overlapping, uniform maps. Along the way, he erected a number of historic milestones. He participated in the naming and the mapping of America, and he devised a new method – a 'projection' – of converting the spherical world into a two-dimensional map. He constructed the two most important globes of the sixteenth century, and the title of his pioneering 'modern geography', the *Atlas*, became the standard term for a book of maps.

No better example is required of genius arising from turmoil. Mercator was born in 1512 and died in 1594. His world was one of violent conflict, social upheaval, religious revolution – and geographical discovery. He was five when Martin Luther precipitated the Reformation, and ten when the survivors of the world's first circumnavigation returned to Seville in their leaking caravel. He knew poverty, plague, war and persecution. He was imprisoned by the Inquisition, yet patronized by an emperor. His life was one of brilliant breakthroughs and abrupt reversals.

In its telling, this is the story of the poor boy made good, the pauper who embraced the world, found fame, faced death, yet triumphed through fortitude. Variously described by his peers as honest, calm, candid, sincere and peaceable, Mercator wore – and wears – an aura of beatitude in troubled times. His attitude to his geographical calling was described by his friend and neighbour, Walter Ghim, as 'indefatigable'. Mercator's extraordinary longevity, eighty-two years, makes him an unusual biographical subject. Surviving for twice as long as many of his contemporaries, he was able to mature through two consecutive life-spans. In his first life, he stumbled through learning, experimentation, acclaim and catastrophe; in his second life, he withdrew to his Rhenish sanctuary and concentrated with single-minded endeavour on the works that would bring him timeless fame.

Writing Mercator's 'Life' has been a humbling experience. My motive for daring to trespass into his aura arose in part from a sense of lack. The best full-length biography of Mercator is still the volume by H. Averdunk and Dr J. Müller-Reinhard, published in German in 1914. Beyond the chapter in Karrow's magisterial *Mapmakers of the Sixteenth Century and Their Maps*, Mercator's story has never been told in full, in English. He deserves a broader acknowledgement.

Mercator is the third book in a geographical trilogy that has occupied me for the past ten years. Each of these geographical narratives has a cartographic plot. The first book, *Clear Waters Rising*, describes a 10,000-kilometre journey on foot along the continental watershed of Europe. The route appears on any map of Europe's physical geography as a sinuous chain of mountain ranges running from northern Spain to western Turkey. The second book, *Two Degrees West*, was also derived from a line on a map, the line of longitude which was selected by the Ordnance Survey as the central meridian, following the decision in 1938 to map Britain on the Transverse Mercator Projection. Two degrees west runs for 578 kilometres through England, and the book describes the places and people found along that geographical cross-section. With rectilinear certitude, *Two Degrees West* led me to the father of modern mapmaking and this, the third of these geographical narratives.

In researching *Mercator* I have run up more than a few debts of gratitude. Without the British Library – and in particular the reading room fondly known as 'Maps' – this book would not have been possible. The collections at the Royal Geographical Society and at the National Maritime Museum, Greenwich, have also been invaluable, as have those of the London Library, the National Library of Scotland and the Plantin-Moretus Museum in Antwerp.

Many individuals gave their time and advice. For generously sharing their cartographic scholarship, I am indebted to Peter Barber, Tony Campbell, Dr Catherine Delano-Smith, Francis Herbert and Dr David Munro. (All errors in this book are entirely my responsibility and in no way reflect upon those mentioned here.) For translating essential primary sources, I am extremely grateful to Richard Bryant, Glyn Davies and William Sutton. Their care and expertise has vivified documents for the modern reader. To David Watson I extend enormous thanks for his linguistic versatility in translating disparate and seemingly arcane works relating to Mercator and historical cartography; David paid an additional price for his interest by being engaged as the book's copy-editor. Thanks are also due to Tom Graves for his indispensable help in marshalling the book's illustrations.

Those who played a greater role than might have been acknowledged at the time include David Bannister, Hol Crane, Inge Daniels, Matt Dickinson, Luke Hughes, Colonel Colin Huxley, Simon Jenkins, Peter Van der Krogt, Roland Mayer, Hallam Murray, Stuart Proffitt, Rodney Shirley, Humphrey Stone and Louisa Young. Chris Crane was the first to read a passage from the book, but he was taken by cancer before I could finish.

Once again, Derek Johns was there at the beginning, at the middle and at the end, proving (yet again) that the trials of a literary agent begin with the sale of a promising idea. To Rebecca Wilson in London and to David Sobel in New York, I would like to extend warm thanks for commissioning the book, and to Richard Milner for guiding it through to its printed state. Finally – and most importantly – there is, as ever, one person who has been tested beyond normal limits of endurance. The book's dedication cannot compensate for the lost years, but it does mark the return of a husband and a dad.

NC
London
March 2002

I

A Little Town Called Gangelt

In the summer of 1511, Emerentia Kremer fell pregnant and the harvest failed. Rye rose to its highest price for a decade and plague returned to the lands of the lower Rhine.

Emerentia and Hubert set off towards the ocean. In the Low Countries they could find food, shelter and a place to lay the baby. For several days they travelled, over the river Maas and across the heaths of Kempen, where robbers lurked by brackish meers.

When the sandy tracks gave way to clay, spires began to pierce the sky. On these seeping levels sprawled some of Europe's richest cities and towns: Antwerp and Mechelen, Louvain, Brussels, each of them arterially connected to the sea lanes of the world by the same deltaic river. It was to this river – the immeasurable Schelde – that the Kremers were bound.

On the west bank, where the shadow of a castle darkened the thatched roofs of a riverport called Rupelmonde, they sought Hubert's uncle Gisbert, a priest in the hospice of St Johann.

It was a cold winter. February brought frosts and snow. At daybreak on the fifth day of March, Emerentia gave birth to a boy.[1] The seventh and final child of the Kremers was given the name Gerard.

The end of winter brought floods as snow-melt and rain poured from the continental uplands into the brimming plains of Flanders and Brabant. When the time came for the Kremers to leave Rupelmonde, they recrossed the Schelde and took the familiar route east to the river Maas and the rising land of the duchy of Jülich.

Beyond the river, the road climbed between fields towards a small, walled town on the edge of a treeless plateau. Seen from afar, Gangelt looked formidably contained, an isolated disc of stone and brick punctuated by thirteen bastions whose firing points squinted into the distance. Steep-pitched slate roofs rose from the bastions, catching the light when the sun lay low.[2]

I

Fig. 1. The border town of Gangelt, just inside the duchy of Jülich. A detail from Mercator's untitled map of 1585 (Private collection)

From the blank plateau, the travellers passed through an arched gateway into an interior world seething with humans and animals. The space within the walls was so confined that a man could walk from one side of Gangelt to the other in four minutes. Hovels and byres were wedged into every stinking, clamorous crevice. The only open space lay in the centre of the town, where the road from Sittard met the road to Jülich at the marketplace. Here stood the church in whose sooted interior generations of Gangelt's peasants had raised their eyes to the Host.

Gangelt had been built by farmers. The town stood on the southern edge of a low, broad plateau that was part of the most productive tillage belt on the continent, a long band of fertile loess, glacial dust blown overland at the end of the last Ice Age and swept by meltwaters into a gentle reef along the northern edge of the continental uplands. Below Gangelt, a sea of meadows and trees reached all the way to the dark rim of hills behind Aachen, one day's ride to the south.

The strange stones that occasionally emerged from Gangelt's sanguine soils were tools and weapons left by the earliest hunters and farmers, who had found that the well-drained soil was light enough to be worked with a wooden plough. In the Kremers' day, flint arrow-heads, pot fragments, axe-heads and the blue glass of Celts periodically came up with the clod. Rarer were the spatulas, bottlenecks and figurines dropped by Romans who had settled this sunny vantage point during their short and precarious toehold east of the Maas.

Fourteen hundred years after Tacitus had complained about *Ger-*

mania's bristling forests and festering marshes, the crops beyond Gangelt's walls were being rotated on strip fields to a three-year cycle, from fallow to early planting of oats and then late planting of rye. Animals grazed – and fertilized – the fallow fields. The little river below the bluff had become the Rodebach – the 'Rode' derived from the local word for a place created from cleared woodland – and now formed one of the internal boundaries of the vast and unwieldy Holy Roman Empire, an ill-fitting jigsaw of hereditary princedoms, Imperial free cities and prince-bishoprics which occupied the heart of continental Europe. While Gangelt supposedly lay within the Christian protectorate of the Emperor Maximilian I, it was ruled by the Duke of Cleves and Mark, who had acquired the duchy of Jülich through marriage the year before Gerard was born. The Rodebach defined the border of the duchy of Jülich. On the other side of Gangelt's stream lay the duchy of Limburg. A short walk to the west lay the prince-bishopric of Liège.

To Gangelt's location on a political border and a geographical divide (between the southern forests and northern loess) could be added yet another fault-line, for the town marked the divide between German-speakers to the east and those to the west who spoke in Brabantine Dutch. Living on the edge gave Gangelt's citizens an unusual awareness of geographical, cultural and linguistic diversity, and a deep sense of insecurity.

But the town had an attribute which was shared by few others on the plateau, for the high road between Cologne and Antwerp passed through its walls. (A lower, longer route, involving more river crossings, passed to the south, through Aachen.) Not only were Cologne and Antwerp the economic giants of northern Europe, but they lay on the Hanseatic trade route between northern Italy and the Low Countries. Gangelt was quite used to merchants who might have begun their journey as far away as Genoa and Venice.

Against the backdrop of planting and harvesting, the dry months welcomed carnivals of passers-by: painters and puppeteers, tinkers, gypsies, wandering friars and itinerant teachers, beggars, tooth-drawers and tumblers, clowns who could juggle balls, fence and dance on a rope, charlatans and *quacksalbers* boasting salves and cures. Gerard grew up with the sound of bagpipes, flutes and fiddles; perhaps even heard 'Sour-milk', the professional lute-player, or Jörg Graff, the blind balladeer. Such men knew no borders. Gangelt was a natural way-station for men-of-the-road, who could find half a dozen venues between Cologne and Antwerp. Changing an audience was easier than changing an act.

Some of these passers-by told of strange, faraway lands. Gangelt may well have seen Savoyard organ-grinders and fortune-tellers, showmen with peep-box views of Constantinople, or mountain men with marmots on a leash; maybe a rare bearward from the mountains of Bilé Karpaty, leading a beast with drawn claws and a ring through its nose. Flageolet-players from southern Italy were known to travel as far as the Rhine and Spain. These perennial nomads were regarded with fascination and suspicion. With grave-diggers, hangmen (and sometimes shepherds too), the sons of German players were prohibited from membership of guilds; some were accused of witchcraft; they were *unehrlich* – shady, and of dubious honour. And so also – to some at least – were the hawkers of papal indulgences, agents of Pope Leo x, then touring the countryside selling remission from purgatory.

A couple of days ride to the east, 'Holy Cologne' was a keenly felt presence. With a population of 40,000, this university city was the largest in Germany, a theological bastion and a production centre for leather goods, textiles and metalware. Cologne was also one of the seed-beds of classical revivalism, humanism, the movement which could be traced back over a century to Francesco Petrarca's triumphant return to Avignon bearing the lost manuscripts of Cicero and Quintilian. The subsequent rise of the Italian *umanista*, the 'humanist' teacher whose ancient Greek and Roman texts described the human rather than the spiritual condition, ignited a demand for works of moral philosophy, of history and poetry, grammar and eloquence, of mathematics and astronomy – works which had suddenly become available to a mass audience with the invention of printing. Cologne had reacted with alacrity to the current passion for rediscovered ancient texts. Only a decade or so after a goldsmith called Gutenberg had begun printing in Mainz, Ulrich Zell set up a press in Cologne, and by 1466 he had printed Cicero's *De Officiis*. By 1500, Cologne was a leader among the continent's 250 print centres.[3] The *mercator* – book carrier or dealer – would have been another of Gangelt's frequent callers.[4]

With new translations of Greek and Roman works, and pamphlets attacking the ailing Church and Empire, Gangelt was also familiar with current affairs 'broadsides', printed sheets sold from trays carried by sellers wearing broad-brimmed feather hats with newsprint pinned to the crown. In Latin and German, the broadsides were headed by a three-line motto and (for the illiterate) an explanatory woodcut picture. Among the woodcuts to have passed this way were those of Germany's most acclaimed painter and engraver, Albrecht Dürer. A friend to leading humanists, Dürer had championed the power of the word in a print which showed the 'Son of Man' with flaming eyes, stars and a sword-hilt to his lips.[5] (Two years after the Kremers were driven from Gangelt, Dürer would come this way while travelling from Nuremberg to the Low Countries to confirm his post as painter to the Imperial court; in his travel journal, the artist would merely note that he had 'passed the little town of Gangelt'.)[6]

So Gerard Kremer spent the first five or six years of his life in a bustling, anxious rural town on the road between two of Europe's fastest-growing cities. For a developing mind, it was stimulating, uneasy theatre. Beyond the walls, the landscape he knew was richly patterned, with strip fields of ripening crops rimmed with cornflowers and poppies. On roads and tracks moved carts and coaches, solitary horsemen and alarming *Landsknechte*, mercenary soldiers in looted finery and *Pludderhosen*, bag trousers drawn together at the knees and waist. The towns he knew rose spired above sawtooth walls.

Too young to attend school,[7] Gerard would have been a regular visitor to the church. Steeped in ceremony and sacramental symbols, Mass was an impenetrable spectacle. In puddles cast by flickering wicks, the boy would have heard the disembodied voice of the priest from behind the rood screen, conducting the liturgy in a strange and unintelligible tongue. Gazing up at the Host, he waited for the 'many graces' which William of Auxerre had promised to those who prayed 'while looking at the Lord's body'.[8]

And there was much to pray for. On God's providence depended the Kremer family's survival. Wheat thrived or failed with the weather. Low temperatures reduced yields and so could dry springs, wet summers, damp autumns, long frosts or rainy winters. Storms could flatten a ripening crop. Too much rain would prevent the hay from drying, cause ploughs to stick, and rot the spring sowings. Summers had to be dry and warm, yet not too hot. Heatwaves were as lethal as winter freezes. Yields could quadruple – or halve – from year to year, while a sequence of bad seasons would lead to a famine and the spectre of skeletal peasants grubbing for rape and roots, flower bulbs, grass, leaves, an overlooked turnip.

Fortunately, the Kremers did not rely entirely upon the land. As a cobbler, Hubert had dual means. For a family which carried the migratory instinct, there was a grim symbolism in this source of income. Mobility was the temporal saviour of peasants on the edge, and mobility needed boots, boots which could walk to fields and markets and, in desperation, to new lands. Everybody needed boots. The *Bundschuh*, a rawhide boot bound to the calf with strips of leather, was the peasants' defining article: clumsy, mud-clogged and quickly stitched from untreated skins. The cobbler's craft demanded few tools and elementary skills. But it was one that could make a difference between survival and starvation: the cobbler could carry his skills from town to town, while a commission to make a pair of riding boots might be worth as much as 1 gulden, enough to buy ten geese or 35 gallons of wine.

Much of Gerard's later character can be ascribed to his father's trade: cobblers belonged to a culture that had ancient associations. Since Lucian wrote of the shoemaker-philosopher back when Romans occupied Gangelt's heights, this had been a noble, mysterious craft. Shoemakers and shepherds had much in common. They followed solitary, meditative lives; they were loners, outsiders. They had time for thinking. But there was one crucial difference: where the shepherd in his upland fastness communed with his flute, the urban shoemaker accumulated human nature, and read. Some of the stories and songs sold for coins in Gangelt's streets were popular German ballads in praise of shoemakers, whose *Bundschuh* was also an emblem of insurrection, the word *Bund* being used to describe both a shoe fastening and an alliance of rebelling communes.

If his father imbued the boy with a predilection for reflection, it was surely his mother Emerentia who bestowed Gerard with extraordinary doggedness. Feminine fortitude had a history on the plateau, for it was a Gangelt mother who walked with her sick daughter all the way to Aachen in AD 828, so that she could pray in the cathedral. Her daughter was miraculously cured and Gangelt earned its first written record, in a Vatican manuscript.

But in young Gerard's life there was a third, absent 'parent': the uncle who had escaped from Gangelt to university and then secured a priesthood in Rupelmonde. In Gisbert, the family had proof that there was an alternative to rural poverty. And it was Gisbert's encouragement that must have been behind Hubert's decision to provide the family with a fallback plan should life in Gangelt prove impossible; before leaving Rupelmonde with their baby Gerard, Hubert had taken a lease on a homestead.

For a family of such slender means, Gerard had been a reckless conception. The Kremers already had more children than most.[9] With seven children to feed, Hubert and Emerentia knew that Gangelt offered little hope for the future.[10] They had neither the land nor the skills to provide for so many. Over three-quarters of their erratic income was required for food, nearly half of which consisted of bread.[11] They rarely tasted meat. In Gangelt, they were locked to the fate of the peasant, who was currently enduring rural Europe's transition from an ancient feudal system to a money economy, where the freedom to work for a wage came at the cost of dispossession from the land, as owners consolidated their estates for commercial production. The rising prices of farm produce benefitted the large farmers and estate owners, but crippled the peasants, who were forced to work more, for lower wages, growing crops which were not theirs. As larger farms became more viable, the ancient privileges which gave peasants the wherewithal to live off the land were eroded. A new term emerged, *robotet*, meaning drudge, toil, slave, fag, sweat. The peasant became a wage slave, a *Robot*. To the daily drudgery were added punitive taxes and periodic demands for men and horses to fight the emperor's campaigns.[12]

But of the multitude of forces conspiring to make the Kremers' toehold in Gangelt so precarious, the most alarming was space. Space was disappearing. Since the great pestilence of 1347–51, the population had bounced back into over-productive recoil.[13] In awe, and then alarm, German chroniclers recorded that couples seldom had less than 'eight, nine, or ten children'[14] and that 'landed property and rents for dwellings be become so very dear that they can hardly go any higher ... rather all villages are so full of people that no one is admitted. The whole of Germany is teeming with children.'[15]

Fields that had run to scrub since the pestilence were put back under the plough, and houses reappeared from the hummocks of crumbled villages. The best land reclaimed, colonists moved to the margins. The land, recorded one chronicle, 'has been opened up more than within the memory of men; and hardly a nook, even in the bleakest woods and on the highest mountains, is left uncleared and uninhabited'.[16]

As space was consumed, those who owned it began to take a keener interest in its dimensions. Landowners seeking to maximize their rents and dues consolidated their holdings. Surveyors appeared, pacing boundaries, sketching perimeters, making maps.

While Gangelt's hard-pressed peasants scrabbled for space, tales of extraordinary new lands filtered along the road between Cologne and Antwerp. Broadsides and hearsay told of voyages made by men of Spain and Portugal, and of a new, fourth continent blessed with riches beyond belief. The first description in German of these lands of plenty passed through Gangelt a decade before Gerard was born. By the time Gerard was a boy, the four-page printed letter that had been written by Christopher Columbus in his caravel off the Canary Islands had reached most towns in Europe.[17] Dated February 1493, 'The Admiral's' letter began by announcing that he had sailed west for twenty days[18] and 'reached the

Indies', where he found 'very many islands filled with people without number'. Claiming and naming a succession of islands, Columbus described the 'numerous rivers, good and large', the 'many sierras and very high mountains ... most beautiful, of a thousand shapes'. He wrote of trees that 'seem to touch the sky', of nightingales and 'little birds of a thousand kinds'. There were lands rich for sowing, for livestock and building, and natural harbours 'you could not believe'. One of these islands was 'larger than England and Scotland together'. To his king and queen, Columbus promised gold and spices, cotton, gum mastic, aloe wood, cinnamon, slaves and rhubarb. Gangelt's incredulous peasants also learned that 'all the islands of India' were patrolled by the canoes of those 'who eat human flesh'. Contrary to expectation, Columbus had 'so far found no human monstrosities', although he'd been told of people born with tails, and an island whose inhabitants were entirely bald. Another island was populated only by heavily armed women dressed in plates of copper. 'In all these islands', he concluded, 'it appears ... that the women work more than the men'.[19]

In 1505, an even more amazing voyage appeared in German. Another printed letter – this time from a raunchy Florentine navigator – told of 'a new world ... a continent more densely peopled and abounding in animals than our Europe or Asia or Africa'.[20] From Lisbon, Amerigo Vespucci had sailed 'seeking new regions towards the south', on a voyage that had taken him beyond the equator to the 'torrid zone', where 'both sexes go about naked' and where the women, 'being very lustful, cause the private parts of their husbands to swell up to such huge size that they appear deformed and disgusting'. This remarkable feat was achieved, according to Vespucci, 'by the biting of certain poisonous animals'. Apparently the women could not resist 'the opportunity of copulating with Christians'. Everything Columbus had done, Vespucci could do better: his letter claimed that he had discovered new stars in the sky and a new method of navigation, while his voyage south down the coast of the new continent had taken him 'to within seventeen and a half degrees of the Antarctic circle'. Where Columbus had merely *heard* of cannibals, Vespucci had *met* a man who had eaten three hundred human bodies. Vespucci's 'unknown world' was not just bountiful, but 'rich in pearls', with gold in 'great plenty' and a climate that was 'very temperate and good'. So marvellous was his continent that Vespucci dared to suggest that, 'if the terrestrial paradise be in any part of this earth ... it is not far distant'. A German engraver quickly translated Vespucci's imagery into a lubricious woodcut of naked 'New World' natives eating human limbs while the explorer's ships peeped in from the horizon.[21]

With incredible letters from men of the sea, Gangelt's book peddlers also circulated *Das Narrenschiff*, the first book by a German author to penetrate the continent's bloodstream. Written in 1494, two years after Columbus' voyage west, *The Ship of Fools* was an allegorical fable about a vessel sailing for the fool's paradise of 'Narragonia'. Its author was the editor of Columbus' letter, the satirical poet Sebastian Brant, and it contained the earliest reference in literature to the discoveries beyond the 'Western Ocean':

> To lands by Portugal discovered
> To golden isles which Spain uncovered
> With brownish natives in the nude
> We never knew such vastitude.[22]

In rough, accessible verse, Brant used his ship as a moral vessel to present over a hundred sample fools selected to represent the shortcomings of the day. Immoral monks and drunkards shared the pages with criminals and crooked judges, robber knights and rude cooks. Each chapter was headed by a three-line motto, and a woodcut, some of them by Brant's friend Albrecht Dürer. Brant believed that his German blood was that of the chosen people. God had rewarded German nobility, magnanimity and orthodoxy with spiritual and worldly leadership – through the Holy Roman Empire – of the Christian commonwealth. But for the degenerating manners, morals and faith of the common people, and the disloyalty, fear and indecision of the nobility, the Empire would not be disintegrating.

Year by year, life on Gangelt's blessedly fertile plateau became less tenable.

The floods were so bad in the spring of 1515 that passage through the Rhineland was impossible. Year's end brought a winter in northern Europe that was cold enough to freeze the river Thames in England. The next winter was nearly as severe, and was followed by a hot, wet summer that rotted the harvest. Lawlessness followed.

The trials of Gangelt's increasingly impoverished peasants were reflected in letters written at the time by the Low Countries' most celebrated humanist, Desiderius Erasmus, whose travel plans had been disrupted by the crisis in the countryside. In June 1516, Erasmus reached the Flemish border after a troubling journey from Basel to Antwerp: 'I ran into soldiers everywhere,' he wrote, 'and saw the country people moving their belongings into the nearest small town.'[23] Rumours had reached Erasmus that soldiers were about to attack Lorraine. Nobody was sure who had sent them, but the humanist suspected them to be mercenaries discharged by the emperor and 'looking for someone to pay them wages instead of him. We are,' added Erasmus, 'in such a state of turmoil, playing dice all the time, and yet we consider ourselves Christians ...'[24] In July 1517, a rampaging mob of dismissed soldiers called the Black Band sacked Alkmaar.

By autumn 1517, rye was unaffordable and prices of barley, wheat and oats were soaring. Harvests were worse, and prices even higher than they had been when the Kremers left Gangelt in 1511. Millennial pre-monitions that the 1500s would bring despair on mankind were beginning to be fulfilled.

On All Saints' eve 1517, the unthinkable occurred, and the Christian commonwealth turned against the Pope. The movement began in Wittenberg, where a Dominican called Tetzel had been selling indulgences to raise funds for the building of St Peter's in Rome. A local monk, Martin Luther, circulated a list – in Latin and German – of ninety-five discussion-points concerning indulgences. Satirical woodcuts appeared, mocking the Church, and news came of uprisings in the south. Revolution was in the air. The daily carnival of travellers passing through Gangelt now included unfrocked clergy, landless peasants in search of work, semi-literate artisans, students, twitchy *Landsknechte*.

As a cobbler, a maker and mender of the *Bundschuh*, Hubert Kremer may have felt exposed through his association with the symbol of insurrection. Recent outbreaks of peasant unrest had frequently been led by the Kremers' class, the artisans and craftsmen, the blacksmiths, cobblers and tailors who found themselves the focus of rural life.

If 1517 was bad, 1518 was worse. Writing in March 1518 from Louvain, Erasmus fretted restlessly about the looming danger: 'I am off either to Venice or to Basel, either of them a long and dangerous journey, especially through Germany, which beside her long history of robberies is now exposed to the plague ...'[25] One week later, Erasmus referred to Germany again, as 'in a worse state than hell itself, for you can neither enter nor leave.'[26] He blamed the rulers: 'God help us, what tragic works these princes have in hand! The sense of honour is extinct in public affairs. Despotism has reached its peak. Pope and kings regard the people not as human beings but as beasts for market.'[27]

The situation in Germany worsened week by week. Still in Louvain, Erasmus was unwilling to 'creep into Germany' for fear of 'Those criminal scum ... Already there are stories everywhere of the murder of travellers.'[28] Cursing the 'idiot theologians' who were driving him to leave Louvain, Erasmus wrote to Thomas More in England that a botched parley by the Duke of Jülich, the Duke of Cleves and the Duke of Nassau with the Black Band had led to a thousand victims being 'cut to pieces'.[29]

The bad harvests, rising prices, overcrowding and general lawlessness were ideal for disease, described in May by Erasmus as 'a new kind of plague ... spreading throughout Germany, which attacks a great many people with coughs, headache, and internal pains and kills a number of them...'[30]

For the Kremers there was only one direction. To the east stretched the darkening lands of Germany. To the west lay the lands of the Burgundians; Flanders; Rupelmonde; Gisbert and a leased homestead in Hubert's name.

Forty years earlier, the Flemish diplomat Philippe de Commynes had kept a chronicle during his years serving the dukes of Burgundy. A seasoned European traveller who had seen the city republics of northern Italy, de Commynes had come to regard the Burgundian territories in a unique light. He wrote of a land like no other, where wealth and peace 'overflowed' and where 'meals and banquets were larger and more sumptuous than in any other place'. Here there was money to spend, and luxurious clothing. 'Therefore,' he concluded, 'it seems to me that these lands, more than any other principality on earth, could be called the promised lands.'[31]

It was to these promised lands that the Kremers now turned.

2

Promised Lands

With their children and possessions, the Kremers passed from Gangelt's chaotic interior onto the open plateau.[1] Reduced by distance, Gangelt shrank into a neat impression of towers and walls against the empty vellum of loess.

Down the slope from Gangelt, the departing family passed through Süsterseel – where Albrecht Dürer would breakfast in two years' time – and then Sittard ('a pretty little town', noted the artist).[2] Dürer was an experienced traveller, a veteran of two trips to Italy and several crossings of the Alps, and he had chosen to take the more hazardous, direct route from Cologne to Antwerp, rather than the longer, busier road which passed south of the Kempen through Diest and Mechelen. The Kremers would have taken the same, direct route.

The road was busier than when Hubert and Emerentia had travelled the same way six years earlier. Now everybody seemed to have a reason to be on the move. There were more landless peasants seeking urban poor relief, more pilgrims, more disenfranchised knights, more merchants, more convoys of heavy-axled carts bound for the boom towns of the northern littoral. Some would have been headed for Mechelen, others for Brussels, or Hondschoote, the town that had won the race to mass-produce serge.[3] Most, however, were bound for Antwerp.

For a small boy who had been removed from the only home he had known, the experience can only have been intense. Gerard had no conscious recollection of the migration that he had taken shortly after his birth. Now he was old enough to know of mercenaries and robbers; old enough to recognize masks of desperation. Among those on the road that summer was Erasmus, who had eventually summoned the nerve to leave Louvain for Basel. When he returned north in September, the worst part of the journey was the leg from Aachen to Maastricht, along the low road south of Gangelt's bluff.[4] In Aachen, he had been poisoned by 'raw' *Stockfisch* and been unable to move before putting his fingers down his throat. These lands east of the Maas held especial horrors for the well-travelled humanist: 'nothing', complained Erasmus, 'is more uncivilized or dreary or barren than that country, such is the fecklessness of the inhabitants; so I was the more glad to get away. The danger of robbers – which was very great in those parts – or at least the fear of them was driven out by the distress of my illness.'[5]

Crossing the Maas at Stocken, the Kremers climbed again to the wasted heathlands of Kempen, passing the nights in roadside hostels, crammed perhaps (as Erasmus was near here) into hot, stinking rooms with up to sixty other rowdy, beer-sodden travellers. Like Erasmus, the Kremers would have been fed on wind-dried, pulped *Stockfisch*, bread and warm beer. At Uilenberg the family parted with ill-afforded *pfennigs* or groats to pay the Brabant land toll, and then took the rutted roads towards the Schelde.

Everywhere lay water, in the slow rivers, in peat diggings and mill leats, in puddles and on polders. The land looked like an immense cracked mirror, refracting the sun from countless wet fragments. Men shrank beneath the weight of the sky. On the thin horizon, unimpeded breezes stirred windmills and lazy sails, which slid through the grasses towards freshwater wharves. One of those wharves was Rupelmonde.

Here was the peace and wealth which de Commynes had promised.

Above the flatness of river and floodplain loomed Rupelmonde's immense castle, with its seventeen steepled towers and formidably high walls. Besides its military role, the castle had been used by generations of Flemish counts as a prison to detain – and to eliminate – those who threatened the well-being of Flanders. Built where the river Rupel joined the Schelde, Rupelmonde ('the mouth of the Rupel') had been a strategic site since the time of the Romans. A century before the Kremers arrived, it had been a thriving centre of nine hundred dwellings, with its own markets, fairs and town walls. But in one of the battles which followed a rebellion by Ghent in the early 1450s, the town had been devastated. Viewed beside its enormous, battered castle, the resurrected Rupelmonde seemed disproportionately small.

Close to the castle stood a new mill, of a scale and construction unseen in rural Jülich. Commissioned by the Count of Flanders and finished a year or so before the Kremers arrived, the mill was powered by the tide and built of brick and Doornik stone with a cathedral roof that doubled its height. The basalt grindstones had come all the way from Andernach in the uplands of Eifel.[6]

Behind the castle, a wide, rutted way led up past cottages roofed with

Fig. 2. Rupelmonde, the view south-east towards the river Schelde, from Sanderus, *Flandria illustrata*, 1641–4 (British Library 177 h.11)

river reed to the marketplace and church, where Rupelmonde's thousand souls came together to pray.[7] Beyond the marketplace the road continued, and a short way along here stood the homestead Hubert had leased since Christmas 1511.

Rupelmonde and the county of Flanders were recovering from a run of bad seasons: in 1515, heavy autumn rains had broken the dykes, flooded the *moer* and led to a rise in grain prices. The winter of 1516–17 had brought gales and more floods, followed by a freeze that had turned the land to stone and frozen the Schelde. Mortality rates had soared.[8] Then 1517 began with a drought.

But by 1518 the worst had passed. Rye in Antwerp had dropped to half its 1516 price. Oats were down too, from 27 groats per *viertel* to 23. At St Bavo Fair, the price of cheese had fallen from its 1517 price of 189 groats per *wage* to 163 groats in 1518. The cost of firewood was down, and interest rates were falling.[9]

Hubert cleared the debt on the land he had leased, and paid off three years of the debt on the homestead, deferring the remainder with the intention of paying it off the following year. Flanders was familiar with Hubert's type. The weavers and spinners, carpenters, diggers, flax workers and cobblers were an expanding breed of specialist 'hand-workers' who supplemented their subsistence plot with a skill which could buttress a family's needs. Hubert had his plot, and his shoes.

The riparian life of Rupelmonde was quite unlike the cyclic rusticity of landlocked Jülich. The river sang with calling boatmen and creaking blocks; the thud of spar and rumbling mill; the mews and cries of waders and waterfowl. Rupelmonde could offer a peasant heavy clay or the sandy loam of the rise on which the village had been built. The earth was either wet and heavy, or acid. Rupelmonde's men netted fish, caught eels in the mud, crewed boats, worked the Schelde's quays, took the bailiff's pay, maintained dykes. Some took pay in the local brickyards; others repaired the great flat-bottomed barges – the *pleiten* – pulled up from the river on timbers laid in the mud of Rupelmonde's shore.

Along the Schelde glided a ceaseless flotilla of shipping, small boats being rowed or poled, larger barges sliding with the current, ships with masts and sails and black-eyed cannon. Few waterfronts in the Low Countries offered such a view. The convergence of the Rupel and the Schelde, brought together streams of river traffic from the Leie and Dender, the Zenne, Dilje, Demer and Nete. Between the interior and the sea, this was one of the busiest junctions in Flanders; the confluence that connected Antwerp to the inland wharves of Ghent and Aalst, Brussels and Mechelen. Passing Gerard's eyes was wheat from Picardy, peat for heating the homes of Antwerp and barges bringing Hemiksen brick and building stone from Brabant. Firewood came down from the Ardenne forests by way of the Demer and Rupel; silk and satin from Mons and Quesnay. Heading upstream from Antwerp was salt and herring, soap, chalk and lard. And luxuries from other lands, like wine and figs and currants bound for Dendermonde, where wagons would carry them on to Hainaut. Way upstream of Rupelmonde, flat-bottomed riverboats pushed deep inland, carrying butter and cheese to places like Béthune and Cambrai, where the rivers were scarcely as wide as the hulls. From Béthune they brought back paving-stones. Many would have moored at Rupelmonde, meeting beer boats and ferries and market-day barges, the *marktschepen*. Rising above the barges were larger vessels, sea-going cogs crewed by men who spoke in other tongues, sailing inland as far as Mechelen and even Brussels.

The river trade spun its own tales. Some were captured in the accounts of the bailiffs: a peasant of Moerkerke fined for illegally loading madder onto a boat for England; a Scotsman smuggling honey and herring; a German fined for selling a parrot that had already been seized by the bailiff. Some boatmen smuggled pigs, others grindstones.

The river was also a conduit of overseas news. On its banks, a boy could meet men who had sailed through the Pillars of Hercules and bumped against the ice floes of the East Sea. Boatmen on the Schelde were connected by calling to crews who had sailed off the edge of their sea-charts; who were reshaping the face of the earth. The printed accounts of Columbus and Vespucci were merely the prelude to continuing geographical revelations. Rumours must have reached the Schelde concerning the Cabots, father and sons who had twice sailed west for the English king, Henry VII. Where Vespucci had turned south down the coast of what had now been termed (by the Italian chronicler Peter Martyr)[10] the 'New World', the Cabots had turned north. So important were their discoveries that the second son, Sebastian, had recently entered service as a cartographer to Ferdinand V of Spain. From Portuguese ships on the Schelde spread the stories of Vasco de Gama, whose astonishing (and unpublicized) voyage of 1497 had taken him around the tip of Africa and all the way to Calicut – and India. Columbus and Vespucci had returned with unfulfilled promises, but Gama had delivered, spectacularly. In circumventing the Genoese and Venetian merchants, and the Levantine monopolies, Gama had opened a trade route between Portugal and the mercantile capitals of India. Navigators from the western seaboard of Europe were extending the points of the compass. By Gerard's eighth birthday, four years had passed since Martyr had published the bloody adventures of Vasco Núñez de Balboa, who had crossed the Panama isthmus to reach the 'Southern Sea'. Stripped naked to ford swamps, Balboa's men had been confronted by snakes, poisoned arrows and a tribe dressed as women, forty of whom were 'torn to pieces'[11] by Spanish fighting dogs. In another encounter with natives, Balboa's men slew six hundred 'like butchers cutting up beef and mutton for market'.[12] More recently, Hernán Cortés had invaded Aztec Mexico, exhorting his *conquistadors* to fight like Romans.

While the river nourished Gerard's geographical imagination, it was his uncle Gisbert who provided the boy with spiritual guidance.

Gisbert had proved that Latin and learning could break the shackles of poverty. Having left his birthplace, Gangelt, long before his brother Hubert, Gisbert had won himself an education at the university of Louvain,[13] and it was his faculty acquaintances who must have helped him to secure the clergyman's post at the hospice in Rupelmonde.[14] Such hospices had once provided lodgings for poor travellers, but over the years they had turned their attentions to the sick and elderly paupers of their own communities. Most had an altar, or a chapel, where the priest could take Mass. Typically, the appointments procedure required the bishop (a graduate of Louvain) to propose three or four candidates for the town's authorities to choose between. Once accepted, Gisbert was protected by his clerical post from the worst adversities, while he also wielded considerable local influence through his control of poor relief. The handouts included staple foods and – significantly for Gerard's father – shoes. As a member of the clergy, Gisbert also received a tax remission of 50 per cent, a saving which would soon be of considerable import.

Gisbert's own faith would not appear to have been conventionally Catholic. Gerard's uncle was almost certainly a sympathizer of Gerard Groote, a charismatic fourteenth-century mystic from whose teachings on the secluded banks of the Ijssel had sprung a religious order called the Brethren of the Common Life. Groote and his twelve disciples had devoted themselves to imitating the life of Christ, preaching, teaching, striving to revive the faith and morals of congregation and clergy. Emphasizing inner strength and educational improvement, the Devotia Moderna, the New Devotion, flowed across lands that had been ravaged by war and pestilence, poverty and the spiritual dues of the papacy. A monastery and numerous brethren houses were established, where the virtues of piety, chastity, humility and charitability offered the lay population a route to practical Christianity. Because they attached little importance to ritual, and took no vows, the Church was unable to attack them over issues of clerical organization and dogma. To mendicants steeped in simony – the trading of ecclesiastical privileges – the high ideals of the Devotio Moderna were a threat, and yet the Brethren made a chimeric target; attacking them cast a brighter light on their moral crusade. In this, the Devotio Moderna was an uncontroversial revolution. The reflective spirituality of the Brethren had been captured by Thomas à Kempis in his *Imitation of Christ*. Kempis had been born in Kempen, not far north of Gangelt. On the banks of the Schelde, his book was second in popularity to the Bible.

Gerard's relationship with Gisbert, and with Groote, Kempis and the Devotio Moderna, could be read in his name. In the dialogue that had finally converted Groote to his life of ascetic reform, William de Salvarvilla had argued that Gerard (Groote) had been born with gifts of nature which would lead to a better life. Even his name spelled a future beyond terrestrial glory.[15] 'Gerardus', wrote Salvarvilla, was derived from *gerens ardua*, or 'aspiring to, or accomplishing, lofty things'.[16] 'Gerardus', he added, was also *gerens ardorem*, 'having the ardour of love for God and his fellow men'.[17] God, claimed Salvarvilla, had spelled out the mystic's acronym as *Gratus Haeres Aeterni Regis Amorem Rerum Divinarum Utiliter Sapiens* – 'Beloved heir of the eternal king, profitably occupied with love of the things of God'.[18] Once Thomas à Kempis had repeated Salvarvilla's wordplay in the *Vita*, the name 'Gerardus' entered the aspirational lexicon of every follower of the Devotio Moderna. From the Ijssel to the Schelde, 'Gerard' was the most revered of Brethren names. Gerard Kremer had been baptised into the family of Gerard Groote.

Through reading and contemplation, the Devotio Moderna raised issues that were as pertinent in the early 1500s as they had been a century earlier. Educated, pious and discreetly unconventional, Gisbert invested his nephew with equal measures of certainties and doubts.

Hubert's youngest son took his first classroom seat shortly after arriving on the Schelde.[19] Unlike rural Gangelt, where the elementary school's teachers may have had to share their educational commitments with the rhythms of the local harvests, Rupelmonde's school was more likely to have been a dedicated affair, benefitting from its proximity to Antwerp and from diverse sources of local income. Teachers at the school may have been clerks or the local priest.

In a large galleried room equipped with tables and benches,[20] Gerard found himself sitting with pupils whose first language was Flemish and whose backgrounds ranged across the social spectrum.[21] Although Gerard must have picked up the Schelde dialect relatively quickly, his German background caused an early isolation.[22] An incomer from a dialectically foreign land, the boy's predisposition toward introspection was fostered at school.

Learning began with the common European language, Latin, the *lingua franca* of the Church, of law and medicine, of commerce and diplomacy, and of education. He began with collections of adages which had the dual purpose of teaching basic Latin and imparting sound morals. Perhaps Rupelmonde's school was sufficiently advanced to have acquired the *Adages* of Erasmus, but Gerard was more probably exposed to the *Dicta Catonis*, a collection of Christian ethical mores familiar to generations of schoolboys, chanted until its hexameters and pentameters were imprinted in every young skull. Over one hundred of Cato's two-line *disticha* advised on everything from anger to inaction and the ethics of poverty: 'Since sons you have – not wealth – such training give / Their minds that they, though poor, unharmed may live'.[23] Cato also contained interminable *sententiae*, one-liners whose remorseless repetition was intended to guide and inspire, and to trip the wayward with a snatched recollection of childhood wisdom: 'Great crises foster deeds enshrined in thought'[24] ... 'If short of best, then emulate the good'[25] ... 'Learning is pleasant fruit from bitter root'.[26]

Gerard also learned Dutch and Latin versions of the Lord's Prayer, the Creed, the Ten Commandments, psalms and an elementary catechism or two. The Bible itself would have been kept on a high shelf, too open to misinterpretation by those unable to distinguish the literal from the allegorical; question-and-answer catechisms were more susceptible to order.

With the beginnings of Latin, Gerard was also taught the 'reckoning' skills required by merchants, and reading. He began to learn writing with the spherical O, the letter from which the roman alphabet took form. Inscribed within a square to keep its concentricity, the schoolboy O was practised until it reappeared without its frame. Once mastered as the intuitive template of all letter forms, the O's semi-circular variants – the Q and the C, the D and the G – came naturally. So too did the symmetrical curves of the other non-linear letters: the B and the J, the P, R, S and the U. Of the linear letters, the most difficult was M, and that also took its form from the sphere. Roman miniscule revolved even more around the o, as linear letters like the l and the e were reduced and rounded into the wave rhythms of the written line. In mastering the Roman O, Gerard had shaped the graphic form of his life's passion.

By the time Gerard was seven, he was speaking and reading Latin. Over the following years, he would embark upon the *trivium* – Latin grammar, logic and rhetoric, the skills of oral and written presentation. Once equipped with the *trivium*, he would be eligible to progress to university. His two eldest brothers were set to follow Gisbert into the Church, and the same destiny must have been expected of Gerard too.

Gerard was still at elementary school when the people of Flanders learned that their lands had been gathered into an empire of unimaginable scale.

Every schoolboy knew that the story of the world had been foreseen in the Book of Daniel. In chapter seven, Daniel had dreamed that four great beasts had come up from the sea. He had asked in his dream 'the truth of all this'[27] and been told that each beast was a king. Later generations of Christian theologians understood that the four beast-kings were the four successive empires that were known to have ruled the world.

The first beast, the winged lion, had been plucked of feathers, raised onto two feet and given a man's heart. That was the Babylonian empire.

Daniel's second beast, the bear with three ribs in its mouth, was the Persian Empire.

The third beast had been like a leopard, with four wings, and four heads: the empire of Alexander.

The fourth beast, 'dreadful and terrible, and strong exceedingly',[28] with great iron teeth and ten horns, was the empire of the Romans. The beast's ten horns gave rise to ten kingdoms, one of them the Holy Roman Empire.

In 1519, the fourth beast spawned its latest emperor, a nineteen-year-old from Ghent. Barge-jawed Charles was the eldest son of Philip the Handsome, the Duke of Burgundy, and mad Joanna, the Queen of Castile. After a rigged election which sidelined the French candidate, Charles became

> King of the Romans; Emperor-elect; semper Augustus; King of Spain, Sicily, Jerusalem, the Balearic Islands, the Canary Islands, the Indies and the mainland on the far side of the Atlantic; Archduke of Austria; Duke of Burgundy, Brabant, Styria, Carinthia, Carniola, Luxemburg, Limburg, Athens and Patras; Count of Habsburg, Flanders and Tyrol; Count of Palatine of Burgundy, Hainaut, Pfirt, Roussillon; Landgrave of Alsace; Count of Swabia; Lord of Asia and Africa.[29]

The young emperor exerted direct rule over his family lands, the Low Countries, Burgundy, Naples, Spain and the Spanish discoveries in the New World. Over Germany, his imperial influence depended upon satisfying the princes and cities that their interests were shared. Charles was the most powerful man in Europe; 'God', crowed his grand-chancellor, Gattinara, 'has set you on the path towards world monarchy.'[30]

That year, an expedition sailing in the name of Charles v left Seville with the intention of finding a route to the Spice Islands from the west. If Ferdinand Magellan succeeded in reaching the Moluccas, he would – in theory – be able to return to Spain using the Portuguese trade route around the Cape of Good Hope. For the monarch of a world empire, the first circumnavigation of the globe would be a prize of immeasurable symbolism.

Nowhere in the Low Countries demonstrated the emperor's global reach more vividly than Antwerp. A two-hour walk from Rupelmonde, the city seethed with mercantile expectation. Had the youngest Kremer been in the city on the Sunday after the feast of Assumption in 1520, he would have seen a procession that triumphantly encapsulated Charles' world monarchy. It was recorded in breathless detail by Albrecht Dürer.

Dressed in their best and marching beneath the signs of their trades and professions were the city's guilds – the goldsmiths, painters, stonecutters, embroiderers, sculptors, joiners, carpenters, sailors, fishermen, butchers, leather workers, clothmakers, bakers, tailors, shoemakers, the shop-keepers and merchants, marksmen with their guns, bows and crossbows. Accompanied by pipes, drums and trumpets, the carnival took two hours to pass and included footsoldiers and horsemen, various religious orders and – wrote Dürer – 'a great troop of widows', dressed in white linen robes, 'very sorrowful to behold'.[31] Twenty bearers held aloft the image of the Virgin Mary and Jesus, 'adorned in the richest manner'. The procession, scribbled Dürer excitedly, continued with 'many wagons drawn with stagings of ships and other constructions'. These included a company of Prophets, scenes from the New Testament and the three kings 'riding great camels, and other strange beasts'.

'Last of all', scribbled the artist, 'came a great dragon', followed by boys and girls dressed 'in the costumes of many lands'.

In Antwerp, Dürer concluded, 'they spare no cost in such things, for they have plenty of money'.

But in Rupelmonde, the Kremers were broke.

Rumours of impending planetary conjunctions in the sign of Pisces had suggested that the better times could not last. A deluge had been anticipated, and a mathematician in Tübingen had even provided a date: February 1524.[32]

Sebastian Brant sensed the end too: in 1520, nearing death, and suffering from acute pessimism and a nervous breakdown, the author of *The Ship of Fools* penned his last poem:

> Such confusion will arise everywhere,
> such horrible happenings,
> as though the whole world were to be destroyed.
> May God help holy Christianity![33]

In Toulouse, President Aurial built himself an ark.

And Providence played into the hands of the astrologers: on the Schelde, the wet summer of 1520 was followed by famine. In November, Dürer's wife Agnes had her purse cut from her body in Antwerp's cathedral, losing two florins, some keys and – noted the impecunious artist – 'the purse itself'.[34] By the end of 1520, the price of grain had driven housewives to riot in Mechelen and Louvain. Imperial troops were put on the streets to restore order.

Gerard's ninth birthday passed in circumstances of imminent need. Heavy snowfalls had blanketed Flanders through January and February, 1521. Firewood reached the highest price since records began, over a century earlier. The weather on the Schelde was windier and damper than it had been back in Jülich, and when the snows came, the drifts piled against the dykes then thawed back into the *moer*, leaving the land even more saturated. Costs soared of renting peat pits.

The snows of early 1521 were followed by a dry summer and a bad harvest that pushed the price of grain to new heights. Famine tightened its grip and in Mechelen, Louvain and Vilvoorde the granaries of convents, burgesses and merchants were looted by women in search of food. In Antwerp there were riots and farmers were robbed of grain they had taken to market. The following winter, the snow returned. And in 1522, so did famine.

Worse still, the war that had erupted in 1521 between the Habsburgs and the French crown had led to massive tax demands by a Habsburg administration needing to finance its army. By 1522, taxes were nearly three times their 1470s level. The war crippled trade and for two years, through 1522 and 1523, no ships from Portugal, Spain or Italy arrived in Antwerp. Shipping on the Schelde dwindled. So did the revenue passing through Rupelmonde's toll.

The crises struck hard at the handworkers of Flanders. Bad weather could ruin a smallholding for a season, but war and unrest could erode markets for years. As recession bit at Antwerp, the demand declined for tapestry, linen, furniture – and for Hubert's shoes. In Mechelen, just up the Rupel from Rupelmonde, the fullers and weavers went on strike. Protests followed in Antwerp, Louvain and Brussels and then in June 1525, the weavers and shoelace-makers of 's-Hertogenbosch turned on the town magistrate. Mostly illiterate, desperately poor, and possessing little or no land, the handworkers were primed for revolt.

In Antwerp, economic connections with Germany and waves of immigrants from the east helped to kindle and spread Lutheran fervour, which was stoked in the autumn of 1520 when Luther called for a reformation of the Church, burning his bull of excommunication and the papal constitution. Augustinians who had settled in Antwerp several years earlier began to circulate Lutheran literature and in a vain attempt to douse the flames, the Edict of Worms in May 1521 banned *'famosos et pestilentes libros'*[35] and then burned at the stake all such books that could be found, together with paintings and pictures which offended the orthodox faith. In Antwerp, four hundred 'Lutheran' books were publicly burned, and in Bruges, a picture of Luther was thrown to the flames. The following year, the Augustinian prior Jacob Proost was arrested. Proost escaped to Wittenberg, but two monks, Hendrik Voes and Jan van Esschen, were executed. By 1522, Antwerp printers (publishing anonymously) had issued twenty-three editions of Luther's work in Latin and in Dutch. Bans followed on books, by Luther. The monastery was destroyed and Erasmus warned a correspondent that 'in Flanders all is war and turmoil'.[36]

The Tübingen mathematician had been right about calamity, but he was wrong about the date and the deluge. In 1524, the rains failed, but one year later, revolution finally engulfed Germany. Word quickly reached the Schelde of the lootings and burnings, of the castles and monasteries put to the torch. Gangelt was spared, but for two weeks in April, Cologne burned. By the summer of 1525, copies of 'Twelve Articles' listing the grievances of the Upper Swabian peasants were circulating in Antwerp. Already printed in twenty-five editions, as many as 25,000 copies of the Articles were distributed through the Holy Roman Empire. Having reserved his judgement earlier, Luther now condemned the rebels. By the end of the year, 100,000 peasants had been slaughtered.

There must have been times when the Kremers wondered whether their prayers were being heard, for the difficulties that had driven them from Jülich appeared to have followed them to Flanders. The crises of the 1520s were the last disaster. Years of poverty, of forced itinerancy and grinding manual labour had taken their toll. In 1526 or 1527, after a decade of struggling to make a living in Rupelmonde, Hubert died. He cannot have been more than forty-five.

Just into his teens, Hubert's youngest son passed overnight into adulthood. Years of hardship had turned the cobbler's boy into a physically strong, resourceful individual. A migrant baby, he had migrated again before school age. He had survived drought, flood, plague and famine. Now he had to overcome the greatest crisis that could befall a child of the time. For a boy of Gerard's age and background, the loss of a father invariably led to the poorhouse door. Fortunately for Gerard, he knew the man *at* the poorhouse door. Gisbert, the kind uncle who had left Gangelt a generation before the Kremers, became Gerard's guardian.

Hubert's death coincided with the arrival in Antwerp of the first printed Bible to contain a map.

At the front of the Bible, before the beginning of Genesis, was the Holy Land seen from heaven. Few on the Schelde had seen such a woodcut: miraculously, the Old Testament had been transformed into a single picture, a picture of places, each positioned so that the viewer could see how one related to the other. Towns were sprinkled across the two pages so that the reader could see how Sidon lay far from Gaza and that Tiberius seemed a short way from Jericho. Mountains rose from the page like terrible waves, bearing down on Damascus, tearing themselves into a maelstrom beyond the land of Judah. The Nile spouted like a storm drain into 'Das Gros Meer'.

The closer one looked, the more one saw: there were lines on the land that were rivers, and others that were the borders of the twelve tribes. Canaan was crowded with towns but there were parts of the map whose very blankness could only be the wilderness. One feature above all drew the eye. Like a road, it turned and twisted across both pages from the parted walls of the Red Sea to the Jordan. It was the route of the Exodus. From the roof of Rupelmonde's mill, it was possible to see the distant, tapered spires of Antwerp's churches, but whoever had drawn *this* view had been able to see *everything*. Even the drowning Egyptians.

The text of the new Bible was also imbued with novelty, for most of it came from the pen of Martin Luther, the monk who had given voice to the anti-papal revolution a decade earlier. The map could be traced back to the godfather of Luther's first son, the Saxon artist Lucas Cranach. Back in the days before his friend achieved papal notoriety, Cranach had cut a wall-map of the Holy Land. It was this map that a Zürich publisher called Christopher Froschauer had reduced, recut, and then bound into the Lutheran volume that reached Antwerp in 1525.

Antwerp was one of the first places in Europe to receive Froschauer's edition. Within months, the printer Jacob Van Liesvelt had produced a Dutch edition prefaced with Cranach's reduced map; 'Das gros meer' became 'Die groote zee'.

For men like Gisbert Kremer, Luther's scriptural picture not only confirmed the Holy Land as an actual place, but related it to the meadows and polders beyond the village. Were not the ebb currents which ribboned the Schelde like those of the Nile? Were not Liesvelt's ships on Die groote Zee the same as those off Rupelmonde? Galilee sat as placidly as the inland waters of the Low Countries. To an immigrant family on the Schelde, none of the images was more universal than the long, winding road from Egypt to Canaan, by way of the wilderness.

3

To the Water Margin

Wanting the best for his brother's son, Gisbert decided to send Gerard to the Brethren of the Common Life in 's-Hertogenbosch.

Erasmus had been schooled there, and its serving Brothers included a rising humanist playwright (and friend of Erasmus), Georgius Macropredius. With the sons of other paupers in the Brethren's *domus pauperum*, Gerard would be able to absorb the mores of the Devotio Moderna, while his Latin could have no better guide than an Erasmian humanist like Macropedius.[1] 'S-Hertogenbosch had also nurtured an outstanding city school whose teachers would be able to share with the Brethren the task of preparing the boy for university. That the son of an immigrant cobbler, sponsored by an uncle of meagre means, could take advantage of such an opportunity, was a measure of the egalitarian nature of Europe's most advanced educational system. It also confirmed Gisbert's faith in his bright young nephew. One of the four principal towns in Brabant, 's-Hertogenbosch lay at least five days' walk away, to the north-east of Rupelmonde.[2]

As Emerentia watched her youngest child leave, she must have wondered whether she would see him again. Gerard was about fifteen. This was the third long journey that he had made since his birth in Flanders, and each had been brought about by difficulty. Now he was leaving his family and his home, leaving the places he knew best. For safety, he would have joined a group of travellers, merchants maybe, or other students.

The road north took the boy back to the heaths of the Kempen. In one day, Gerard travelled from the most densely populated part of Brabant to the emptiest. Those who could wrest a living from the sand and gravel lived in isolated *kampen* and hamlets. Where there was soil, there were fields of rye and buckwheat. Flocks of small sheep – prized in Antwerp for their flavour – grazed the scrub.

A few years earlier, Gerard would have met tides of fattened oxen being driven south from Drenthe and Overijssel, Friesland, Groningen and even Holstein, bound for the weekly cattle markets of Brabant.[3] But the wars north of the Maas had severed the trade and Brabant's smaller markets were being squeezed to extinction: a couple of years before Gerard left Rupelmonde, the magistrates in Diest announced that revenues from the livestock market had collapsed and that there were insufficient funds to pay poor relief. Many of the newly destitute headed for the fabled poor-tables of 's-Hertogenbosch.

The faces along the road were as wretched as any Gerard had seen in his life.[4] Across Brabant, peasants were on the move, drifting from country to town, from town to town. In Turnhout and Tilburg, the two towns on his route, conditions were dire.[5] Like Diest, Turnhout had depended on its cattle market for revenues. The waves of migrants and rising poverty among artisans had hit small towns harder than cities. Antwerp's golden aura did not reach as far as the heaths of northern Brabant.

Beyond Tilburg, the road passed through hummocked sands that had once been shaded by a great forest hunted by the dukes of Brabant. Where the sands ended, half the rivers of northern Brabant came together to pour into a swamp.

Across the swamp writhed the braided lower reaches of the Rhine and Maas. Slow and silt-laden, these wide, meandering serpents were the necks of drainage basins that reached south to the Alps and east to the edge of Bohemia. They were 'rain rivers', prone to spates and catastrophic floods, and where they ran parallel along Brabant's northern border they formed a watery gulf as wide as the channel between France and England.

The rivers had been a frontier for a long time. At the end of the last Ice Age, they marked the limit of glacial advance, and the Romans had turned them into a natural vallum along the Empire's northern shore, patrolling their banks, building sanctuaries. In the river clay lay bronze cicadas and glass bowls, coins and beads, likenesses of Bacchus and Caracalla, gold foil embossed with dolphins swimming with souls to the Lower World. Where the Styx ran in so many braids, the strange treasures that came to light contributed to the delta's sense of mystery.

These same rivers still formed the most dangerous internal border in the Low Countries. Beyond the Maas lay Gelderland, the lair of Duke Karel. In his continuing quest to sabotage the Habsburgian hegemony of Emperor Charles, Karel had chosen the year that Gerard left home, 1527, to invade Utrecht. Unpredictable and violent, the Gelderlanders

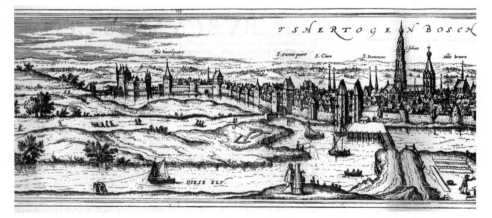

Fig. 3. Watery 's-Hertogenbosch, on the northern border of Brabant, from *Civitates Orbis Terrarum*, 1572 (Private collection)

were a perpetual, invisible threat, interrupting trade and perennially poised to swarm across the Maas to waste Brabant.

So Gerard Kremer had reached the most formidable internal border in northern Europe. Here, he stood on the edges of perception; at the northern limit of the world which he would regard as carrying familiar labels. The Rhine and Maas divided the Low Countries as effectively as the Alps separated Germany from Italy.

As a cultural Rubicon, the Maas had been a cartographic expression for at least a century and a half – ever since the little map that separated the student 'nations' of 'Anglicana' to the north of the Maas, from those of 'Picardia' to the south.[6] In the early 1500s, the population density beside the two rivers was still as close to zero as that of Zeeland's islands.

The swamp could be crossed at 's-Hertogenbosch.

The name came from the woods – *bosch* – of the duke – *hertog*.

From the south, 's-Hertogenbosch looked like an island in the polder, its walls, turrets and spires prickling a horizon that ran flat and wide to the three rivers. Water encircled the walls and threaded meadows where horned cattle grazed and women spread linen.[7] Sailing barges nuzzled the quay on the Dieze while two men on the towpath hauled a floating cargo towards the Maas and Holland's tidal archipelago. Dürer had been this way in the autumn of 1520. A 'great storm of wind'[8] had overtaken him on the Maas and he'd ridden with his companions into 's-Hertogenbosch on hired farm horses, without saddles. 'Herzogenbusch', he noted in his travel journal, 'is a beautiful city and has an extremely beautiful church and a strong fortress...'[9]

Upstream of 's-Hertogenbosch, the dyke along the south bank of the Maas had two *overlaten* cut from its wall to allow floodwater to spill onto the marshes. When the rains came, and when the Rhine and Maas were swollen with snow-melt, the water poured through the *overlaten*, to mix with the waters of the Aa and Dommel as they attempted to drain northern Brabant. Seen in winter, 's-Hertogenbosch seemed from a distance to be a ship on glass, sublimely reflective, ark-like upon its private ocean.

The perversity of its site spoke of a higher ideal. The elevated, ruled horizon divided swamp from sky, earthly labour from celestial awe, hell from heaven. Catholic dogma found potent symbols in this uncompromising landscape. Here were bleak visions that could play on fatalism and remind the men of the clay that the forces of evil would never cease to strive for eternal damnation. Morality and beauty stalked the city walls. Below lay black peat and stinking mud; overhead loomed a vast sky poised to unleash famine and flood. Between the two stretched a fragile plane of river and meadow, moved upon by miniaturized figures, by barges and by slowly turning post-mills.

These were the empty wetlands that Hieronymus Bosch had populated with grotesques. Bosch had died in 's-Hertogenbosch ten years before Gerard arrived and among the extraordinary works he left was a triptych depicting the history of earthly folly.[10] On the exterior of the triptych, Bosch had painted a murky primordial orb at its moment of transformation into the world. Inside, the three panels cavorted with figures from the garden of Eden, the earthly paradise, and hell. Across the central, paradisiacal panel, mankind debauched under the spell of an enchanted landscape of lake, meadow and mountain. Dominating this false paradise were four rivers flowing inward towards a central 'sea' whose focal point was a tiered structure rising from a plated sphere pierced at water-level by what appeared to be a drain-hole to the underworld. This symmetrical quartet of polar rivers, and their 'drain', would emerge from Gerard's consciousness in later years.

Gerard Kremer went to Hinthamerstraat, the long, reeking street which ran from the marketplace through the city's poorest district. About halfway between the market and the town walls, he turned into Schilderstraat. Almost at the end of the street, he found the *domus pauperum scolarium*, the Brethren's poorhouse. Here, he would lay his head for the next three and a half years.

Fig. 4. The playwright Georg Macropedius, Mercator's teacher at 's-Hertogenbosch (1486–1558). From Joannes Franciscus Foppens, *Bibliotheca Belgica*, 1739 (British Library 679 f.10)

Schooling had improved in the forty years since Erasmus had com-plained that the Brethren's 'chief purpose, if they see a boy whose intelligence is better bred and more active than ordinary, as able and gifted boys often are, is to break their spirit and depress them with corporal punishment, threats and recriminations and various other devices – taming him, they call it – until they make him fit for the monastic life'.[11]

Instead of cowled maniacs, Gerard was greeted by the very remarkable Joris van Langhveldt, known by his Latinized name as Macropedius. Born to a noble family at the castle of Langhveldt, one day's walk to the south-east of 's-Hertogenbosch, Macropedius had been schooled by the Brethren in Louvain, and then become a priest, a teacher (he was a renowned grammarian) and the leading Latin dramatist in the Low Countries. He had been with the Brethren of the Common Life for twenty-five years. An engraving showed Macropedius as if cut from stone, sat square to the artist, head and shoulders hooded but for the symmetrical oval of his implacable face. Below the hairless, domed forehead stared enlarged pupils, pooled by high eyebrows and low, weary sockets. Doubt had nowhere to hide. Above the cleft promontory of the chin rode an incipient smile, or a memory of amusement. Macropedius was the only detectable humanist at the Brethren house in 's-Her-togenbosch. He had just turned forty.

By the time Gerard arrived in Schilderstraat, Macropedius had written six or so plays which he had reworked from the Latin comedies of

Terence and Plautus.[12] Vulgar and moralistic, these 'Latin school dramas' were acted by his schoolboys, who had to play the female roles too.[13] In one of the more raucous works, *Aluta*, the audience were warned against halitosis and flatulence, and then subjected to heavy drinking, vomiting, undressing and a comedic exorcism. Macropedius excused himself for what might have been regarded as excessively smutty for a Brother of the Common Life by explaining that *Aluta* was a play intended only for boys. Images of debauchery were sound educational material, as long as purity and truth prevailed by the end of the play. Macropedius took the Ciceronian view that vice and virtue were reflected in the mirror of life; what better way to stiffen morals than to dramatize the punishment of evil and the reward of goodness? In the epilogue of *The Rebels* he reassured his audience that 'this play is safe ... for one whose mind is pure'.[14]

Macropedius must have been aware of how potently he was arming his schoolboys. Humanist texts gave him the facility, without recourse to heresy, to equip his schoolboys with the tools required to investigate the truths of opinion. Through prosody, he was providing the techniques a young man could employ for versifying his thoughts; for self-expression.

The school theatre may also have earned Gerard money. At performances staged for the public (for example at Shrovetide), money taken at the door was distributed by Macropedius to his poorer boys, the boys the playwright addressed in the prologue to *Bassarus* as 'impoverished, deformed and suffering fellows ... lazy, scabrous and beardless wretches'.[15] The audience looking down on 'this plebeian crowd'[16] of spotty wastrels was told that if they spent twice as much for the performance, in keeping with their 'rank and station',[17] the playwright would give them a higher, better seat in the future. So the meagre funds provided by Gisbert to educate his nephew may have been augmented by Gerard's theatrical skills.

A 'foreign', fatherless pauper, Gerard had reason to adopt a disguise, but it was Macropedius who taught him how to be inscrutable.

For Gerard Kremer and the boys in the *domus pauperum* on Schilderstraat, daily routine was divided between the Brethren house and the Latin school in the centre of town. In the Brethren house, one brother supervised administration and discipline, while another undertook the education of the boys. Latin was spoken at all times, and personal property, alcohol and women were forbidden.

Before six each morning, dressed in hooded grey, the boys left the *domus pauperum* for the Latin school. If it was a clear dawn, the sun would warm their backs all the way up Hinthamerstraat to Torenstraat, where they crossed to the chilly shadow of St Jan's. Above on the buttresses crouched tiny bagpipers, workers, supernatural grotesques, and flights of demons from the hands of the master mason, Allart du Hameel, a friend of Hieronymus Bosch.[18] For two centuries the church had been rising. Inside were chapels and altars dedicated by guilds to their saints, and du Hameel's font, with a great finial of climbing figures which made the lid so heavy that a cast-iron lever was required to lift it. With the schoolboys and townsfolk who came to St Jan's each day were miracle-seekers paying their devotions to the Sweet Lady, whose elaborate chapel was lit by the stained glass, and a chandelier and crucifix by Bosch. The school lay on the other side of the church, in the angle of Kerkstraat and Peperstraat.

The '*groote school*' in 's-Hertogenbosch was the largest in the city, famed for its Latin, and one of the finest secondary schools in Europe.[19] Largely due to the influence of Erasmus' headmaster at Deventer, Alexander Hegius, old scholastic regimes were catalysing with the 'new learning' of humanism. The most influential educationalist of his age, Hegius had succeeded in reconciling the ascetic ideals of the Brethren with the humanist quest for knowledge and eloquence. In the eyes of Erasmus, Hegius had risen to the stature of the ancients. Schooling was at its most alchemic.

At the Latin school, Gerard embarked upon the full *trivium*, the grammar, rhetoric and logic that would form the foundation of his university studies.[20] The basis of Gerard's humanist education was outlined in Erasmus' educational treatise *De ratione studii*,[21] in which he listed some of the principal authors that a boy at secondary school should study. 'First and foremost,' instructed Erasmus, came 'the Greeks and the ancients', Plato, Aristotle and Theophrastus serving as 'the best teachers of philosophy'.[22] Beside the Bible, the theological authors to read were Origen, Chrysostom, Basil, Ambrose and Jerome. Poets should be studied for their ability 'to flavour their compositions with knowledge drawn from every quarter' – especially mythology, Homer being 'the father of all myth', while Ovid's *Metamorphoses* and *Fasti* were 'of no small importance'.[23]

Erasmus completed his reading list by drawing specific attention to a subject which would ignite in Gerard an extra-curricular passion. 'Geography', wrote Erasmus, 'must also be mastered.' Geography was 'useful in history, not to mention poetry'. As far as sources were concerned, geography had been 'dealt with most succinctly by Pomponius Mela, most eruditely by Ptolemy, and the most comprehensively by

Pliny'. The 'main object' of geography, continued Erasmus, was to learn 'which of the vernacular words for mountains, rivers, regions, and cities correspond to the ancient'. Erasmus saw geography as an etymological and locative prelude; only by constructing a mental map could a schoolboy begin to locate the moral, political and philosophical coordinates of humanism.

By the time Gerard reached 's-Hertogenbosch, fifteen years had elapsed since *De ratione studii* had been lifted from a Louvain press by Thierry Martens. A lifetime spent travelling to and fro across Europe, on horseback, by boat and carriage, had given Erasmus a rich sense of place and a fondness for maps. Strabo, Mela, Pliny and Ptolemy supplied the geographical allusions which Erasmus habitually tossed into his letters. Thus 'the Indies' became a place where – in a letter to Dürer's friend Willibald Pirckheimer – 'my stone and your gout could together be sent packing'.[24] Once Erasmus had learned from Pomponius Mela that the Orkneys lay 'at the farther end of Britain',[25] the islands reappeared in his correspondence as a symbol of 'the remotest of the remote'.[26]

So the incidental geography that Gerard absorbed at 's-Hertogenbosch was likely to have been limited to place-names recorded by a Greek, two Romans and an Alexandrian, all of whom had lived in the century of Christ. Besides informing the boy that 'Barbaria' was the traders' name for the south-east coast of Africa, and that the Straits of Hercules allowed Venetian galleys to enter the Western Ocean, the ancient authors added form and detail to the imaginary world that Gerard was already constructing with his own eyes and ears. Up until this time, his absorption of geography had been unconscious. The occasional geographical texts at 's-Hertogenbosch turned a curiosity into a conscious discipline.

Lessons at the Latin school ran from 6 a.m. until 8 a.m. and then resumed after a morning Mass, which was taken back at the Brethren house. Each day, the boys sang with the chapter choir, and after school at 6 p.m. they attended High Mass, led by a Brother following the Roman rite. On Sunday afternoons they listened to a sermon in the Brethren house.

It was in the silence of the Brethren house that Gerard practised his handwriting. Before the days of the printing press, the Brethren had specialized in copying manuscripts. Together with the liturgical texts and theological books which they copied for their own use, they generated income for the houses by accepting commissions for bibles, missals, prayer books and illuminated lives of the saints. The Brethren wrote for an hour, on two evenings every week, and the house library was lined with books reflecting the fruits of their labours, St Jerome, St

Chrysostom and Boethius, rubbing spines with Thomas Aquinas, Duns Scotus and Valla – their pages thick with spiked palisades of gothic black letters.[27] It may have been here that Gerard picked up his 'bastard' gothic, a cursive version of 'black letter' that had evolved from chancelleries where a faster form of handwriting was required. By his teens, Gerard could express himself in a variety of hands, from roman to gothic to everyday 'secretary hand'.

Outside the walls of the Brethren house and the *trivium* lay the observable world. Schilderstraat continued across the last undeveloped meadow in 's-Hertogenbosch, to the city walls. To the left, a short walk along the walls led to a vantage point over the glistening Dieze and beside it, the rutted road north to the Maas, the Rhine and the Ijssel towns.[28] Under the walls a timber quay had been erected for river trade, and from here, boats left every week for Amsterdam and Delft, The Hague, Schiedam and Rotterdam. This view north could be as bleak as any in the Low Countries, and colder. Towards the rivers, the wetlands were darkened by *kommen*, the back-swamp lands that had formed where meanders had been breached to leave depressed basins of alder carr and scrub.

There were moments during Gerard's sojourn in this Brabantine outpost when the citizens of 's-Hertogenbosch had feared that the troubles to the north would pour across the Maas and swamp the city. Following Duke Karel's invasion of Utrecht, his notoriously brutal commander 'Black' Maarten van Rossum had bludgeoned his way across Holland and pillaged The Hague. A violent counter-attack had forced Karel into handing Utrecht and Overijssel to Emperor Charles, while retaining Drenthe, Groningen and – in sight of 's-Hertogenbosch – Gelderland. Increasingly, the wetlands were a barrier to be breached; English sweating sickness had struck the city within months of Gerard's arrival, and then plague. And the water itself brought death: it was a fever caught on the stagnant waterways of the Low Countries that had recently killed Albrecht Dürer, aged fifty-four.

Turning from the city walls, Gerard could observe a crammed, diverse urban geography. The population of 's-Hertogenbosch had virtually doubled since 1400, from 9,000 to 17,000, and it was now the largest city in the north of the Low Countries; twice the size of Deventer, larger than Utrecht and Amsterdam, dwarfing Rotterdam. Roaming its districts revealed how the architects of high-value space had distinguished themselves from the victims of poor space. Landlords had multiplied their rents by subdividing their properties and squeezing hovels into courtyards and gardens, while the wealthy had withdrawn to new properties on the 'clean' side of town. In 's-Hertogenbosch, the money stayed on the west side of town, around Orthenstraat and Vismarkt, and in the southern blocks around Vughterstraat and Vughterdijk, Weverplaats and Kerkstraat. The grander dwellings on the marketplace included the large stone home built by the father of Hieronymus Bosch. Seen from the heart of 's-Hertogenbosch's poor zone off Hinthamerstraat, the grand houses on the windward (and therefore odour-reduced) side of town were of another place.

Gerard's mother Emerentia died while he was away in 's-Hertogenbosch.

Eighteen, parentless, homeless, penniless, Gerard Kremer assumed a new identity. Like Erasmus, Macropedius and countless others, he dropped his family name for a humanist tag.

At issue was the name to choose. Erasmus had begun as 'Herasmus Rotterdammensis' and changed to Desiderius Erasmus Roterodamus, the last word probably embellished with a proparoxytone to become Rot*ero*damus. Macropedius may have been playing with 'Paedagogus', from the Roman plays of Terence and Plautus, where the word was used to describe a slave who accompanied a child to school and who acted as a domestic governor, a spiritual guide, a provider of learning, a diviner of life. Johann Reuchlin had taken the Greek word for 'smoke', Kapnion. His nephew Philipp Schwarzerd misinterpreted the etymology of his German name and became Melanchthon, the Greek word for 'black earth'. Most name changes revealed a pretension for high learning. (An exception was the Polish humanist Nicholas Vodka, who became Abstemius.) There was no regulatory body to control the fashion for Latinization, no central index of taken names, so duplication sometimes occurred. Several prominent Germans had name-changed to 'Agricola', the lionized Roman commander who wrote the life of his son-in-law Tacitus, the author of *Germania*, the book which put the German tribes on the map of Europe. There was also a Finnish Agricola, a bishop of Turku who published a biblical prayer book that included notes on astronomy and hygiene.

'Kremer' and 'Cremer' being the German and Dutch words for 'merchant', Gerard chose to become a Latinized merchant, *Mercator*. In doing so, he also associated himself with the *mercator* who peddled books from town to town, and with the merchant who had become the global citizen, the self-made, multi-cultural opportunist operating across the borders of Church and Empire, speculating, wholesaling, striking deals at fairs and on quaysides, patronizing arts and fleets. Through Macropedius, Gerard would also have learned – perhaps performed – Plautus' popular comedy *Mercator*. And Macropedius, who liked to quote Pliny in his plays, presumably introduced his pupil to chapter III of *Natural History*, where the word *mercator* had been employed in a way which would have appealed to a young man seeking a new image. Referring to the Tiber as 'Mercator placidissimus', the Roman writer had described Italy's most important river as 'the most placid entrepreneur of products from all over the world'.[29] It may have been Pliny's symbol of the busy Tiber as a benevolent medium of global exchange which most appealed to the imagination of a gentle boy who'd grown up on the Schelde.

So Gerard Kremer became Gerard Mercator. But the formulation of his new identity did not stop there. In common with Erasmus, Mercator demonstrated his keen sense of place (and his preferred roots) by adding a geographical agnomen to his name. His full name would read 'Gerardus Mercator de Rupelmonde'. Thus did Gerard forsake his German ancestry – and repay the uncle on the Schelde who was funding his education.

4

The Castle

After three and a half years on Brabant's water-margin, Mercator's uncle Gisbert, now his sole guardian, 'then transferred him thence to the famous university of Louvain'.[1]

It was summer 1530. Travelling by day, Mercator took the Antwerp road again, hitching himself to fellow voyagers to cross the Kempen. The chapters of Mercator's life were coming to be divided by migrations across the vast, uncultivable core of the Low Countries. Each bleak crossing of this interior waste was a reminder of dearth; a journey back to a landscape unmarked by the multiplications of detail that man was adding to the collar of fertile alluvium that looped around the Kempen from Antwerp to the mouth of the Maas. And each traverse drew a line between geographical polarities, a cord connecting the crowded and familiar, with the blank and unwritten.[2]

To a mind sensitized to spatial extremes, the Low Countries were far from uniform. A traveller in these sedimentary lands quickly developed an instinct for relating place with place. And in common with many migrants, Mercator had accumulated a rich collection of topographies. He had been the victim of his involuntary treks, but he would also be the beneficiary. Already his bank of recollections contained many more images of diversity than those of a young man bred securely within the protective walls of a city or rural estate. The troubled loess of Gangelt was unlike tidal Rupelmonde; metropolitan Antwerp and ethereal 's-Hertogenbosch were more remote from each other than their miles implied.

As Bosch had shown, this was the terrain of well-muscled imaginations. The deltas and their windy interfluves were a flat stage for souls to toil beneath an over-arching cosmos. Mercator's perception took form on these plains. The views he knew were invariably level, verticality was rare, scales were horizontal. There were no chasms or cliffs that closed out the sky and shadowed the land. Or peaks that could offer another

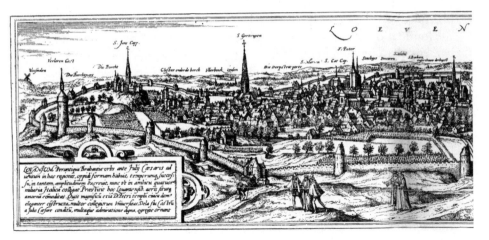

Fig. 5. Louvain, from *Civitates Orbis Terrarum*, 1572. Castle College stood near the Fish Market, on low-lying ground between the churches of Saint Gertrude and Saint Peter. (Private collection)

dimension. The Low Countries existed in only two dimensions. The world Mercator knew was a plane. It was a world as flat as a map.

The city rose well from the banks of the Dyle. Where the river had cut the Brabantine sand, a slope had formed above a small isle. It was bound to have a human allure. The shallow valley was well-drained and sheltered, and crossed by one of the roads from the Rhineland to Flanders. Boatmen from the Schelde could reach this crossing and Caesar was said to have camped here. Then Vikings. After Arnulf of Karinthe fought the Normans here in 891, the first Duke of Louvain built his castle on the island, and the castle became the kernel of a settlement that swelled then encircled itself with walls, then poured over its battlements. A second circle of walls was built, enclosing an area seven times as large as the first. By the early 1300s, the city was home to 20,000 people, producing 756,000 ells[3] of cloth a year. Then Louvain committed civic suicide, turning on itself in a murderous feud between citizen and patrician. The weavers fled and so did one-quarter of the population.

But reason – in the form of a university – returned to the crippled city. Across Europe, from Salamanca to Krakow, from St Andrews to Salerno, universities already thrived. France had seventeen and England two. Scotland three. The Germans had around ten. The Low Countries had none. For young men in search of higher learning, the closest professors were at Cologne, Paris or Trier. Only in Scandinavia and Ireland did a prospective student suffer such a lack of choice. Brussels had rejected the foundation of a university on the grounds that its presence would be hazardous to its daughters' virtue. Urged by Louvain's magistrates and the chapter of St Pierre's, Duke John IV of Brabant petitioned the Pope, who granted a Bull of Foundation in 1425.

Four years before Mercator emerged from the Kempen, Louvain university had celebrated its first centenary with a Mass in St Pierre's. The 'Belgian Athens' had risen so fast that it could already outbid every university on the continent for professors. Only Paris could compete in terms of numbers and prestige. The last decade had seen the establishment of three new colleges: on the Fish Market, the classics lectures at the innovative Collegium Trilingue had drawn – according to Erasmus – 3,000 students in one year; on Meiersstraat, Pope's College was educating a new generation of theologians, while St Jerome's had been established to concentrate on philosophy. Students undertook trans-continental treks to matriculate on the Dyle. They came from Portugal and Spain, Italy, Hungary, Norway, Sweden and Denmark, Scotland and Ireland, even France. The greatest number came from England.[4] To peripatetic humanists like Erasmus and the Spanish philosopher Juan Luis Vives (who'd left Paris disgusted by its scholasticism) Louvain was the hub of northern humanism. Even the Italians were having to concede that there was intelligent life north of the Alps. Nobody, Erasmus had written, 'could graduate at Louvain without knowledge, manners and age'.[5]

Entering the outer walls by the Aerschot Gate, Mercator passed fields and orchards, and hovels of wattle and thatch, and then the road became squeezed by houses and twisted away from the river, running as straight as a rule through a gate in the inner walls to a maze of tight streets packed with shops and colleges, grand houses, markets, churches, every thoroughfare filled with students and townsfolk. In the centre of this maze he came to an open space filled with light. Rising above the surrounding roofs was the church of St Pierre, its west front demolished thirty years earlier to make way for three towers, intended by their designer (Joos Metsys, the brother of the Antwerp painter Quinten Metsys) to come closer to heaven than any other man-made structure. Beside the church, the town hall rose like a giant stone reliquary, intricately carved with countless niches and pinnacles.

Mercator's journey ended at a modest building near the Fish Market. The teaching house known as the Castle opened onto the street which ran north towards the Mechelen Gate. At the rear of the building, a garden ended at the Dyle, which formed a loop around the Castle and its immediate precinct. Again, Mercator found himself surrounded by water.

With the Lily, Pig and Falcon, the Castle was one of four teaching houses – or *pédagogies* – in Louvain which offered the two-year Faculty of Arts course. Described by one of Mercator's fellow students as the university's 'leading and most distinguished school',[6] the Castle had attracted a number of bright, well-connected students who could expect to graduate from Arts to one of the university's 'higher' faculties: Medicine, Canon Law, Civil Law or Theology.

Among those sharing the Castle with Mercator were Andreas Vesalius[7] and his friend Antoine Perronet de Granvelle, who had arrived at the Castle two years earlier. As 'rich students', in private rooms, both men would have been set apart from Mercator and his fellow paupers in the Castle's dormitory. Mercator and Perronet nevertheless formed a binding friendship during their days in the Castle. Both originated from humble stock: although Perronet's status implied a noble past, his father, Nicholas Perronet de Granvelle, had started life as a blacksmith's son and risen to be the keeper of the seals for the emperor. Another 'rich' student studying in the Castle between 1530 and 1532 was Georg Cassander. One year younger than Mercator, Cassander had been born – like Mercator – in Flanders. In Vesalius, Perronet, Cassander and Mercator, the Castle was fostering four distinguished men of the future: an anatomist, a politician, a theologian and a cosmographer.

Elsewhere in the city were others whose names would become familiar to Mercator long after he had left Louvain. The future botanist Rembert Dodoens of Mechelen had matriculated three weeks before Mercator. Beyond St Pierre's, Mercator would have found fresh-faced Andreas Masius among the new arrivals boarding in the Lily, still a magnet for those of Erasmian inclinations. Masius would also reappear later in Mercator's life.

Having reported to their teaching house, prospective students had two weeks in which to appear before the university rector for matriculation. Always a celibate churchman, the rector was the university's highest official, taking the head of processions with his two beadles, and passing judgement on student miscreants, who could only be arrested on his authority.

On 29 August 1530, Mercator stood before Pierre de Corte and his beadle. Resplendent in his fur-trimmed purple gown, de Corte had taken office a few months before Mercator arrived in Louvain. A friend of Erasmus since their days together in the College of the Lily, de Corte had contrived to remain loyal to Christian humanism while defending

Catholic orthodoxy, most publicly by preaching against Luther in St Pierre's.

Having confirmed his familiarity with Latin, sworn to obey the university statutes, and paid his registration fee, Mercator was enrolled among the *pauperes ex castro*, the poor students of the Castle. In sprawling secretary-hand, the new student's name, origin and social status were inscribed in the *Liber intitulatorum*. Confirming his new identity, the boy born on the banks of the Schelde described himself as 'Gerardus Mercator de Rupelmunda'.

The academic year opened with a Mass on 1 October, the feast of St Remigius and St Bavo. University statutes were recited and orations delivered from a professor from each of the five faculties.

Rising at dawn, Mercator's days within the Castle's walls were prescribed by the rigid house timetable. Breakfast was taken after Mass and then classes were attended until lunch, which was followed by a rest in the garden. Further classes and private study in a cell continued until dusk, dinner and bed.

Although he wore the standard, undecorated ankle-length tunic and plain bonnet, Mercator was reminded daily of his social standing. With the Castle's other poor students, he took his meals at the lowest table, at the far end of the hall from the nobles. He may also have suffered ritual beatings, ambushes and extortion meted out by older students, many of whom regarded newcomers as beneath contempt.

For students of limited means, the classroom was the only refuge. It was also the only route to a better future. As they had done for generations, students still gathered in a *schola* or *familia* around their master. Acceptance into a *familia* could be more troublesome than the more formal requirements for university entry. It was also selective, gathering students of similar social backgrounds or geographical roots. Introductions helped, and so did patronage.

The regard humanists held for Louvain did not always extend to the Faculty of Arts course. Despite the Castle's reputation as the university's leading teaching house, its masters still adhered to the traditions of scholastic theology.[8] Of the seven liberal arts, Mercator did not progress far beyond the grammar, rhetoric and logic of the *trivium*. He was taught arithmetic, and possibly elements of music, but had to educate himself in astronomy and geometry after graduating. Neither was the method of teaching any more advanced than it had been at 's-Hertogenbosch. Learning by rote, Mercator and his fellow students listened to readings from set texts. If they were lucky, the master provided interpretations and glosses where necessary, and at daily disputations they could raise questions and debate the responses. And those set texts were mildewed with tradition. Mercator and his fellow *logici* spent their first nine months sitting through lectures on logic, the authority being Aristotle. Porphyry's fourth-century *Introduction* and *Commentary* on Aristotle's treatises on substance, quality and quantity – the *Categories* – would also have been read and debated. For the later, physics, element of the syllabus, Aristotle was again the authority, and for mathematics, Boetius' *Arithmetica*.

But for most of the two-year course, Mercator knelt before Aristotle, 'the master of those who know',[9] the philosopher whose separation of the constantly changing earthly sphere from the serenely circular progress of the stars supported so neatly a biblical heaven and earth. Louvain's university regulations were explicit: at all times 'you will sustain the doctrine of Aristotle', they instructed, 'except in cases which are contrary to faith'.[10] There could be no space for intellectual manoeuvre: 'No one may be allowed to reject as heretical an opinion of Aristotle, which the Catholics have diligently defended, unless the opinion has been previously declared heretical through the Faculty of Theology.'[11] Doubters of Aristotle could expect to be ejected from the faculty, and only the most humiliating recantation would facilitate readmission: in 1486, Marsilius de Craenendonck only managed to return to his studies after a formal assertion of Aristotle's repute and an admission that he'd acted out of frivolity.

Much of the frustration felt by Mercator's student contemporaries was directed at theologians who misrepresented 'the Philosopher'. Vives had written for many young minds when he had penned *In pseudodialecticos*, his attack on the study of philosophy: 'They do not even know who Aristotle is,' raged the Spanish humanist. 'They have no first-hand knowledge either of his natural or moral philosophy, or even of his logic, which they shamelessly profess to teach without having laid eyes on one of his books of logic.'[12] Too much of the curriculum was devoted to learning logic 'for its own sake', rather than 'going on to the other sciences'.[13] And in words which must have struck a personal chord with the son of a shoemaker, Vives wondered who could tolerate 'a cobbler who does nothing but sharpen his needles and his awls and his various knives, and twist, wax and add horse bristles to his thread'.[14]

Within the Castle, Mercator had to endure the same 'empty and senseless babbling' that had driven Vives from 'the Cimmerian darkness' of the university of Paris. One of the Castle's masters stepped straight from the pages of *In pseudodialecticos*. A theologian, he was remembered by Vesalius for introducing 'his own pious views' to commentaries on Aristotle's *De anima*, insisting that the brain 'was said to have three ventricles'.[15]

Aristotle was equally misleading in his notion of the earth's five parallel zones, whose balanced disposition reflected the symmetry that had been observed elsewhere in nature. *Antarkticos* must exist, because the lands of the Arctic required a balancing antipode. Aristotle's whole 'habitable world' was confined to the zone between the northern 'frigid zone' and the uninhabitably hot 'torrid zone'. Below the torrid zone was another 'temperate zone' (which Aristotle refrained from inhabiting) and the Antarctic frigid zone. Although 'squint-eyed' Strabo had accepted the same five zones in his *Geography*, Aristotle had an early doubter in the first century geographer Pomponius Mela, who had asserted that *both* of Aristotle's temperate zones were inhabited, the southern one being the home of *antichthones*, antipodeans; it was only the terrible heat of the intervening torrid zone which made the *antichthones* inaccessible.

Mela was something of a celebrity at Louvain. Eleven years before Mercator matriculated, a young German humanist called William Nesen had announced that he would be giving a series of lectures on Mela's geographical compendium, *De Situ Orbis*. The first printed edition of *De Situ Orbis* had been published in 1471, in Milan, but an improved edition was long overdue and Nesen's lectures would provisionally address that need. But Nesen had not sought permission from the university authorities to lecture privately. Neither had he even matriculated. He had simply fixed a notice to the door of St Pierre's informing students that his geographical lectures would take place in the Convent of the Augustines. The Faculty of Arts were outraged by Nesen's impudence and the Senate convened an investigative committee. Nesen struck that night, calling at the rector's house with three cloaked accomplices armed with swords. Hiding his face with his sleeve, he handed a letter to the rector's terrified *familiaris* and the four ran off. The following morning the letter was read aloud at a meeting of the Senate convened in the Upper Chapter Room of St Pierre's. The unidentified courier was threatening a riot should the Mela lectures be prevented from continuing. The Senate responded by placing a notice on St Pierre's threatening dire penalties upon those who refused to reveal within three days the identities of the letter-carrier and his three accomplices. Nesen was implicated, but his three accomplices were never identified. For a short while, the Senate thought that they had a culprit in the local printer Rutgerus Rescius, who was ordered by the rector to be arrested. Freed after the intervention of Erasmus, Rescius took his revenge by suing the rector.

Cast from Louvain, Nesen continued his vendetta against Louvain's theologians from Frankfurt, where he fell under the spell of Luther, and then moved to Wittenberg, where he lectured on geography.

Four years after his celebrated prank in Louvain, Nesen drowned in the Elbe. By then, 1522, Mela was the subject of official lectures being given in the city by Vives, and Aristotle's zones had been conclusively breached by humble seamen who could claim to have seen through the folly of the ancients: 'It appears to me,' Amerigo Vespucci had written to his patron, Lorenzo di Piero Francesco de' Medici, 'that by this voyage of mine the opinion of the majority of the philosophers is confuted...'[16] Far from being too hot for human habitation, the 'fresher and more temperate' air of the torrid zone had attracted such a large population that more were 'living in it' than 'outside of it'.[17] Such revelations shone as beacons in the dark halls of the Faculty of Arts: 'Rationally,' winked Vespucci, 'let it be said in a whisper, experience is certainly worth more than theory.'[18]

Mercator would have been one of many who took advantage of the free instruction at the new Collegium Trilingue on the nearby Fish Market. Inspired by Erasmus, the Trilingue's lectures were delivered in Greek, Hebrew or Latin. Through comparison and experimentation, students criticized and tested the old authorized texts; sources were questioned rather than accepted. For students already exploring territories unknown to the Church fathers, this was learning at its most adventurous and critical. Louvain's conservative students demonstrated in vain with cries of 'Out with the Fish Market Latin!'

Mercator completed his first year in the Castle, graduating to the level of *baccalauréat* by the end of 1531. Ahead of him stretched four months of metaphysics and ethics, and three months of *repetitiones* – the practicals and revision that would lead towards his final exams. At the end of October 1532, he was awarded his *magisterii gradum*, the master's degree. While friends like Masius and Vesalius set their sights on one of the four higher faculties, Mercator walked out of the Castle to freedom.

For the first time in his life, Mercator found himself beyond the confines of home, dormitory and curriculum. Whatever intentions he once held of a future in the Church now receded fast as he embarked upon a journey of his own making. Alone and unsupervised, he 'read philosophy privately',[19] allowing the winds of curiosity to carry him to places far beyond the Faculty of Arts. It was a journey which rewarded the twenty-year-old with 'great pleasure'.[20]

Mercator's solo voyage revealed as many contradictions as it did truths. In the course of long, silent hours of reading, the views of Aristotle became increasingly difficult to reconcile with 'the most true and holy history'.[21] Dictated by the Holy Spirit, the Bible revealed all truths, to which nothing should be removed or added. Let the Bible, Mercator wrote later, 'in its complete state, with everything remaining, reveal its hidden mysteries'.[22]

And the mystery before all was that of Creation. Comparison of Aristotle with the Bible appeared to pitch Moses into a confrontation with 'the master of those who know'. How could Aristotle, for example, claim that 'change' required pre-existing matter, when God had created the world out of nothing? By any terms, the creation of the world was a substantial change, but none of Aristotle's theories allowed for his 'unmoved mover' to pull a stunt such as that described in Genesis. When God 'created the heaven and the earth', and placed 'a firmament in the midst of the waters', the Bible was clearly stating that He was producing matter where there had previously been none. Living things had been the subject of changeless creation too: 'Let the earth bring forth grass, the herb yielding seed' described the miraculous rooting of living vegetative matter which could not be explained by Aristotle's assertion that plants must always come 'from seed'.

Among theologians, the problem was hardly new. Three hundred years before Mercator had been born, Thomas Aquinas (a keen sympathizer of Aristotle) had side-stepped the Aristotle-Moses-Creation problem by allowing that the philosopher's theory of change was partially acceptable when applied solely to the sublunary world. New to unsupervised reading, Mercator's curiosity must have been sparked by a closer look at Aquinas, whose *Summa Theologiae* he would have known from his 's-Hertogenbosch days.[23]

The words of the Prophet and the Philosopher were irreconcilable. And for Mercator, they raised questions about the entire basis of his education: 'But when I saw that Moses' version of the Genesis of the world did not fit sufficiently in many ways with Aristotle and the rest of the philosophers,' he wrote later, 'I began to have doubts about the truth of all philosophers.'[24]

Doubting Aristotle was dangerous practice, especially in Louvain, where his views were regarded as canonical.[25] Undeterred, Mercator began 'to make investigations into the mysteries of nature'.[26]

'It was amazing,' he confessed later, 'how the contemplation of nature pleased me.'[27] From nature, he added, 'becomes known the cause of things, from which all knowledge is obtained'.[28] As the pages turned and the weeks passed, Mercator found himself drawn towards the structure and description of God's remarkable creation. 'I took particular pleasure,' he noted, 'in studying the formation of the whole world.'[29] It was the suspended orb of the earth, he observed, 'which contains the finest order, the most harmonious proportion and the singular admirable excellence of all things created'.[30]

Painstakingly, Mercator began to write a gloss on the first chapter of Genesis. Consulting sources from Plato and Parmenides to Pliny, Mercator believed that he could add flesh to the bones of Moses' account, and create an entirely plausible explanation which accorded with contemporary knowledge of the 'earthly machine'.[31] It was, however, a project which was likely to draw the attention of Louvain's theologians, and soon enough he found that his investigations proceeded 'in such a way'[32] that living in Louvain became untenable.

With the benefit of hindsight, Mercator may have come to regret such an early conspicuity. Once noticed by Louvain's hardliners, an individual became a suspect. And the distinction between a suspect and a heretic was vague and flexible.

Later, Mercator recalled how he 'went voluntarily, alone, from Louvain to Antwerp'.[33] By implication, this was a journey undertaken on foot. The distance could be covered in a day, passing through Mechelen half-way. Solitary, resolute, Mercator walked briskly up the rubbish-strewn cobbles from the Fish Market towards the portal of light in the town walls.

5

Triangulation

The road led Mercator to a Pauline revelation. Leaving Louvain as a wavering philosopher, he would return as a committed geographer.

'Antwerp was like a world,' another fugitive would recall, 'one could lie concealed there without going outside it.'[1] The pumping heart of the global economy had become a vast diffusion chamber of cultures, classes and ideas. Every estuarine view and cramped street trembled with activity: on the choppy Schelde were Genoese galleys, single-masted English cogs, sleek caravels out of the Canaries and huge three-masted caracks from Portugal bristling with cannon. Between thickets of spars and the city walls scurried pack carriers, wheelbarrowmen and carters shifting sacks of verdigris, ginger and galingale from Spain, Rhenish wine, Flemish salt, Hungarian steel, Sumatran cubeb. And in the streets behind the quays stalked the merchants and factors and migrants who had helped to swell the city's population from 33,000 to 50,000 in only forty years.[2] Blessed with the geography of success, Antwerp lay near the mouth of a navigable river that was watered by the most densely populated region of Europe. Only Tuscany – Europe's other pole of population – came close to matching the human densities of the Low Countries.

By now, the port on the Schelde had also overtaken Deventer as the most important print centre between the Rhine and the Seine. Over half of the 4,000 or so books printed in the Low Countries since 1500 had been lifted from presses in Antwerp, which had also become the art capital of northern Europe.[3] South of the Grote Markt, in and around the Lombardenvest, Kammenstraat and Steenhouwersvest, Mercator would have found a thriving, subversive community of painters and engravers, cartographers, woodcutters, dealers, printers and bookbinders. 'Access to the city is good', an incoming French printer would enthuse;

the many different nations to be seen in the market square there testify

46

Fig. 6. Antwerp in winter, from the west bank of the Schelde, by Joris Hoefnagel (1541–1600). (Stedelijk Prentakabinet, Antwerp, Belgium/Bridgeman Art Library)

to that. And in Antwerp all the materials so necessary to the art of printing may be found. Manpower enough to train in any of the crafts can be found without any difficulty ... and finally there flourishes the University of Louvain, outstanding for the learning of its professors in all subjects and whose learning I reckoned to turn to profit for the general well-being of the public in manuals, textbooks and critical works.[4]

Of necessity, Mercator's investigations remained discreet. The mutterings of his detractors still reached him from Louvain, where certain theologians appeared keen to provoke him into revealing his sympathies in print. But the continuing disapproval of those he'd originally sought to elude only served to encourage his investigations. Affected 'with weariness' by these 'futile conversations', Mercator 'feigned an excuse'[5] to prolong his absence. And in doing so, he was drawn by degrees towards the branch of knowledge that was best suited to explaining the structure and mystery of God's creation. That discipline was geography.

'Since my youth,' he remembered later, 'geography has been for me the primary object of study. When I was engaged in it, having applied the considerations of the natural and geometric sciences, I liked, little by little, not only the description of the earth, but also the structure of the whole machinery of the world, whose numerous elements are not known by anyone to date.'[6]

One man on Mercator's road was particularly well placed to educate a philosophical refugee on the true state of knowledge concerning the fabric of the world. He lived in Mechelen, midway between Antwerp and Louvain.[7]

Franciscus Monachus (Francis Monk) had entwined his name with his calling by becoming a Franciscan monk. At a time when most geographers were theologians, Monachus had been one of the first in the Low Countries to break with the synthetic mix of biblical cosmogony and Aristotelianism that had characterized orthodox geography for centuries. Educated at Louvain university, Mechelen's bold monk practised a geography which drew on investigation; on experience and observation.[8] His terrestrial globe had been the first to be constructed in the Low Countries.[9]

Monachus belonged to a revived geographical movement which had begun to gather momentum over a century earlier, when a Greek copy of Ptolemy's *Geōgraphikē hyphēgēsis* – 'Guide to Drawing a World Map'[10] – arrived in Florence from Constantinople. A native of Egypt, an astronomer and a geographer, Ptolemy had written his *Geography* in Alexandria during the second century AD. Once the *Geography* had been translated into Latin and printed in Vicenza, further editions followed in Bologna, Rome and Florence, and Ptolemy's 1,200-year-old cartographic treatise became a template for modern mapmakers. Mercator's own destiny would become so entwined with the Alexandrian's legacy that some would come to regard Rupelmonde's son as the Ptolemy of their age.

In its opening lines, the *Geography* distinguished two forms of cartography. First, there was *geographia*, or world cartography, which was 'an imitation through drawing of the entire known part of the world together with the things that are, broadly speaking, connected with it'.[11] Then there was *chorographia*, or regional geography, 'an independent discipline' which 'sets out the individual localities'.[12]

Underlying Ptolemy's treatise was a grid of geographical coordinates which could be stretched like a fishing net over a globe, or laid flat on a map. The north–south lines represented longitude, and the east–west lines, latitude. Each line had a number, counted off from the Fortunate Isles in the case of longitude, and from the equator in the case of latitude. It was a uniform, universal system, as applicable to sixteenth-century Europeans as it had been to second-century Egyptians.

In the main parts of the *Geography*, Ptolemy provided a topographic description of Europe, Africa and Asia, he outlined the role of astronomy in gathering geographical information, and he described how to depict the world on globes and on maps. He also illustrated how to create a 'graticule', the flat grid of meridians and parallels which could be used to plot geographical locations. Each variety of graticule was derived from a projection, a mathematical formula which allowed the cartographer to transpose coordinates from the spherical, three-dimensional surface of the world, to the flat, two-dimensional surface of a map. To help his readers with this difficult concept, Ptolemy provided two sample projections, one 'inferior and easier' and the other, 'superior and more troublesome'.[13] In a gazeteer, Ptolemy listed the coordinates of around 8,000 places. Most of the early printed editions of the *Geography* carried twenty-seven or so maps constructed using these coordinates.

By the time Ptolemy reached the German print cities, with the Ulm edition of 1482, a succession of mathematicians and astronomers residing in the geographic heart of the Holy Roman Empire were already preparing the way for Monachus and the Low Country geographers. One of the earliest of these pioneering Germans was Johannes Regiomontanus, who set a standard in 1472 by installing a press in his Nuremberg house and announcing a forthcoming programme of maps and books, among them a new translation of Ptolemy's *Geography*. From the university in Tübingen (where his students included Melanchthon), Johann Stöffler issued waves of astrological texts. In Nuremberg, Johann Werner (a friend of Conrad Celtis) translated Ptolemy and Euclid, wrote extensively on mathematics and astronomy, and in 1514, published a work on map projections which would give Mercator a platform for his own cartographic career. Also living in Nuremberg was 'the ingenious Erhard Etzlaub',[14] the compass-maker whose sundials were 'sought after even in Rome',[15] and whose *Romweg* of 1500 had been the earliest printed route map.

West of the Rhine, in the mountains of Lorraine, the remarkable Martin Waldseemüller launched his cartographic career in 1507 with a book – *Cosmographiae introductio* – in which he (or his collaborator Matthias Ringmann) suggested that the continent visited by Columbus should be named after Amerigo Vespucci. 'Since another fourth part [of the world] has been discovered by Americus Vesputius,' proposed the *Cosmographiae*'s author, 'I do not see why anyone should object to its being called after Americus the discoverer, a man of natural wisdom, Land of Americus or America, since both Europe and Asia have derived their names from women.'[16] *Cosmographiae introductio* was accompanied by a globe (on which was marked the continent 'America') and a large map of the world 'containing the islands and countries recently discovered by the Spaniard Americus Vespucius in the western sea'.[17] Waldseemüller followed this astonishing debut with a treatise on surveying and perspective, a booklet on using globes, a map of Europe and then – in 1513 – a new edition of Ptolemy. With its twenty-seven woodcut maps, tables of coordinates, index of over 7,000 place-names and Latin translation of Ptolemy's text, were twenty modern maps: 'We have confined', explained Waldseemüller in his address to the reader, 'the Geography of Ptolemy to the first part of the work, in order that its antiquity may remain intact and separate.'[18] In separating Ptolemaic geography from his *Supplementum modernior*, Waldseemüller had become the first to publish a systematic collection of modern maps. Waldseemüller collected 'firsts'. Not only was he associated with the first use of 'America', but he was the first to show its continent. His world map was the first to cover 360 degrees of longitude, and his was the first printed map to show the entire coastline of Africa.

Meanwhile in Nuremberg, Albrecht Dürer's experiments with space had led inevitably to mapmaking and a collaboration with an Austrian mathematician and poet-laureate, Johannes Stabius. (The laurel had been received from Celtis, for a Sapphic ode which Stabius had written for Emperor Maximilian I.) On their 'imaginary orb'[19] of 1515, the artist and the poet had plotted a perspective view of the world's surface geography, as seen from the heavens. Effectively a picture of a globe, seen from one side only, Dürer and Stabius had created the first map of the world projected as it were onto a solid geometric sphere. That same year, Johann Schöner of Karlstadt published his first terrestrial globe: 'A very clear description of the whole earth, with many useful cosmographic elements and a new and more accurate form of our Europe than earlier.'[20] Two years later, Schöner followed it with a celestial globe, and in 1523 – following the return to Europe of the survivors of Magellan's circumnavigation – he produced a new terrestrial globe updated to include the 'islands and lands recently discovered upon the instigation of their most Serene Highnesses the Kings of Castile and Portugal'.[21]

The name 'America' may well have been introduced to the Low Countries on a world map which had been bound into a 1520 commentary on Solinus. Copied from Waldseemüller at a reduced scale, the map had been created by a bright young Saxon mathematician, Peter Apian. Since Apian – or his editor Lorenz Fries – had deliberately omitted Waldseemüller's authorship, most in the Low Countries would have considered the map to be the work of Apian, an unknown prodigy. Four years later, the graduate of Leipzig and Vienna justified his reputation by reviving the pursuit of sound theoretical geography with his *Cosmographicus Liber*. Inspired by Ptolemy, Apian's measurement-based 'Book of Cosmography' would be a major influence on Mercator's life. As it was understood by Apian, 'cosmography' was less of a specific discipline than an umbrella term for the study using descriptions, maps and diagrams of the whole universe. Astronomy was involved, and geography and theoretical cartography. Apian's cosmography included diagrams that illustrated the earth's climatic zones, methods for calculating latitude and longitude and the application of trigonometry for distance measuring. Varieties of map projection were included, and an elaborately printed world map fitted with no less than four volvelles revolved to illustrate latitude and longitude, the zodiac and the rotation of the earth on its polar axis. Part II contained the corrected latitudes and longitudes of nearly 1,500 places, from Antioch to Oxford, and a scattering of new discoveries in 'Insulae Americae'.

Cosmographicus Liber had been printed in Antwerp the year before Mercator matriculated, and its circulation was one of several factors which tilted Europe's centre of geographical excellence from Germany towards the Low Countries.

Leading the Low Country geographers was Franciscus Monachus, the globe-maker of Mechelen. On that first globe and in its accompanying booklet, the monk had refuted 'the nonsense of Ptolemy and other early geographers',[22] by assimilating as many recent sources as he could get his hands on. These included the work of Schöner and Apian and a report from Francisco de Hoces, who in 1526 had found land in the southern hemisphere at 0 degrees longitude, 52 degrees latitude.[23] Monachus also had access to a copy of *De Moluccis Insulis*, the printed letter written by the diplomat Maximilianus Transylvanus after he had debriefed the survivors of Magellan's round-the-world voyage.

Monachus had not departed entirely from the traditional geography of ancient Greece. The Aristotelian view, that the world's southern pole must be occupied by a landmass which balanced the northern Arctic, seemed to conform to the information provided by Magellan and de Hoces (and Schöner). Linking the land seen by Magellan south of his storm-bound straits to the land found by de Hoces, the monk had ruled a conjectural coastline around the pole, to which he attached the legend: 'This part of the world not yet discovered by our navigators.'[24] The map in Monachus' booklet was one of the earliest to recognize the rounding by Magellan of America's southern tip, and to confirm the existence of the southern continent. Elsewhere, Monachus was more controversial.

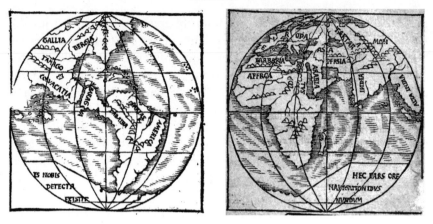

Fig. 7. The world view of Franciscus Monachus, *De orbis situ ac descriptione ... c.* 1527 (British Library C107.bb17.(1))

Departing from Waldseemüller, he had connected 'America' to Asia with a tapered waist of land, a configuration that he'd derived from the pages of Marco Polo, Sir John Mandeville and Odoric of Pordenone. Midpoint, the 'waist' was cut through by a channel wide enough to suggest a sea route between Europe and India.

Monachus may also have shown Mercator a map of the northern lands that he had derived from *Inventio fortunata*, a book which described the voyage two centuries earlier of an adventurous Franciscan monk from Oxford who had sailed to Greenland and maybe even Markland – Labrador.

Unlike de Hoces, or Oxford's intrepid Franciscan, Monachus did not sail for his source material. He was a navigator of libraries; an explorer in search of hidden texts. From Monachus, young Mercator could have learned that the quest for cartographic veracity depended less on certainties than on weighing probabilities.

While the monk may have introduced Mercator to the principles of geography, the man who became his mentor was a thin, brilliant mathematician whose name had been taken from his place of birth and from the Latin for gem.

Four years older than Mercator, Gemma Frisius had been born with crooked legs to impoverished parents on the edge of Friesland's windy marshes. Orphaned, the boy had shuffled on crutches until he was six, when his stepmother had taken him one feast day to the shrine of St Boniface in Dokkum. The pilgrimage cured his legs and the small,

Fig. 8. Gemma Frisius, Mercator's mathematics tutor (1508–1555). (Science Photo Library, London)

frail boy was educated at Groningen and then Louvain, where he had matriculated in 1526 as a poor student at Lily College. Unlike Mercator, Gemma stayed on after graduation, studying arithmetic and astronomy. By the age of twenty, he was in print, having made a few alterations to Apian's textbook on cosmography and taken it to Monachus' Antwerp publisher, Roeland Bollaert. On the title page, readers were informed that the German mathematician's work had been 'carefully corrected, and with all errors set to right, by Gemma Frisius'.[25] Any buyer who troubled to compare Gemma's edition with Apian's original would have had difficulty finding any textual corrections. But in its appearance, Bollaert had masterminded a thoroughly modern impression, substituting Apian's inky gothic typeface with a rounder, sharper roman font. And while the forty or so illustrations had – with a couple of significant exceptions – been materially copied from Apian, efforts had been made to reduce the ink-load by lightening decorative motifs. Gemma had also tinkered with the world maps at the front of Apian's book, changing the Indian peninsula that Apian had updated from Contarini back to a Ptolemaic island. And where the German had shown America as a single, crinkled crescent straddling the equator, Gemma deferred to Waldseemüller and pushed the crescent south of the equator, and added a second continent to its north. The southern continent he labelled 'AMERICA'. The northern one he left nameless.

Gemma's Apian appeared in 1529, and this was followed a year later by 'a geographical globe with the most important stars of the eighth celestial sphere'.[26] This ingenious all-in-one terrestrial/celestial globe was engraved by a Louvain goldsmith, Gaspar Van der Heyden, and on sale by the end of 1530. It was accompanied by a three-part booklet whose full title, *Gemma Phrysius de Principiis astronomiae & cosmographiae, deq[ue] usu globi ab eodem editi. Item de Orbis diuisione, & insulis, rebusq[ue] nuper inuentis* (Of the Principles of Astronomy and Cosmography, with Instruction for the Use of Globes, and Information on the World and on Islands and Other Places Recently Discovered) obscured the breakthrough it contained. *Principiis astronomiae* appeared in Antwerp, just as Mercator was matriculating. Bollaert had recently died and Gemma's publisher and printer was now Bollaert's old printer on Antwerp's Lombardenvest, Johannes Grapheus. In a 42-leaf volume that was compact enough to slip into a large pocket, a student could learn the principles of latitude and longitude, of winds and meridians, of poles, eclipses and zodiacal signs; to tell an isthmus from a peninsula. The second part, *De usu globi*, described chapter by chapter how to use a globe for investigating astronomical and cosmographical matters. The third part, *De orbis divisione*, took the reader around the world from Babylonia to 'Bresilia', with a cast that included American cannibals and Prester John, and flora that ranged from cinnamon to rhubarb trees. The island where Magellan had died with a spear through the throat nine years earlier was listed, and the booklet contained one of the first descriptions of America. Again Gemma rebutted Monachus: 'Many', insisted the Frieslander, 'connect this part of the earth [America] with Asia and say that it is one single continent, but their arguments are not valid.'[27]

But the jewel in this slim volume was slipped into the tail-end of *De usu globi*. In chapter XVII, Gemma tackled the longitude problem.

The Spanish, reminded Gemma, had discovered new lands in longitudes that were 'uncertain or completely unknown to us'.[28] Estimating longitude by referring to lunar eclipses was unsatisfactory, since the eclipses were too infrequent and suffered anyway from parallax and the latitude of the moon. Gemma agreed with Ptolemy that the 'winding paths' of a sailing voyage made lunar fixes too uncertain for practical use. 'Therefore,' he concluded with just a trace of presentiment, 'let me attach something I thought of myself...'[29]

That 'something' was no less than the solution to the longitude problem. In half a page, chapter XVIII described how a mariner sailing with a portable clock which had been set to the time of the port of departure could always calculate his longitude as long as the clock never

stopped. As Gemma pointed out, 'small clocks of ingenious construction'[30] already existed. Such clocks were 'little burden to a traveller
on account of their small size' and would 'often run with a continuous
motion for up to 24 hours, and if you wish, they will run with an almost
perpetual motion'.[31]

Propelled into circles of influence which reached far beyond Louvain,
the orphan from Friesland was spirited off to the Imperial court at
Brussels by the Polish ambassador, Johannes Dantiscus, who then tried
to lure Gemma back to the Vistula to collaborate with another gifted
mathematician, Nicolas Copernicus. According to Dantiscus, Copernicus had just completed a revolutionary work on astronomy but was too
nervous to publish. After several vacillations, Gemma moved back to
Louvain, where he resumed teaching and turned his mind to plotting
terrestrial space.

Surveying had barely progressed since the Egyptians began measuring
fields by standing three men in a line with their fingers in the air.
Vitruvius had helped with his first-century BC 'hodometer', a wheeled
device that measured distance by releasing a pebble into a box with every
revolution, but a surveyor was still a man with a rope and a 'pole made
of wood ... the length onely of one perch'[32] who 'muste other ryde or
goo over, and see every parcell thereof, to know howe many acres it
conteyneth.'[33] That worked in an approximate fashion for fields and
small estates, but was impractical for surveying larger areas. Maps of
regions were compiled by counting unequally spaced mileposts or wheel
revolutions along roads which were never straight, then applying the
general trend of the road to a series of compass bearings. Periodic latitude
fixes helped to correct north–south errors, but with longitude impossible
to compute accurately, east–west errors could be enormous.

In Italy, the humanist architect Leon Battista Alberti had touched on
the use of triangles as a systematic method of survey, but had failed to
elaborate on the idea, or to put it into practice. Then a German Hebraist
and theologian, Sebastian Münster, crept closer to the solution, by using
sighting lines to form triangles during a survey of the Heidelberg district
which he published in the form of a small map in 1525. Münster had
measured the angles from Heidelberg to the towns on his map, then
estimated the distances to them by means of land-traverses. This was
such a laborious method that he had appealed to scholars to help him
extend the survey area, since 'it would be too heavy and costly a burden
for one man to observe and describe Germany properly'.[34]

Gemma's nineteen-page *Libellus de locurum* was included in the second
edition of his *Cosmographicus liber Petri Apiani*, printed in Antwerp in

1533, at about the time that Mercator arrived in the city. In it, Gemma described how to survey an area of any size on the basis of a single ground measurement. All his readers required was a makeshift instrument: a flat piece of wood inscribed with a graduated circle, from the centre of which revolved a pointer – an alidade – fitted with pins, for sights. By holding the 'planimetrum' level and orienting it with a compass so that its north–south line was parallel with magnetic north, the surveyor could rotate the pointer until its sights were aligned with the landmark in question. The bearing was read from the graduated circle. By taking bearings from two landmarks, the position of a third could be fixed. Gemma suggested that the surveyor began by ascending a suitable high-point, such as the tallest tower in a town, taking a series of bearings in a circle, then adding them to a circle drawn on paper. *Libellus de locurum* then described how to repeat the process from a second tower, using the intersecting sight-lines to fix each landmark. Gemma finally pointed out that a third set of bearings would resolve any problems caused by two sight-lines converging in a straight line. Coasts and rivers, he added, could be mapped in the same way.

To draw a map to a known scale, continued Gemma, the surveyor needed to create a base-line by measuring the actual distance between two of the centres. He gave Mechelen and Antwerp as an example, and described how to replicate their relative positions at a reduced scale on the map. Distances between other places on the map could be calculated using similar triangles. (In walking from Louvain to Antwerp, Mercator had effectively followed Gemma's instruction that a base-line could be measured 'by walking over this distance'.[35] A measured base-line from Louvain to Antwerp would have been sufficient to embark upon the first mathematical survey of the region.)

Using Gemma's method, surveyors could accurately map an entire area by taking sightings on prominent points. And on the plane surface of the Low Countries, the Church had already provided the Trinity of every triangle, in the form of towers. The two-dimensional geography of these floodplains permitted unimpeded sight-lines, while the long straight roads were ideal for measuring base-lines.[36]

In covering the land with imaginary triangles, Gemma had also given geographers a startling new means of simplifying the surface they were trying to represent. At its surveying stage, triangulation created a mimetic map, a pictograph laid over the landscape at a scale of 1:1. Seeing maps as a tracing of reality was one of the perceptive warps that would help Gemma's generation of earth-modellers to break free from the imaginary worlds of the Middle Ages.

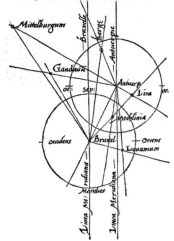

Fig. 9. The use of triangulation for surveying, illustrated by Gemma Frisius in his treatise, *Libellus de locorum describendorum ratione* ... (1533). (British Library C113c3 (1–2))

The concentration of geographical activity along the Louvain–Mechelen–Antwerp axis was compelling.

Monachus and Gemma were theorists whose field of interest extended down through surveyors, cartographers and instrument-makers to publishers and printers. Antwerp's printers had responded with alacrity to the accounts of voyages to distant lands: Willem Vorsterman printing a Latin edition of Vespucci's second and third voyages, and Jan van Doesborgh publishing vernacular descriptions of exotic lands, among them Africa and (in Latin and English) the 'newe landes'[37] across the Atlantic. More recently, Michiel Hillen van Hoochstraten had produced an account describing the campaigns of Fernando Cortes. The letter accompanying Monachus' globe had been printed in Antwerp by Maerten de Keyser. Printers and publishers were vying for a slice of travel and cosmography: as soon as Joannes Grapheus printed Apian's cosmography, his Lombardenvest neighbour Peter de Wale published a copy of Peter Apian's world map.

Grapheus had also printed an appeal for better treatment of Lapps and Ethiopians.[38] This had been written by Damião de Gois, a thirty-year-old Portuguese nobleman and occasional diplomat for João III of Portugal. De Gois was typical of the kind of character that Mercator could expect to meet in Antwerp's print shops in the early 1530s. An ardent Erasmian, poet and musician, De Gois had come to Antwerp in 1523, following his appointment as secretary to the Portuguese factor. He soon met Grapheus' brother Cornelius (recently released from prison after serving his sentence for exercising his views on Church reform), and Dürer, and various others who found the city a congenial location for exchanging unorthodox views. Mercator walked into Antwerp soon after de Gois had returned from Wittenberg, where he had met Luther and opened a friendship with Melanchthon. The year before Mercator graduated, de Gois had matriculated at the university of Louvain, where he fell in with leading Erasmian humanists such as Conradus Goclenius and the printer Rutgerus Rescius. An impulsive activist, de Gois would reappear later in Mercator's life, as the saviour of Louvain.

At around this time too, an acquaintance of Gemma's called Jacob Van Deventer, a *lantmetere* (surveyor), may already have begun mapping Brabant using triangulation.[39]

Money from trade, exploration and diplomacy was being lavished on globes, cosmographies, maps and instruments. Geography was no longer restricted to describing the appearance of the world created by God. In Germany, Luther's reformist geographers were more interested in how God's world actually *functioned*. Mercator found himself between two polarities. He was committed to the world plan outlined by God in the Bible, yet moved by the spirit of Ptolemy's mathematical geography. When Ptolemy wrote that it was 'necessary to follow in general the latest reports that we possess',[40] he was inviting his readers to forsake the old authorities for critical observation and exploration. Yet the Bible seemed to sanction such investigations. 'In wisdom,' stated the Book of Psalms, 'hast thou made them all: The earth is full of thy riches'.[41] The Book of Psalms opened the door to Mercator: 'Wisdom', he wrote later, 'is to know the causes and the ends of things, which can be known no better than from the fabric of the world, most splendidly furnished and drawn out by the wisest architect according to the causes noted in their order.'[42] God and Ptolemy appeared to be working towards the same, wise end.

Geography was irresistible for other reasons too. For anyone who could master the essential mathematics, this burgeoning field offered money, recognition and patronage, attractions which an impoverished, immigrant cobbler's son could not ignore. Philosophy held no such promise. With apparent reluctance, Mercator accepted that his philosophical studies 'would not enable him to support a family in the years to come, and that he would be put to even greater expense in pursuing them before he could reach the point where he would be able to count on a comfortable income for himself and his dependents from this source.'[43] Exerting pressure on the young idealist may have been his great-uncle Gisbert, whose generosity had presumably subsidized his two additional years of reflection. Mercator may also have felt that he had betrayed his uncle's expectations that he would enter the Church.[44]

The financial pressures on Mercator were exacerbated by the economic crisis that had been clamped across most of Europe since 1527. By 1533, the Low Countries had suffered a six-year spiral of harvest failures, famines and epidemics whose effects had been worsened by European wars, financial disasters and disruptions to international trade. Demand for goods had fallen, and so had wages. Unemployment had swollen the continent's caste of beggars, whose persecutions now ranged from Church bans in Augsburg to the stocks and whipping in England. Beggars in Flanders and Brabant could choose between emigration and the galleys.

As a mathematical discipline, geography offered a source of income, from teaching, from instrument-making. Perhaps even from map-making. By 1534, Mercator had made up his mind.[45]

Short of money, unmarriageable and unpublished, Mercator 'abandoned the study of philosophy' and decided to follow the example set by the inestimable Gemma. With the absolute conviction of a convert, he 'threw his mind into mathematics'.[46]

6

The Mathematical Jewel and
Other Suitable Tools

Back in Louvain by the end of 1534,[1] Mercator presented himself at Gemma's lectures on the 'theories of the planets'.[2] The eager 22-year-old failed to record how long it took him to realize that he was wasting his time and money.

Ill prepared by his university curriculum, Mercator found that Gemma's teaching passed over his head. Later, Mercator recalled that the lectures had borne 'little benefit' since he had 'never studied geometry before'.[3] But Gemma clearly took pity on his student, and provided Mercator with a reading list.

Days of esoteric contemplation over, Mercator adopted the principles of Juan Luis Vives: 'A tool should be acquired quickly,' Vives had advised, 'and no more care should be given to this preparation than is suitable for carrying out the task.'[4] Foregoing food and sleep,[5] Mercator restricted himself to the subjects which applied to his ultimate goal. 'I have', he recalled, 'completely subordinated mathematics to cosmography.'[6] His aim was less to attain 'the level of perfection'[7] than to master that which was sufficient to 'analyse the principles of geography, the disposition, dimensions and organization of the whole machine of the world'.[8] No time or energy would be wasted. 'I chose from these subjects', he wrote, 'only what I considered necessary for my needs.'[9]

Some elements of cosmography proved easier to grasp than others. In astronomy, he concentrated on 'the apportionment of worlds, the proportions of movements, the distances and sizes of stars',[10] but admitted to mastering these 'with difficulty'.[11]

In geometry, Mercator studied 'only that which allowed me to measure the positions of places, to draw chorographical [regional] maps, to establish the dimensions of regions, the distances and size of celestial bodies'.[12] On Gemma's advice, Mercator 'studied at home'[13] the *Elementale*

Geometricum of Joannes Vögelin, referring to the Frieslander whenever he found himself stuck. (Vögelin was assimilated 'very quickly',[14] a result Mercator ascribed to the author's 'simple method'.)[15] Having 'acquired the rudiments of geometry without much difficulty',[16] Mercator then tackled the *Elements* of Euclid, mastering the first six books 'within a few days'.[17] Again, he studied selectively, isolating only the Euclidian propositions 'which appeared to be some use to my profession'.[18] Among these were the definitions and properties of lines. For practical applications of mathematical theory, Mercator turned to works by the brilliant young French mathematician, Oronce Fine.[19]

Vives had been right. With the principles of mathematics committed to memory, 'the way forward ... for the understanding of matters concerning the sphere' suddenly cleared and the 'understanding and evaluation of the movements of the planets' was bathed in 'a surplus of light'.[20]

At last, Mercator could see where he was going. Not only that, but he quickly discovered that by applying mathematics 'to the composition of the universe', some 'extremely agreeable *contemplationes*'[21] emerged.

The conversion of Mercator to a mathematician included the adoption of a new 'mathematical' style of writing. *Cancelleresca* had been imported from Italy, where copyists who had grown frustrated with the multiple pen-lifts demanded by separate, upright, roman letters, had created a cursive derivative, with linked letters, a rightward lean and olivary curves. Fast and informal, the new hand had spread from the chancelleries of the Vatican to those of Florence, Ferrara and Venice, where it was converted to type by Aldus Manutius. It would take twenty years for the new sloping hand to reach the Low Countries, and when it did, in 1522, the first printer to try it was Dirk Martens of Louvain.

By the time Mercator had left university, Italian *cancelleresca* had become the calligraphic signature of humanism, but to the Faculties of Law and Theology it was a frivolous distraction. Erasmus – who had long since forsaken spiky secretary hand for flowing *literae latinae* – had slipped a sub-treatise on handwriting into his hugely popular book *De recta pronuntiatione*. To Louvain's famous humanist, *cancelleresca* was not only the clearest and quickest form of handwriting, but the *sole* form of lettering suited to the Latin language: 'Write a speech of Cicero's in gothic letters', argued Erasmus in *De recta*, 'and even Cicero will seem uneducated and barbarous.'[22] Invoking St Paul as the unofficial patron of clear cursive, Erasmus reminded his readers that the apostle of the

Gentiles had 'sent his Letter to the Galatians all in his own hand-writing',[23] a biblical precedent reiterated by Dirk Martens, who had chosen the *Epistles of St Paul* for his first demonstration of sloping type. Other controversial advocates of northern *cancelleresca* included Cornelius Grapheus, the brother of Gemma's publisher. A Latin poet, an admirer of Luther and a friend of Erasmus, Dürer and More, Cornelius wrote in a clear, relaxed cursive which became even more eloquent after his incarceration by the Inquisition in 1523.[24] Printers also grasped the advantages of *cancelleresca*, using a type-set version to complement upright roman type. Shorter passages such as introductions and quotations could now be separated from the main body of roman text.

For the measurers of Louvain, *cancelleresca* was the script of the future. Its clarity, compression and openness to decorative flourishes made it ideal for instrument notation, maps and globes. Apart from its visual value, *cancelleresca* had a particular appeal among men who worked in the immutable world of mathematics. Despite its apparent fluidity, *cancelleresca* was constructed along mathematical principles, with precisely defined rules concerning the form of each letter and the angle of slope. (When Mercator came to write his own guide to *cancelleresca*, he would stipulate that principal strokes should slope at 3.75 degrees from the vertical.)

Gemma had been one of the first in Louvain to have seen the cursive script compressed onto a map, for Apian's 1524 edition of his *Cosmographicus liber* had employed a tentative *cancelleresca* for regional names on a map of Greece, leaving 'GRECIA' distinguished by capitals. A close look at the most successful lettering on the map – the word 'Macedonia' – would have instantly revealed to Gemma the space-saving clarity of cursive lettering.

Outshining Apian's example of geographical lettering were two books which must have arrived in Louvain by the early 1530s. The first had been published in Rome in 1527. *Antiquae Urbis cum Regionibus Simulachrum* by Marcus Fabius Calvus included nineteen city plans which had been given a dramatic sense of spatial order by the systematic use of roman capitals for major features and italic for minor. On the fourth map for example, the capitalized 'AMPHITHEATRUM' was categorically separated from the diminutive, precisely lettered '*castra peregrina*' next door.

Equally striking was Benedetto Bordone's *Isolario*, first published in Venice in 1528.[25] Originally a miniaturist and illustrator from Padua, Bordone had published a highly successful book of island maps, which he labelled to great effect using – like Calvo – a combination of roman capitals and flowing cursive. Bordone was thinking along the same lines as Gemma, for his miniaturist mind had also discovered the pleasures of 'a world map in the round form of a ball, a new thing, and moreover of marvellous utility'.[26] With its susceptibility to compression, italic was ideal for lettering globes, where space was even more limited than it was on maps.

Around this time too, the new lettering was creeping into the work of two of the continent's foremost cosmographers. In 1531, Sebastian Münster published his *Horologiorum*, a treatise on sun-dials, with complex illustrations whose detailed annotation was achieved through tight cursive lettering. Two years later Gemma produced the second edition of Apian's cosmography for the printer Johannes Grapheus. It was an edition which displayed several startling modifications: the second volvelle, and the celestial hemisphere of the *Cosmographicus liber*, had been re-engraved with cursive lettering, and the introduction was now set in Aldine-inspired type, as was the introduction of the book's second part, *Libellus de locorum*. Indeed, every illustration in Gemma's surveying treatise was lettered in the new cursive. The lettering was rudimentary, and Grapheus – whose controversial brother had been writing in exquisite italic for years – must have known that there was room for improvement.

Apart perhaps from the printed cursive of Bordone and Calvo, none of the map lettering Mercator can have seen came close to realizing the true potential of *cancelleresca* as a formal, mathematically precise complement to capitals. In acquiring the new hand, Mercator had two likely sources of instruction: the writing manuals of Ludovico Cicentino (Arrighi) and Giovanniantonio Tagliente, both of which had been published in the 1520s. A third book, an anthology by Ugo da Carpi, may also have been Mercator's primer. And two of these authors may already have been known to Mercator through other interests: Tagliente had once illustrated a book on surveying (with a title page showing Ptolemy wielding a pair of compasses at a globe), while da Carpi's *Thesauro de Scrittori* contained several pages of mathematical tables.

How long did it take Mercator to learn the new hand? Arrighi reckoned that a novice could learn his *cancelleresca* style in 'a few days'.[27] Erasmus had pronounced 'that in handwriting style and practice bring speed but that speed and practice never bring style'. Both were correct. Mercator might have picked up the mechanics of italics in a few days, but it would be many years before he ceased improving.

*   *   *

Mercator was quickly repaid for his propaedeutic rigour. The university granted him permission to give students private tuition in mathematics, and he was soon supplementing his teaching fees by constructing mathematical instruments. These included 'spheres and astrolabes ... astronomer's rings, and similar apparatus in bronze'.[28]

Of the Louvain craftsmen from whom Mercator may have learned instrument-making, the most likely was the engraver and globe-maker Gaspar Van der Heyden.[29] An accomplished artist who had begun his career making copper seals and silver chains for the city authorities, Van der Heyden had constructed for Monachus the first printed terrestrial globe in the Low Countries.

Sawing, filing, drilling, polishing, engraving, Mercator's hands blackened with oxidized copper and brass as he sweated to turn lumps of dull metal into intricate, shining instruments. The glitter of golden dust cast an aura over the measurers of Louvain. These men were making tools that could test the truth of the cosmos. Gemma's circle of mathematical practitioners operated beyond the bounds of faculty and theology. The abstract hardware being assembled in Van der Heyden's *atelier* was intended for revolutionary applications.

Some of the devices taking shape on Van der Heyden's benches had never been seen before. Others, like the astrolabe, had been around for a long time. But the astrolabe had proved an enduring, versatile tool: it could be used for determining latitude, for compass bearings, for telling the time, for calculating the heights of mountains and for compiling horoscopes. A mariner's version of 'the mathematical jewel'[30] had sailed with Columbus.

One of the prototypes to emerge from Van der Heyden's workshop was Gemma's 'astronomer's ring'. Described by Gemma in a short paper which he completed in February 1534, this was not the simplest of instruments to construct, and consisted of three rings fitted together to form a hollow sphere that could nest in a cupped hand.[31] The inner ring had to revolve within the outer two, and each ring had to be marked off with scales such as degrees, hours, months and weeks, tangents and signs of the zodiac. Assembled, Gemma's ring could perform any angle-measuring task.

The ring, admitted Gemma, was 'not entirely a discovery of mine'.[32] He did however claim to have 'augmented the ring so much that from simply showing the hours of the day and the four directions it now rivals whatever mathematical instruments you will'.[33] The proposals of others 'here and there in long treatises on quadrants, cylinders, and astrolabes is brought together into this single ring'.[34] The ring was another example of Gemma's ability to create a unifying concept from pre-existing fields of knowledge. Gemma's view, that it was 'an estimable thing to add to discoveries and disseminate them',[35] became one of Mercator's guiding principles.

Of the varied instruments which were shaped by Mercator's hands at this time, one device in particular would serve to make his name.

Although it had been Strabo who first identified the globe as the best method of representing the known world, the use of globes over the intervening millennia had been restricted by the cost and expertise required to reproduce the earth's features on a sphere. Sometimes the geography had been engraved on metal balls, others times painted onto wood, or paper. No two globes were entirely alike. Printing made possible multiple copies of identical, less expensive globes. Again, German ingenuity had paved the way, Waldseemüller's printed globe of 1507 being followed by Schöner's celestial and terrestrial pairs of printed globes. It was 'insufficient copies of the work of Schöner'[36] that had prompted Jean Carondelet, a privy councillor and special adviser to Queen Maria of Hungary, the regent of the Low Countries, to encourage Van der Heyden to produce a printed celestial globe. Carondelet had referred to the Louvain goldsmith as 'the most capable engraver of world globes known to him.'[37]

Among the instruments being fabricated in Van der Heyden's workshop, globes were the most complicated, most prestigious and most expensive to construct. Printed globes – whether celestial or terrestrial – were painstakingly assembled from papier mâché hemispheres, which had to be weighted then glued then coated with plaster to form a perfect sphere before the printed paper segments, or 'gores', could be pasted onto the exterior. The geographical surface of the globe was coloured by hand, then varnished. The finished globe had to be mounted on a stand of some complexity. Typically, a terrestrial globe would be equipped with a horizon ring, and a graduated meridian ring which could be used to set the globe to a specific geographical latitude. On the meridian ring would be an 'hour circle' and a pointer. Oblique to the meridian ring would be a *circulis positionis* which could be used to locate places on the globe. The globe might also be provided with a spherical set square, a *gnomon sphericum*, which could be used for determining latitude.

By far the most complex element of a printed globe was the engraving, printing and pasting of the paper gores to the exterior of the sphere.

Pioneered in Germany by Waldseemüller and Schöner (and described in a 1527 treatise by Henricus Glareanus), the technique required the globe's cosmographical information to be engraved onto twelve flat, separate, tapered segments, which would (theoretically) fit together when applied to the spherical surface of the globe. Since features such as coastlines and rivers would inevitably cross one or more gores, the potential for misalignment was great. Engravers learned to confine the lettering of a place to one gore. Once printed, the gores would be carefully matched and pasted to the globe's surface, a process which required judicious stretching to avoid misalignment and wrinkling.

As Strabo had pointed out, no other form of terrestrial representation could compare with a reduced, spherical model of the earth. And as the ancient, symmetrical form of the world dissolved into an apparently random distribution of land and ocean, those whose interest it was to know the earth's geography were turning to the most sophisticated model available – the globe. In the Low Countries, the leaders in this field were Gemma and Van der Heyden.

Since 1531, the two men had been talking of a printed terrestrial globe to supersede the model that they had completed in 1529. The pair had protected their proposed work by taking out a series of Imperial charters, first for ten years, and then a revised version for four years, later extended to ten. The Imperial wording of the final version, dated December 1535, referred to their intention 'to publish a globe or sphere of the whole world, on which the recently discovered islands and lands will be added, and which will be improved and enriched and more beautiful than their earlier [globe]'.[38] The charter also referred to a second, celestial globe, 'for the general use of enthusiasts'.[39]

Emperor Charles V ('by the mercy of Divine Grace, Emperor of the Holy Roman Empire and King of Germany, Spain, both Sicilies, Jerusalem, etc., Arch-Duke of Austria, Duke of Burgundy, Brabant, etc., Count of Hapsburg, Flanders, Tyrol, Artois, etc.')[40] had high expectations. The new terrestrial globe would be out of the ordinary, a symbol of the Empire's technical prowess and temporal reach. The globe – proclaimed the charter – would 'make mathematics more illustrious ... keep alive the memory of old kingdoms and events and ... make known to coming generations our time and our realm in which (by the clemency of God) very many islands and areas unknown in earlier centuries have been discovered and more than just a few of which, happily, are being introduced to the Christian religion ...'[41] Gemma's new globe would define the Empire as earthly custodian of the past and future.

In return for moulding the earth to an Imperial standard, Charles would protect his two cosmographers from copyists. A fine of ten silver marks would be levied against absolutely anyone ('all printers and sellers of books and all directors, employees and personnel in the book trade as well as everyone else, subject to us and our realm')[42] who published without permission any globes or associated manuals bearing the name of Gemma or Van der Heyden. Since the charter specified that copying was prohibited 'in the same or any other form',[43] the two globe-makers were affirming intellectual copyright over their forthcoming material. Gemma and Van der Heyden could exert their privilege until February 1539.

To fulfil the charter's wording, Gemma had to construct a sphere the likes of which nobody in the court of King Charles would have seen before; a globe of exquisite beauty, crammed with the latest geographical data: new coastlines, hundreds of place-names, legends, geographic descriptions, principal stars...

Such a globe could not be printed from wood-blocks. They were just too crude; too inflexible. The recent works of Monachus and Apian had both suffered from chunky, imprecise capital lettering, over thick line work, ink blots and hatching that looked like badly stacked straw. The new globe would have to be engraved on copper. A relatively new medium, copper plates were very expensive but produced remarkable prints. Larger plates could be used (which also made copper an ideal medium for maps) and its malleable surface permitted smoother curves and finer lines to be incised. Engraving errors were much easier to correct on copper, by 'hammering out' the mistake from behind, and then repolishing the surface. Copper also facilitated smaller, neater lettering. Were Van der Heyden and Gemma to succeed in producing a copper-engraved printed globe, they could claim a first – and so could the emperor.

But the emperor's expectations (and Gemma's ambition) exceeded Van der Heyden's skills. By 1535, Gemma had altered the original charter twice, and still not delivered the 'improved', 'enriched', 'more beautiful' terrestrial globe. Although Van der Heyden was the most experienced constructor of globes in the Low Countries, and although Gemma had the mathematical expertise, their combined talents were insufficient for the project in mind.

Maybe they had been delayed by geographical uncertainty, Gemma unsure whether to repeat his assertion that 'America is not connected to Asia'. Maybe they were waiting for an engraver who had mastered Italian *cancelleresca*. Maybe they were waiting for a patron who could authenticate a Magellenic globe. Whatever the reason for the delay, the two men had accepted by the end of 1535 that the globe required a third collaborator. That collaborator was Gerard Mercator.

7

Neither Known Nor Explored

The three names destined to be engraved onto the proposed globe were joined by a fourth.

Authoritative, well-travelled, courtly and erudite, Maximilianus Transylvanus was best known in the world of exploration and cartography for having written *De Moluccis Insulis*, the report on the first circumnavigation of the world. By the mid 1530s, Transylvanus was an old man, retired to his palace in Brussels after a lifetime of Imperial diplomacy. For Mercator, this would have been a memorable encounter: Transylvanus had actually met the survivors of Magellan's expedition; talked to them; taken down their tales and observations.

From this respected member of the Imperial chancellery, the young engraver would have heard first hand how Magellan's expedition had ended. How, on 6 September 1522, the battered caravel *Victoria* had sailed out of the Atlantic and up the river Guadalquivir towards Seville bearing the eighteen emaciated survivors of a fleet that had once numbered five ships and 265 men. *Victoria*'s foretopmast had gone. Her timbers weeped for want of caulking. Her sails were patched and stained. At a fork in the river known as La Horcada, the caravel had been met by a skiff from Seville, and with extra hands on the ropes, she had crept around the river's final bend, beneath the octagonal Torre del Oro to the city waterfront. Watched by a curious crowd, the spectral caravel had fired her bombards across the greasy water. The following day, Sebastian Elcano had led his diseased men along the Mole. Barefoot, limping and bent, each man had carried a flickering candle to the Franciscan convent. Before the shrine of Santa Maria de la Vitoria they had prayed for the soul of Magellan, the captain general who had died with a spear through his throat. They had prayed for their lost companions; for their emperor. Then they had crossed the pontoon bridge to Seville's cathedral and the chapel of Santa Maria de Antigua, 'The Ancient One'. Before the altar of the Virgen de la Rosa, they had fallen once again to their knees.

Radiantly gold, with a rose in her right hand and the Christ Child in her left, the Madonna had gazed down at the redeemed seamen. Five days later, Elcano had been summoned by the emperor to the Spanish court at Valladolid. With Albo the pilot and Bustamente the barber, the Basque had taken the long road north to the capital of Old Castile. At the court of inquiry, Elcano and his gaunt companions had described how they had eventually found the hidden straits at the tip of Vespucci's 'America', and how they had sailed for ninety-eight blistering days to reach the furthest side of that vast new ocean. The court heard of battles and shipwrecks, and of Magellan dying to save his men in the bloody surf of a faraway island. And they heard of the Moluccas, the Spice Islands.

Among those at Valladolid who had listened with awe was Maximilianus Transylvanus, then a secretary to the emperor. Already a veteran of Imperial assignments to England and Italy, Flanders and Germany, the diplomat interviewed Elcano, Albo and Bustamente. Transylvanus' account, *De Moluccis Insulis*, had been printed in Cologne in 1523, the year that a cobbler's son from nearby Gangelt had passed his eleventh birthday.

Twelve years later, the elderly diplomat and the cobbler's son had been brought together by a globe. Before he died, Transylvanus had information to bequeath, information that he'd gathered years earlier in Valladolid from a boatswain, a pilot and a barber. Mercator was to share the engraving of the globe with Van der Heyden. The combination of copper-plate printing and Mercator's *cancelleresca* promised a feat of cosmographical compression new to globes.

But before they could commit burins – engraving tools – to copper, Gemma had to corroborate his geography. This exhaustive exercise in correlation required the mathematician to pull together reports and printed accounts, maps and Ptolemaic coordinates. Achieving a best-fit world had commercial implications, for the more accurately the continental outlines could be fixed, the longer the new globe would stay in print.

Only one continent attracted any level of consensus. The Portuguese caravels that had been plying the trade routes to India for forty years had created a clear image of the African coast, an image which had been refined by Ruysch's copper-plate engraving of 1506 and then by Waldseemüller's wood-block of 1513. No longer did Africa extend parallel-sided to meet a contiguous 'Terra incognita' girdling the southern reaches of the planet. Ptolemy's super-continent had been severed at around 35 degrees south, and tapered. Waldseemüller had been able to mark over three hundred coastal locations around the perimeter of the continent.

So Gemma's Africa would assume its revised form, with a coast labelled by the Portuguese and an interior populated by the ancient tribes of Ptolemy: the Trogloditae by the mountains of the east coast; the Nigritae and Garamantes of the north. The identity of the two African islands – Madagascar and Zanzibar – would stay true to the great nomenclator of the east, Marco Polo. Illustrating that their learning spanned the gulf between the authority of the ancients and the evidence of modern navigators, the globe-makers added to Madagascar the label 'vel S. Laurety', the name by which the island was known to more recent, Portuguese visitors.

Every other continent posed Gemma with irreconcilable evidence.

In Europe there was a problem with the northern and southern shores. Only three years earlier, the Bavarian cartographer Jacob Ziegler had published a new map of Scandinavia which depicted an extra gulf beside 'Finlandia'. The same map portrayed a non-Ptolemaic Scotland which protruded northwards from England, instead of taking the traditional kink eastwards towards the Danish coast. Although Ruysch had shown a similarly aligned Scotland as long ago as 1507, Gemma stayed with Waldseemüller and Apian – and Ptolemy – and showed a right-angled Scotland. But he liked Ziegler's new gulf.

Conveniently, Europe's southern limit had always been defined by the Mediterranean, but the sea of the ancients had been subject to controversial shrinkage. The recent appearance of vast landmasses between Europe and the Indies had meant that Europe and Asia had to be compressed in width if there was to be enough space left on the globe for the new continent. Since the time of Ptolemy, the Mediterranean had occupied a width which totalled over 60 degrees of the earth's longitude, or one-sixth of the circumference of the globe at that latitude. Oronce Fine, whose 1531 map marked a 56-degree-wide Mediterranean, was one of several cartographers who were convinced that Europe occupied too much of the earth's circumference. But a consequence of reducing the width of the Mediterranean was that Spain became smaller, a development which may not have been greeted with enthusiasm at the court of Charles v. In Paris, Fine had no need to concern himself with Imperial sensitivities; the Frenchman had reduced Spain to less than half the width of Sylvanus' Ptolemaic version twenty years earlier. Gemma felt unable to disagree with Ptolemy.

If Africa formed the meridian of cognizance, the longitudes to east and west were meridians of increasing speculation. Work on Asia was still in progress. Now that the Portuguese had followed Arab traders to the coast of India, this western peninsula of Asia had solidified as a triangular spit tapering to the equator. East of here, the Asian mainland was being redrawn every year, although few reports had been as influential as that of *De Moluccis Insulis*, which had made it impossible for Ptolemy's great limb to extend half-way to the pole from south-east Asia. Like Monachus and Fine, Gemma lopped off the limb to leave a stump. Transylvanus had also confirmed the non-existence of the world's biggest island: 'For where Ptolemy, Pliny, and other geographers placed Taprobane', Transylvanus had written, 'there is now no island which can possibly be identified with it.'[1] As it happened, Ruysch (relying on the Portuguese) had got there first. In 1507, he had towed Taprobane across the Indian Ocean and moored it off Malacca, reduced in scale to a floating landmass a little larger than Madagascar. Monachus, Waldseemüller and Fine had followed. And so did Gemma.

The southern pole had been variously covered with a continent bigger than Europe, and by ocean. The unexplored landmass that Monachus had marked had probably been derived from the 'Brasilia regio' that Schöner had thought lay south of America. Apian had followed Schöner, with a landmass suspended between Magellan's straits and the southern pole. Then Fine placed his vast heart-shaped 'Terra Australis' over the pole, complete with several mountain ranges and a 'Brasielie regio' on the part of his continent that lay south of India. Although Waldseemüller and Münster had failed to acknowledge a southern polar continent, Gemma followed Monachus and Fine, and sketched in a vast landmass, parted from the tip of America by the tight straits that Magellan had found. In repeating the Antarctic continent, Gemma was rejecting the views expressed by Elcano's men, who 'thought that there was no continent, but only islands, as they occasionally heard on that side the reverberation and roar of the sea at a more distant part of the coast'.[2] Without such a continent to balance the weight of the landmasses of the northern hemisphere, the world would fall to disequilibrium.

America itself had by now settled as a fairly reliable presence. From Magellan's straits, the 'New World' widened in a roughly triangular fashion to its broadest northern limit in the latitude some way north of the equator. And it was up here that the problems for European cosmographers were most taxing.

Monachus and Fine understood America and Asia to be a contiguous landmass (although Monachus had divided the two by a strait just wide enough for the emperor's ships to reach the Indies). Such an interpretation kept Asia where Columbus had believed it to be, just a relatively short voyage west of Europe.

The Germans thought otherwise. Waldseemüller and then Apian and

Münster had prised the New World away from Asia to create a vast ocean floated upon by Marco Polo's 'Cipangu' (or in Apian's case, 'Zipangri').[3] All three cosmographers supported the notion of a sea route past Waldseemüller's 'Terra ulteri incognita' (or 'Terra de Cuba' as it had become on Apian and Münster's more recent maps) to the Indies. No European court wished this to be true more than that of Charles V. Having crossed the Atlantic to Isabella, Spanish ships could strike north past Cuba and then cut through the undiscovered north-west passage to Cipangu and the Moluccas.

Gemma allied himself with the Germans. Their new globe would show a slim continent that was no more than 30 degrees wide, separated from the Arctic landmass by a narrow north-west passage, and from Asia by a tapering ocean. Their conviction was based upon accounts unknown to – or ignored by – Monachus when he published his terrestrial globe with Van der Heyden a decade earlier. The accounts came from the Corte Reals and Cabots.

When John Cabot had sailed west from Bristol in 1497, he had sighted land which he took to be Asia. Twelve years later, his son Sebastian had set off to the north-west to search for the mouth of the anticipated passage. At around the same time – in 1501 and 1502 – a pair of Portuguese brothers, Gaspar and Miguel Corte Real, also found land on the far side of the Western Ocean. Both brothers disappeared. (A third brother, Vasco Annes, was refused permission by the king to search for Gaspar and Miguel.) The Louvain globe would indicate that the Corte Reals had reached the coast of Asia.

Finally there was the question of the north pole. Where Fine (and Ruysch before him) had depicted four islands orbiting an aquatic north pole, Gemma insisted that the entire polar region was a vast landmass connected to Asia and Europe by two arms which surrounded a gigantic inland 'Mare glaciale'. The image of those four islands would return to Mercator.

As the cobbler's son picked up his burin to cut the names of new lands onto the globe, he knew that he was signing up to a world removed from the boot-sucking mud of his boyhood.

The sheer number of place-names which Mercator fitted onto the gores must have amazed Gemma. Where Waldseemüller had managed to fit fifty American locations onto a wall-map as wide as a man was tall, Mercator reduced sixty onto a sphere whose diameter was two hand-spans.[4] By the time he had finished, better-known coasts like those of 'Bresilia' and Africa were so closely packed with names that space ceased to exist between them. All these names had to be squeezed onto the land-surfaces, so that they would not obscure the more important seaward side of the coastline. (The corollary of a clear seaboard was an overcrowded hinterland, where the name of a port might find itself competing for space with a river, a mountain or a star.)

To help the globe's eventual users distinguish between the various types of information, Mercator established a calligraphic convention: capital letters for regions; roman for place-names and cursive for star-names and geographical descriptions. In Europe, where there was no space to fit the names of major towns and cities, Mercator engraved numbers which referred the reader to keys in the south Atlantic.

Various legends emphasized the globe's use of the latest reports. Against the Arctic gateway to the Moluccas, Mercator engraved a ref-erence to the Cabots: 'Arctic straits or straits of the three brothers, by which the Portuguese tried to go to the East to travel to the Indies and Molucca'.[5] Thus, the north-west passage existed because navigators had *tried* to find it. (And it had a printed existence because globe-makers had tried to plot it.) To substantiate their conviction that the Corte Reals had pushed past the Arctic to Asia, the globe-makers created a stubby peninsula part-way down the Asian coast, which Mercator labelled 'Promontorium Corterealis'.

The Louvain globe-makers also confirmed that there was land – as yet unexplored – between Florida and the 'Arctic straits'. In Ziegler's book *Quae Intus*, they had seen that the east coast of Greenland – or 'Terra Bacallaos', 'the land of the cod' – appeared to correspond with this unknown coast, so 'Baccalearum Regio' became the top end of the northern continent. In the absence of a generally accepted name for this new land, they labelled it after Nuño Guzman's recent gold-seeking expedition: 'Hispania major a Nuñno Gusmano devicta anno 1530'.[6]

Gemma and his partners had now given the continent that was Amer-ica's northern neighbour a more specific form than Waldseemüller had managed. But the western seaboard of the continent was completely unknown. There had been no reported landfall between Hispania nova and Baccalearum Regio. Against this vast and very conspicuous zone of absolute conjecture, Mercator placed a disclaimer: 'These coasts,' he engraved, 'from the port of Matonchel[7] to the mouth of the Arctic straits, as well as the interior of the area, are neither known nor have been explored.'[8]

Work on the globe ran into the winter of 1535, then through to the new year.

One of the last jobs to be completed was lettering the decorative cartouches. Having dedicated the globe to Maximilianus Transylvanus, the man who had contributed to some of the globe's most significant data, the three globe-makers appeared in a second cartouche beneath the Tropicus Hybernus. The work had been 'described' and given 'this form' by 'Gemma Frisius, doctor and mathematician ... from various observations made by geographers'.[9] The last part of the cartouche seemed to suggest that it was the team's youngest member who had cut the most copper: 'Gerard Mercator of Rupelmonde engraved it with Gaspard Van der Heyden, from whom the work, a product of extraordinary cost and no less effort, may be purchased.'[10]

Never had anyone in Louvain seen such a globe. Gemma had pinned his continents beneath a net of numbered meridians and parallels. America and its northern appendage were fixed between the Euro–African block and Asia. Antarctica was independent. Anybody used to woodcuts would have been astonished by the clarity and minuteness of Mercator's cursive lettering. It was a pretty globe too, with sailing ships and sea life, a sprinkling of stars and a bunch of Vespucci's over-sexed cannibals inland from Brasilia. The globe also had a thrilling topicality: the New World settlement of 'S. Michaelis' had been established by Pizarro only three years earlier; the reference to 'Nuñno Gusmano' related to the Spaniard's expedition of 1530; and the Imperial eagle that floated over the roofs of Tunis symbolized the city's capture by Charles v in July 1535, a matter of months before the globe's gores were printed.

Mercator's name had been engraved upon the future of mapmaking. But he – and perhaps *only* he – knew that his lettering had been far from perfect. The couple of years he'd had to practise had been insufficient to develop a natural, personal style. He still had trouble with his *m*s and *n*s, which had a habit of leaving their line and of adopting irregular slopes, and he knew that his *f*s and *c*s didn't look quite right.

In putting his name to the globe, Mercator had also committed himself to a particular configuration of God's creation. Mapmakers had always been able to conceal their uncertainties by choosing projections which placed the known world – Europe, Africa and western Asia – in the centre of their sheet. Zones of the unknown were marginalized to the map's peripheries where vague continental presences could be melded into cartouches and decorative borders. A globe-maker could cast no such mists over lesser-known landmasses because a sphere had no periphery.

8

Celestial Maidens

In the two years since his flirtation with philosophy, Mercator had secured an income, a reputation, and a girl from Louvain.

Barbara Schelleken was the daughter of Jan Schelleken and Johanna Switten. Like Mercator, Barbara was one of seven children, and like Mercator, she had lost her father too early. Jan had died in March 1528, a couple of years before Mercator arrived in Louvain. With her sister and five brothers, Barbara had been raised – presumably with some difficulty – by the indefatigable Johanna.[1]

In the Schellekens' home, Mercator refamiliarized himself with family life. He clearly found Barbara delightful company. A girl of 'chaste morals',[2] she was extremely capable about the house and had been born to a mother whose child-rearing record suggested that Barbara too would make an excellent mother.[3] But there was another reason why Mercator may have been drawn across the Schelleken threshold. While Johanna retained an allegiance to the traditions of the Catholic Church,[4] her daughter Barbara shared Mercator's enthusiasm for the Gospel.

To Barbara, Mercator could promise with some degree of certainty that he now had the means to provide for a family.

The couple married at the beginning of September 1536. Mercator was twenty-four. Still a man of slender means, he was nevertheless a tireless worker, devout and loyal, and his name was emblazoned on a work which promised further commissions. After years in paupers' dormitories, he can only have craved the bosom of a family home. In the hard-working daughter of Johanna he not only had an intimate who could share his love and faith but a bride who could cook and mend and bear his children. By December, Barbara was pregnant with the couple's first child.

It was a time when confidences were best kept between bedroom walls.

On the streets of Louvain, a culture of suspicion prevailed. Since the religious persecutions of the 1520s, the learned class had lost some of its enthusiasm for the new doctrines. Surviving dissenters had been driven underground. To Louvain's conservative theologians, Luther was the root of all evil and dissidents of any complexion were regarded as *luther-iaenen*. A careless word, an overheard conversation or a misplaced letter could lead to charges of heresy.

Hazards for Louvain's secret evangelists had become more extreme with the recent return of the university's most feared theologian, Jacobus Latomus. A tiny, chilling, man with thin lips, dark, bagged eyes and a limp, Latomus had been studying then teaching at the university since 1500, but had left to take a post in Cambrai a few months before Mercator matriculated. The champion of the Catholic assault on Luther, he was also a vitriolic opponent of Erasmus, who had once referred to the theologian as 'Hephestion', the *Iliad*'s crippled metal-smith with the comical limp.

Almost as prominent as Latomus was Ruard Tapper, a theological professor who had once taught Aristotelian physics and logic at Louvain. Threatening, ill-mannered, badly dressed and plagued by a speech impediment, Tapper had risen to be president of the College of the Holy Ghost, a canon of St Pierre's and most recently, chancellor of the university. A disgruntled German geographer had once caricatured Tapper – who had been born in Enkhuizen on the furthest edge of Holland – as a slovenly oaf from 'Paphlagonia', the end of the world. A feared theological assessor with over a decade's experience of heresy trials, Tapper's learning could be called upon by the procurer-general, Pierre Dufief, whose reputation for extracting evidence had been gained by scrutinizing letters for heretical passages, sitting at interrogations, passing judgement, overseeing torture and attending personally to burnings.

Mercator had good reason to avoid Dufief's attention. Most recently, the procurer-general had snared the Englishman, William Tyndale. The year before Mercator married, the translator of the Lutheran Bible to English had been living in Antwerp's English House when he was betrayed to the court in Brussels by one Henry Phillips. Tyndale was arrested and taken to the castle of Vilvoorde, a gloomy 160-year-old stronghold between Louvain and Brussels. In Louvain the news of his arrest spread quickly. Tyndale's books and possessions were confiscated and sold to pay for his upkeep in prison. While Mercator had been busy with Gemma's globe, Latomus and Tapper had sat at the commission of seventeen, under Dufief. Relishing the dialectical contest, the two theologians had tried to trap Tyndale on the central Lutheran issue that

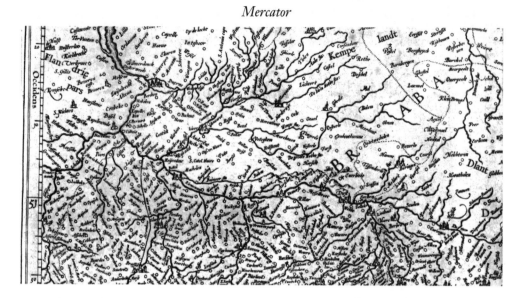

Fig. 10. Mercator's homeland for the first half of his life: A detail from his untitled map of 1585, showing Rupelmonde and Antwerp (upper left), Louvain (lower centre) and Brussels (lower left). Mechelen can be seen midway between Antwerp and Louvain. (Private collection)

salvation could be achieved by faith alone, not works. Latomus kept a record of the debate, which he later turned into three books. In his cell, Tyndale too wrote a book. Close by in Louvain, Mercator and his humanist friends would have been aware of the arguments as a stream of lawyers, theologians and friars rode out to Vilvoorde to substantiate the case against Tyndale. A few weeks after Mercator's marriage, Tyndale was taken from his cell, strangled and burned.

Erasmus had also died that summer, and in dying, he handed his humanist mantle to a generation whose methods of confrontation would be less direct.

Meanwhile, the measurers of Louvain pursued their private investigations. At the centre of this group was Gemma, who had been awarded his licentiate degree in medicine in 1536.

In the autumn, Andreas Vesalius – the intense young student of medicine who had shared the Castle with Mercator – reappeared in Louvain after three years in Paris, a sojourn he had been forced to abandon when Charles v attacked France in July. As inspired by Hippocrates and Galen, as Mercator was by Ptolemy, Vesalius had already begun his own exploration of the human body, and was unenthusiastic about a return to conservative Louvain. In the medical school in Paris, Vesalius had witnessed a couple of dissections, and on the way back from France he had participated in his first autopsy. Although the dissection (of an eighteen-year-old girl who was thought to have been poisoned) had been 'undertaken by a thoroughly unskilled barber',[5] Vesalius' first hands-on encounter with the internal world of the human proved revelatory. In methodically mapping the structure of the body before him, Vesalius had become an anatomist. Back in Louvain, Vesalius fell in with 'that celebrated physician and mathematician Gemma Frisius', and the two of them went 'looking for bones'.[6] Outside the city walls 'where the executed criminals are usually placed along the country roads', they came across a collection of corpses, one of which had been 'partially burned and roasted over a fire of straw and then bound to a stake'.[7] As Vesalius later enthused, this was an 'unexpected but welcome opportunity' since 'the bones were entirely bare and held together only by the ligaments so that merely the origins and insertions of the muscles had been preserved'.[8] Helped by Gemma, Vesalius scaled the stake and removed the femur from the hipbone, returning over successive nights to smuggle the remaining parts of the corpse into the city, where he boiled the bones and reconstructed the skeleton so quickly that he was able to convince everyone that he had 'brought it from Paris'.[9]

Gemma was becoming increasingly absorbed by medicine, and had recently become a father too, so it must have been a relief to Mercator when the Frieslander reconvened the team to construct the celestial partner for their terrestrial globe. Gemma had his own, astrological reasons for modelling the heavens. Besides being used for fixing the positions of new stars, for determining the latitude of a terrestrial location, and for estimating the length of twilight, a celestial globe was also marked off with the signs of the zodiac, and could anticipate the culmination of a star. Celestial globes were essential for selecting the most opportune time for taking decisions and for embarking on a course of action. They were not, however, true representations of the heavens, in the sense that a terrestrial globe was a model of the earth: because it was impractical to model the heavens in such a way that a user could view them as they appeared from a vantage point on earth, modellers plotted the constellations on an artificial sphere. The user thus peered in at the heavens from a viewpoint beyond the universe.

Gemma sought the best source, then repackaged it, with style. That source was the pair of woodcut planispheres of the northern and southern celestial hemispheres, published in 1515 by Albrecht Dürer and the mathematicians Johann Stabius and Conrad Heinfogel. The two planispheres were the first printed maps of the stars. Not only were Dürer's constellation figures beautifully executed, but the celestial mapping was accurate too.

Gemma and his team plotted and then engraved the flat celestial maps as a set of globe gores. The constellation figures were copied, with minor alterations. At the end of the river of stars known as Eridanus, an awkward space on Dürer's original gained a naked maiden who had floated out of a recent edition of *Hyginus fabularum liber.*[10] Winningly she swam, arm over arm, towards the southern pole. As gratuitous as she was gorgeous, Gemma's heavenly bathing beauty was unlikely to harm the globe's sales. Gemma also introduced symbols to represent the six magnitudes of stars described in Ptolemy's *Almagest.*

A duplication job requiring minimal lettering, the celestial globe was completed by 1537. The result was a stunning, extra-dimensional Dürer. The names of the three collaborators were engraved onto the globe, but this time no distinction was made between their roles. Stating that the globe had been made 'by Gemma Frisius, doctor and mathematician, Gaspard Van der Heyden and Gerard Mercator of Rupelmonde'[11] elevated Mercator beyond the mere engraving role he'd played on the terrestrial globe – even though it was probably unjustified: since this was the second of a pair of globes, there was no need to repeat the respective roles, but for those who saw only the celestial globe, the effect of this pared inscription was to promote Mercator and Van der Heyden to a status equal to Gemma's. Mercator had joined the stars.

Barbara gave birth to the couple's first child, a boy, on 31 August 1537. They called him Arnold.

For Mercator, these were days of heady optimism; of fertility, boldness, decisiveness, commitment. Inspired by his 'happy progress in the mathematical arts', he 'set his mind to engraving maps'.[12]

Mercator had reason to feel confident. He had trained himself in mathematics, was familiar with reconciling conflicting geographic sources and had proven himself as an engraver and as a master of cursive lettering. Individually, these aptitudes were not uncommon within the extraordinarily creative radius of Antwerp, but collectively wielded by one man they represented the full complement of mapmaking skills.

And the demand for maps was booming. The increasing reliability of maps and their mass distribution through printing was enabling soldiers, politicians, princes, merchants and navigators to scrutinize regions without stepping outdoors. Situation reviews and courses of action could be tested on a sheet of printed paper. To the up-and-coming Friesland jurist Viglius van Aytta, '*Geographicus chartis*'[13] enabled him 'to know places, regions and people',[14] commenting to his friend Hadrianus Marius that cartographical knowledge 'may prove to be of certain importance should one of us be called upon for this republic'.[15] Among the maps he was curious to obtain were those covering the region then troubled by the rise of Anabaptism. An outspoken opponent of religious liberties, Viglius could observe on his maps the spread of the 'hearth gods and tutelar divinities'[16] which he so passionately detested. In a letter to Hadrianus (to whom he'd offered a Westphalian ham in reward for any maps his friend could send his way), Viglius boasted that he owned 'a map presumably drawn by a sailor ... that I think that Cranevelt shows by no means such a copy'.[17]

In daring to compare himself with Frans van Cranevelt, Viglius was sharing the aspiration of many young humanists in the Low Countries. Long considered the most eloquent man in Louvain, Cranevelt was also a knowledgeable collector of cartography. Now living with his books and maps in a large house in Mechelen, Cranevelt had been educated at Cologne and Louvain and had sat on the Grand Council of Mechelen since 1522. A brilliant linguist, he had been a good friend to Erasmus, to Vives, and to the recently beheaded Thomas More. Cranevelt's long correspondence – in fine italic hand – with Erasmus had covered various contentious issues, from the recovery of Erasmus' imperial pension to Luther's marriage. He had also sided with Erasmus against Louvain's theologians, a favour Erasmus returned by including the councillor in his will.[18] Humanism and love of learning had also won Cranevelt several dedications on the title pages of books.

Mercator may have been introduced to Cranevelt through his old university acquaintance Antoine Perronet, recently appointed Bishop of Arras. Almost certainly, Mercator had met Perronet's father, Nicholas Perronet de Granvelle, Charles v's first councillor and keeper of the seals, who was also a friend of Viglius and of Cranevelt. And Mercator's useful contacts in court also included Adrien Amerot,[19] a grammarian who had won the admiration of Erasmus while teaching at Louvain's Lily College. Amerot had been trusted by Granvelle to tutor his son, Antoine Perronet, and he would soon extend his support to Mercator.

Cranevelt could open many doors for Mercator. A friend of the university rector, Pierre de Corte, and of Gemma too, Cranevelt's court connections included a close friendship with the Polish ambassador, Johannes Dantiscus, and with Nicolaus Olah (Olaus), the long-standing Transylvanian counsellor to the Emperor's sister, the regent of the Low Countries, Queen Maria of Hungary. (Homesick himself, Olah had been one of those urging Erasmus to return from exile to Brabant.) For the last three years or so, Olah's secretary had been an enthusiastic, slightly unreliable calligrapher and mapmaker called Liévin Algoet. Educated in Louvain, Algoet had worked for Erasmus (who found him lacking 'all earnest desire to work at his books, and . . . only good to carry messages')[20] before making a 'navigation chart of the world'[21] for Dantiscus. Since then, Algoet had produced a map of the northern lands. By the time Mercator had committed himself to mapmaking, Algoet was securely lodged at court, as the master of Queen Maria's school for pages. Algoet's patron was another enthusiast of earth-modelling, Marcus Laurin, who owned a terrestrial globe 'so extraordinarily large it is [was] hard to believe such a thing could be made of this [that] size'.[22]

From this circle of court humanists, it was Cranevelt who now stepped forward to support Gemma's young engraver in creating his first map. It would be a map of Palestine, the remote land beyond the Mediterranean where God had first revealed Himself. Its towns and rivers and tribes had been marked on the Bible maps of Liesvelt and his successors, but Cranevelt – who had taught himself Hebrew – had in mind a map which he could hang on a wall, a map which would shine with the purity and precision that only a copper plate could produce.

Mercator knew Palestine better than any place outside the Low Countries. He had grown up with its miracles and revelations. He knew its history. Palestine had been the subject of the first map that most of his generation had ever seen. And like the Bible maps of his boyhood, his would show the route described in the Fourth Book of Moses. The immigrant boy from the Brethren house would begin his mapmaking career by recording the trek taken by the Israelites from bondage in Egypt to salvation in the Promised Land.

9

Terrae Sanctae

The first regional map ever printed had been a woodcut of Palestine, and by the time Mercator and Cranevelt became acquainted, the Holy Land was probably the most printed regional map in existence.

Among geographers, the place where God first revealed Himself was also the first place which should be explored by the children of the Church of Jesus Christ. As printing gathered momentum, the Holy Land appeared as a pilgrimage map by Bernard von Breitenbach, as a Parisian copy of Lucas Brandis' original woodcut, and as a modern map in later editions of Ptolemy's *Geography*.

Among the working classes of the Low Countries, the Holy Land's geography was known through the twin-folio maps folded into Bibles that had poured from Antwerp's presses following the success of Liesvelt's Lutheran edition of 1526. By 1537, Mercator was familiar with Liesvelt's reduced version of Cranach's wall-map, and he had probably seen a second Bible map that had been circulating since 1535, a map that Miles Coverdale had bound into the first complete English-language Bible. Coverdale's title page had been designed by Hans Holbein, who may have contributed to the map – a much neater version of Cranach's.

Holbein – if indeed he was the draughtsman – had marked the location of many more holy places. His carefully shaded mountains, his rushing rivers and rolling seas, his delicately roofed towns and shadowed woods lent an enhanced reality to the biblical landscape. Where Cranach's Exodus road had blundered doggedly through Sinai, Holbein's stony track twisted uncertainly, interrupted by exquisite little vignettes: the Israelites bunched like a market throng between the sheer walls of the Red Sea's parted water; the intersecting lances of Hebrew and Amalekite; tented camps and the Brazen Serpent slung from its T-pole. Blank areas of sea and desert were exuberantly decorated with scrolls, windheads and a magnificent aquatic cartouche eyed by a fanged sea-serpent.

Compared to the globe gores which Mercator had worked on with

Gemma and Van der Heyden, the woodcuts in the Bibles of Liesvelt and Coverdale were crude. Liesvelt's mountain ranges looked like ploughed earth and he had named less than thirty settlements, most of them in the wrong place. Holbein's map had more plausible uplands, but they had been hatched in a style which fused them into the peaks and troughs of the Mediterranean. And although Holbein's map included many more settlements, but the appalling lettering was made to look even less competent by the fabulously worked cartouches and delightful windheads. Holbein had also turned his map 'upside down'. This had the compositional merit that the sea lay at the foot of the map and the mountains at the top, but it caused the Exodus itinerary to travel 'backwards' from right to left.

Where Cranach and Holbein had been concerned with presenting a pictorial *terra cognita*, the German humanist Jacob Ziegler had dreamed of sending surveyors to measure the world. Forty years older than Mercator, Ziegler had spent the last decade of the 1400s at the university of Ingoldstadt, and his circle included Conrad Celtis and the eminent humanist Willibald Pirckheimer, to whom he had outlined his intentions to describe Africa and India, to edit Strabo, Pliny and Ptolemy, to produce a map of the latest discoveries and – as if that wasn't enough – to publish 'the complete Ptolemy in a new edition based on new research and conclusions.'[1] His masterplan was to undertake a complete planetary survey, despatching teams of surveyors to take astronomical readings which he would use to create a new generation of maps. Ziegler added that he would delay the publication of this encyclopedic marvel until the cartographic effect of America's discovery could be determined.

Unsurprisingly, Ziegler ran out of time, but in 1532, two years after writing to Pirckheimer, he published the book which Gemma had picked through for Scandinavian data. *Quae Intus Continentur* also contained detailed descriptions of Palestine and Arabia, supported by seven maps of unprecedented accuracy. In the preface to the 300-page manuscript original of the book (which Ziegler had written in a precise, chancery hand), Ziegler revealed his sources to be 'The entire Holy Scriptures, from Moses to the Maccabees, Saint Hieronymus, together with others, who have described the places populated by the Hebrews: Josefus, Strabo, Plinius, Ptolemaeus and Antonius'.[2] By the time he came to work on the final version of *Quae Intus*, he had also consulted the modern *auctores*, Bernard von Breitenbach and Burchard of Mt Sion. Each place on the map had to be confirmed by at least two sources to merit inclusion.

Although most of Ziegler's data was over one thousand years out of date, and although his plans to use data-gathering emissaries had failed to materialize, it must have been apparent to Mercator that *Quae Intus* represented hundreds of hours of scholarship.

But *Quae Intus* was an unfinished project. Ziegler had gone to the trouble of listing many of the longitude and latitude coordinates of the exodus route, but he had failed to plot its course on his maps. A glance at the seventh and last map in Ziegler's Holy Land series suggested a failure of nerve or lack of time, for the Bavarian had established the first 'scene' in the trek, with a depiction of the Red Sea crossing. Away in the desert, the brazen serpent hung from its pole, a random, isolated symbol of a journey which Ziegler couldn't complete. Neither had Ziegler translated many of the place-names from Roman and Greek. Again there were inconsistencies, Jerusalem appearing as 'Aeeia' on one map, and as 'Ierusalaim' on another.

There were other more superficial failings in Ziegler's maps, failings which would have fired the confidence of a competent young engraver and letterer like Mercator. The crude line work and lack of detail were not unusual for woodcut maps of the time, but they looked antique when compared to the copper-plate engravings coming out of Van der Heyden's *atelier*. And despite having mastered cursive, Ziegler had failed to follow the example of Calvo and Bordone, for all of his maps were lettered with the same stentorian capitals.

If Mercator could plot Ziegler's coordinates onto a Ptolemaic grid, he could make the Bible lands geographically tangible. For the first time, the exodic itinerary would be seen as a mathematical truth.

Mercator's first task was to define the rectangular limits of the region he was mapping and to select an appropriate orientation. Unfortunately, the Holy Land was angled north-east to south-west, so it appeared on the north–south, east–west grid of longitude and latitude as a thin diagonal strip bordered by blanks of sea and desert which occupied half the map area. These were the voids that Cranach and Holbein had filled with exaggerated mountains, fleets of ships and stupendous cartouches.

Mercator devised an ingenious – if un-Ptolemaic – remedy. Framing his mapping area on Sidon and Onuphis (which neatly gave him natural margins of the Nile on the left, and the mountains of Lebanon on the right), he revolved his map so that it was oriented with north-west at the top. The Mediterranean now formed a slim shore along the upper perimeter of the map, while at the foot of the map, the Arabian desert was trimmed too. (When he came to explain himself in the cartouche, he referred to his desire to minimize space on the map which was *inutilia*, useless.) The resulting shape may have been wider than the Golden Mean, but by no more than the maps of Cranach and Holbein had been narrower. Mercator had also created a pleasing symmetry of composition, for the shore at the top could be balanced at the foot of the map by the indentation of the Red Sea on the left, and by a cartouche on the right.

It was an elegant solution, but it did skew the map's mathematical grid, a problem Mercator addressed by marking off the four borders of his map with a herringbone pattern, latitude against longitude. With the aid of two rulers (and a few minutes' practice), coordinates could be set and read as easily as if the grid had been parallel with the map's borders. In the centre of the map, a compass showing the four cardinal points reminded viewers that the map was not oriented with north directly 'upward.'

Palestine framed, he now had to create and populate its landscape.

Mercator began by plotting onto his map the coordinates of known features. In the absence of a triangulated survey, he had the pages of descriptions and locations detailed by Ziegler in *Quae Intus*. But as the 25-year-old began to transpose the data, he began to see that *Quae Intus* contained as many inconsistencies and errors as the exodic itineraries of Cranach and Holbein. Comparisons with a pair of dividers revealed that each of the eight maps in *Quae Intus* had been drawn to a different scale. There were also signs of haste and carelessness. On the last of the series of Holy Land maps in *Quae Intus*, the woodcutter had misnumbered the latitudes on one side of the page. Furthermore, Ziegler's misspellings of place-names suggested an unfamiliarity with Hebrew. More critically, the coordinates listed in Ziegler's text sometimes failed to correspond with their actual location on his maps.

Later, Mercator recalled how he had 'placed the cities, mountains and other places of Ziegler's description in conformity to the proportion of distances', but had found the Bavarian's information to have been 'without proportion and confused'.[3] If Mercator had known how one of Ziegler's sources had mapped Greenland, he might have been less trusting: Ziegler had copied from the Danish cartographer Claudius Clavus a series of place-names which – in the absence of any factual source – had been derived by taking words from a popular Danish ditty. Read backwards, the list of Greenland place-names which Ziegler had compiled for the manuscript prototype of what eventually became *Quae Intus*, formed the following verse:

There lives a man on a river in Greenland
His name is Spjellebod

86

He owes more fell of louse
Than he owes fat of pork.
From the north is the sand drifting.[4]

By the time Mercator had finished picking through Ziegler's text, he had plotted over 400 places on the map, as well as the territories of the twelve tribes of Israel.

Having created a best-fit geography, Mercator now attempted to plot the map's central narrative, the Exodus route. Here – as elsewhere – Ziegler had marked locations that were distant from the coordinates which he had so methodically listed in the text a few pages earlier. Paran, Chaseroth and Sepher formed three random examples on the early part of the Exodus route. Ziegler's map placed them close to the longitude of 64 degrees. Yet the text listed all three a whisker from 63 degrees of longitude – just one degree, but far enough west to have been guiding the Hebrews down the Nile into the Mediterranean. Although Ziegler's description of the Exodus route occupied thirteen pages of text, his bibliographic excavations had not produced locations for all forty-two Hebrew camps listed in the Book of Numbers. Indeed, the only parts of the route which Mercator could plot with some element of certainty were the beginning and the end. Of the eleven camps between Hazeroth and Hashmonah in the central part of the route, Ziegler had provided coordinates for just one. Each side of the camp below Mount Shapher, was a set of five camps which could not be fixed. That left a bewildering number of route permutations. On their smaller-scale maps, Cranach and Holbein had been able to avoid the issue because they made no attempt to mark all the camps. Cranach had depicted the route taking a rather arbitrary zig-zag through the Arabian mountains; Holbein had shown the route crossing itself in a loop. Either was possible. Mercator went for the loop.

Eager that there should be no cloud veiling his method, he made no attempt to provide locations for the unknown camps. Listing them beside his route in two blocks of five names, he added notes that the location of these camps were 'uncertain' or 'unknown'. It was a warning which could be applied to the entire map. Faced with more than one location for most places, Mercator had little option but to best-guess the actual geography.

Having fixed every place to the surface of his map, Mercator could stand back and take a breath before embarking upon the second and final phase, the flowering of his coordinates through the use of symbols into a biblical landscape.

Like all mapmakers Mercator had to imagine his viewers looking down to earth through a celestial window. The higher the viewer, the more compressed the scale and the more remote the image. Mapmakers had learned to reduce this remoteness artificially, by using pictorial symbols which were of a far larger scale than the map itself. This was the illusion which beckoned the viewer towards the landscape. And in Mercator's case, enticing his customers down into the very peaks of Palestine was fundamental to his scriptural brief.

He wasn't the first to pull a perspective trick on a map, but he did it well. By spacing out the mountains in the lower, left part of the map, where the exodus route started, and by cramping the ranges of the Promised Land in the upper right half of the map, Mercator encouraged the eye to follow its exodic itinerary from the barren spaces of Egypt towards distant, teeming Jordan.

Many of the landforms which rose in relief from the blank plane of the map's surface bore a curious resemblance to the *landschaft* vignettes which could be found on the 'world views' of the Flemish painter Joachim Patinir – whose work Mercator must have known well. Mercator carefully engraved symbolic ranges which ascended through gentle foothills and tree-crowns to a teetering peak. A couple of Patinir's rock arches crept in, while the two fish supporting the authorial cartouche were clearly copies of the squamous monster being wrestled to the shore in a Dirk Vellert engraving of 1522.[5] Mercator replaced Ziegler's stiff, artificial rivers and lakes with fluid, Holbeinian waters, while the Bavarian's sea – so tormented and heavily worked that it always threatened to inundate the land – was substituted by patches of placid hatching cruised by ships and improbable fish. Under Mercator's burin, new woodland burst from the shores of Samachonitis and foothills bloomed in Arabia. Of Ziegler's topographical symbols, only his mountains survived. Familiar with the fanged peaks of Bavaria, Ziegler had copied Cranach and turned every Holy range into jaws of toothsome Alps. Mercator – who had never seen a rock outcrop, let alone a peak – embellished Ziegler's ranges into rows of rotten teeth, with forked peaks and weathered stumps. One mountain looked like the gaping jaw of a surfacing whale; another had mushroom overhangs. The effect of the burin point on copper called forth extremes of precision beyond the touch of the carver of wood blocks. Mercator's watercourses were not arbitrary channels but carefully graded in width so that the viewer could separate a stream from a river; the muscular rope of the Jordan from the multi-braided Nile.

Unlike Cranach and Holbein and Ziegler, Mercator did not populate the parted waters of the Red Sea with fleeing Hebrews; he left the parting vacant, inviting the map's viewer to take the Hebrews' place. Elsewhere, the biblical scenes were enhanced by tiny figures. Peering into the ranges, an attentive viewer would be able to pick out Israelite tents west of the Red Sea, and the pillar of cloud blocking the Egyptians' pursuit. Further along the exodic road, a miniature Moses tossed a branch into the bitter pool of Marah; at a suitably vertiginous Mount Horeb, God glowed like a beacon from the summit while a figure far below struck the rock for water. Arrival in Canaan demanded of the viewer even closer attention, for the landscape suddenly became crowded with towns and villages and castles. Not everyone would immediately find the glowing figure 'transfigured before them' atop Mount Tabor, his face shining 'as the sun' and his raiment 'as white as the light'.[6]

Underlying the map's pictorial vignettes and biblical imagery was a systematic depiction of settlements. Drawing on Ziegler's text, Mercator created different symbols for rural settlements, hamlets and forts. Towns were marked with an open circle and major centres such as Damascus and the towns of the Nile were shown as pictorial views.

Mercator's exodus route looked as if it had been graded by Roman engineers. Instead of the ragged paths of Cranach and Holbein, Mercator neatly pricked out a parallel-sided 'road' punctuated with numbered bays whose descriptions could be found in a similarly numbered key set in a cartouche – a technique that Mercator had used on Gemma and Van der Heyden's terrestrial globe, where he had been faced with fitting an overwhelming number of European cities into a small continent.

With the physical landscapes engraved, Mercator added the lettering, explanatory cartouches and various decorative devices which would occupy unused space. Some of Ziegler's place-names were replaced with equivalents which Mercator took from the Hebrew index in the back of a recent Latin translation of the Bible by Sanctes Pagninus.

To help the map-user distinguish between types of location, Mercator used three forms of lettering: roman, cursive and the Gothic-derived *lettre batarde*. More confident than he had been when he lettered the globe, he switched to cursive for place-names, with a relatively sparse use of roman for regions. The virtually unreadable *batarde* was used for the names of the tribes of Israel. The pithy captions appended to land-forms and biblical scenes were largely lifted verbatim from Ziegler's text.

In the central cartouche at the top of the map, Mercator inscribed two passages from the Old Testament. From Micah, he transcribed the plaintive lament for backsliders: 'O my people, what have I done unto thee? and wherein have I wearied thee? testify against me.'[7] And quoting Deuteronomy, he reminded the map's users of the fruitfulness that lay beyond bondage and the wilderness: 'For the Lord thy God bringeth thee into a good land, a land of brooks of water, of fountains and depths that spring out of valleys and hills; a land of wheat, and barley, and vines, and fig trees, and pomegranates ... a land whose stones are iron, and out of whose hills thou mayest dig brass.'[8] Thus did Mercator attribute to the land of no scarcity the source of brass, the raw material for making mathematical instruments. Above the cartouche, Christ peered down upon the Holy Land and a centrally placed compass. But the most ornate cartouche had been reserved for the map's patron: the dedication to 'Francisco Craneueldio Caesaris' occupied a huge garland of fruit placed like a blooming oasis of fertility in the Arabian desert.

The care, scholarship and technical skill which Mercator had devoted to his first solo production were impressive. And the packaging was shrewd. Like Ziegler, Mercator had printed his maps onto a number of sheets, but – unlike Ziegler – he had used a uniform scale. Each set of six map-sections were folded once and slipped into a cover with an additional sheet describing how the purchaser could trim the sheets and paste them together to form a wall-map.[9] Also included were twelve strips of paper printed with a decorative border. Assembled, the finished map was nearly four times the size of Holbein's woodcut in the Coverdale Bible, and much more detailed.

By offering a product which could be kept in a bookcase or displayed on the wall, Mercator was increasing his market. His self-assemble flat-pack was also easier to store and to transport than conventional wall-maps, which had to be rolled or folded in barrels for shipment. The flatpack was also easier to hide. Mercator had taken his first steps as a spatial missionary.

The map was ready for printing before the end of 1537. In spite of Mercator's own reservations concerning its quality,[10] it won 'the admiration of many'.[11]

The 'many' would not have included the likes of Latomus and Tapper, the hardline theologians of Louvain who had recently burned Tyndale. Somewhat provocatively, Mercator had made it clear in the map's title that his intention in mapping the Holy Land had been 'the better understanding of both testaments'. Close inspection of Mercator's map also revealed a compositional kinship with the most famous of the Lutheran propagandist images, *Law and the Gospel*. Typically, the image was presented as two halves, separated by a tree. On the side of the

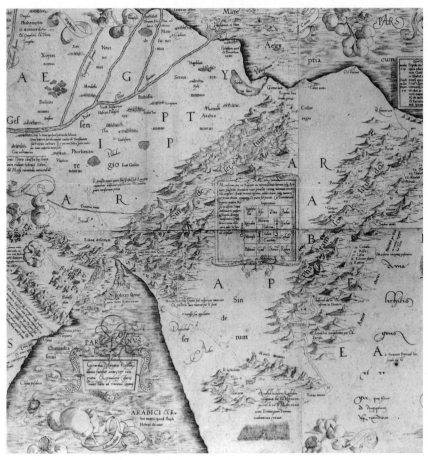

Fig. 11. Mercator's first map: A detail from his Holy Land map of 1537, showing the Exodus route from the Red Sea. (Bibliothèque Nationale, Paris)

Church would be Moses with the tablets of law, and an irredeemable sinner being prodded to the flames by Satan. On the side of the reformers would be the resurrected Christ. Not only had the sorrowing, pierced Christ made His way to the focal point of Mercator's map, but He looked remarkably similar to the Christ shown in a *Law and the Gospel* woodcut produced by Cranach six or seven years earlier. Both Christs rested above a rectangular frame; in Cranach's woodcut it was the rim of the opened sarcophagus; in Mercator's map it was the only rectangular cartouche on the map, lined inside with the passages from Deuteronomy and Micah.

There was even a case for arguing that the general composition of the map had been modelled on *Law and the Gospel,* for Mercator's tilted orientation had placed the land of the Pharaohs on the left, and the

Promised Land on the right. Instead of the dividing tree, Mercator had placed in a line down the centre of the map the figure of Christ, the passages from Deuteronomy and Micah, and the compass rose. Mercator had positioned Cranevelt's dedication on the right-hand half of the map, while his own authorial cartouche was humbly sited over on the left, by the beginning of the exodic itinerary. Cranevelt, on the side of grace, was the kind of Christian humanist whom Mercator might wish to join.

There was another connection with Cranach's propagandist woodcut. For reasons which cannot possibly have been accidental, Mercator had omitted from his exodic route the climactic episode. Where was the vignette of Moses brandishing the tablets of stone? Was Mercator's deletion of Law further proof that his map was a hymn to the Gospels of Matthew, Mark, Luke and John? Or was the lawgiver's absence an apocalyptic warning to those drifting from the Church?

Besides the buried ambiguities, the production of a Holy Land map in Louvain (of all places) raised further doubts. Mercator's *Terrae Sanctae* belonged to a cartographic lineage that had been passed down through the Lutheran bibles of Froschauer, Liesvelt and Coverdale. The map in the bibles of Froschauer and Liesvelt had come from the studio of Lucas Cranach, the godfather of Luther's son, and the reformer's chief propagandist. Coverdale had used Holbein, who had painted his own version of *Law and the Gospel*.

Mercator's association with Jacob Ziegler was undisguised, the opening lines of the main cartouche acknowledging the Bavarian's work. Among the theologians of Louvain, the name Ziegler rang warning bells. Was it not Jacob Ziegler who had defended Erasmus in print? Indeed been *praised* by Erasmus? Who had met with Luther and Zwingli? Who had been offered a chair at Wittenberg, the birthplace of Lutheranism? Who had fought the prelates of Strasbourg? Who had dedicated his book *Quae Intus* to 'Renatæ à Gallia', the reform-minded Duchess Renata of Ferrara, daughter of France's Louis XIII? Who had been heard to refer to the Emperor as that '*Fleming*'? Neither was Mercator's Erasmian dedicatee, Frans Van Cranevelt, especially popular among Louvain's theologians.

Finally, there was that packaging. Mercator had claimed in his address to the reader on the booklet's cover that the format had been intended to reduce the chance of damage in transit, and to limit the retail cost. But what better means of smuggling a wall-map than to ship it as a kit?

And who *was* that tiny lone figure pointing towards Nazareth as he strode from Tabor's foothills? One of the Old Testament prophets? The apostle St Paul en route for conversion (for he is walking and pointing towards Damascus)? A pilgrim? St Jerome? The map's viewer? Mercator himself?

Just one vignette from the New Testament appeared on Mercator's map. For many viewers, the tiny shining transfigured Christ on Mount Tabor carried an unequivocal, unseen caption: 'This is my beloved Son, in whom I am well pleased: hear ye him.'[13] Listen, said the map, to the Scriptures.

In mapping the Holy Land, Mercator had glimpsed his own future.

He had seen it on Ziegler's main map, *Universalis Palestinae*. It took the form of a circle set inside a square. The circle was ruled with the four cardinal points of the compass, north at the top. Pivoted on the centre of the circle was a compass needle. But it was not pointing north. Measuring the angle between the needle and the map's meridian would have revealed to Mercator that Ziegler's needle was being drawn by a source of magnetism some 28 degrees west of the north pole.[14]

In this intriguing discrepancy between 'magnetic' north and 'map' north lay the beginnings of Mercator's investigation into the mysteries of the compass needle – an investigation which would ultimately lead to one of his greatest discoveries.

10

Naming America

With the Holy Land in print, Mercator felt ready to reveal his life plan. He did so on the legend of his second map: a map of the world.

For a statement which would have such extraordinary consequences, it was remarkably low-key; exclamatory rather than declamatory. In it, Mercator announced that he was embarking upon a systematic description of the entire planet.

In abbreviated Latin, the author explained that the map before his readers' eyes 'was a division of the world along broad lines', and that it would be followed 'successively' by individual maps of 'particular regions'.[1]

The announcement was brief, but its meaning was clear: the new world map introduced readers to the general disposition of the planet; the following series of maps would describe each region of the world by turn.

The source of this hugely ambitious project was obvious to anyone familiar with Ptolemy. The *Geography* suggested just such a scheme, a depiction of 'the known world as a single and continuous entity, its nature and how it is situated ... its broader, general outlines (such as gulfs, great cities, the more notable peoples and rivers, and the more noteworthy things of each kind)',[2] to be followed by regional cartography, which registered 'practically everything down to the least thing therein (for example, harbours, towns, districts, branches of principal rivers, and so on).'[3]

This was a project which had been brewing since Mercator's youth, when he had made geography his 'primary object of study'.[4] His early fascination with the form and nature of the world, and his curiosity in those 'numerous elements' that were 'not known by anyone to date'[5] had at last found a means of expression. Mercator would describe – and investigate – the world through maps.

*   *   *

The world map whose legend bore the weight of Mercator's plan had been designed to catch the eye.

Any author contemplating a new world map had two prime concerns: he had to select the most appropriate projection, and he had to acquire the latest geographical information.

All projections sacrificed spatial truth. No matter how a sphere was flattened, distortion would arise. The fact that every projection was a compromise had not dissuaded mathematicians from exploring new methods of transforming three dimensions into two. The last thirty years had seen planar worlds which ranged from Dürer's perspective orb to the separate circular hemispheres of Monachus and a variety of strange, if symmetrical, shapes derived from Ptolemy's 'superior and more troublesome'[6] projection.

Ptolemy's intention had been to devise a projection which preserved the world's mathematical proportions while allowing the eye to rove over a flat, mapped *oikoumene* as if contemplating a spherical globe. It was Ptolemy's partial success in achieving 'proportionality of the parallel arcs'[7] on his second projection that gave his successors, 1,300 years later, the starting point for their own attempts to improve upon the proportions of the planar world map. Four years after Waldseemüller had extended the span of Ptolemy's second projection from 180 degrees to the full 360 degrees, Bernardus Sylvanus in Italy devised a projection on which the degrees on the central meridian were in correct proportion to those of the parallels. Sylvanus' concentric parallels, measured from a point on the central meridian 100 degrees north of the equator, produced a world map which bore a curious resemblance to the human heart.[8] Three years later in Nuremberg, Johannes Werner published a diagram of a projection whose parallels were also proportional to the central meridian.[9] But he had moved the centre for the semi-circular parallels to the north pole, and this time the map was a perfect heart. In France, Fine used the heart-shaped projection for a manuscript map of 1519, and Apian may have copied the Frenchman when he too published a heart-shaped map in 1530. But it was Fine who provided Mercator with the inspiration for his first world map.

Mercator had probably come across Fine's map in the Paris edition of a book of voyages, *Novus orbis regionum*. The book had appeared in 1532. Folded inside was an extraordinary new image of the world.[10] Fine had bisected the globe at the equator and created two maps which would have been circular but for a curved cleft. At the centre of each semi-circular map was one of the earth's poles. The northern hemisphere occupied the left-hand map, and the southern the right. The two maps touched where the meridian of 90 degrees east ran in a straight line between the two poles. All the other meridians curved out from the poles, while the parallels turned concentrically. The impression given was of two maps contrarotating like a pair of cosmic gear wheels.

'We offer you, fair reader,' announced Fine in the map's legend, 'a universal representation of the world . . . with the proportion both of the equator and of the parallels conserved relative to those which come from the centres [the meridians].'[11] Fine's incised semi-circles were a significant mathematical advance on his earlier heart, whose proportional qualities had been compromised by severe geographical distortion in the southern hemisphere. The new map produced a lesser and equivalent distortion across both hemispheres. The geography of the map had retained a truer shape, and you could see more of it. This was, however, a mapmaker's map, and few of Fine's readers can have looked at it without scratching their heads. The two polar regions dominated the field of view, Africa and America were pulled in two, Europe balanced upon Portugal, and Cathay floated with celestial detachment up in the foliage of the map's decorative border. By unpeeling the planet's skin in this symmetrical, bipolar form, Fine had, however, illustrated how the land-masses of the known world were congregated in the northern hemisphere. The largest landmass in the southern hemisphere appeared to be uninhabited.

In his map legend, Fine had drawn the reader's attention to three characteristics that his ingenious double-polar projection possessed. Firstly, it had been plotted 'in accordance with the understanding of recent geographers and hydrographers'.[12] Secondly, it conserved the proportions of the parallels and the equator with the prime meridian. Thirdly, its 'coextended' shape took 'the paired form of the human heart, the left of these comprising the northern part of the world, the right, indeed, the southern'.[13]

Fine's cloven semi-circles may have been mathematically derived, but they were loaded with meaning. As the governing centre of an individual's entire being, the source of 'the issues of life',[14] the heart had a profound symbolic resonance. Fine may also have been alluding to the broken heart of compassion and righteousness, and to the heart that accorded with the humanist emphasis on that which was central rather than highest. The *coeur*, the core, was the essence of being. Perhaps also he was alluding to the Lutheran heart: to Melanchthon, the heart now expressed the nature and purpose of Christian ritual; the word of God entered the ears 'to strike the heart', while the rite entered the eyes 'to move the heart'.[15] Melanchthon's heart was central to experiencing the

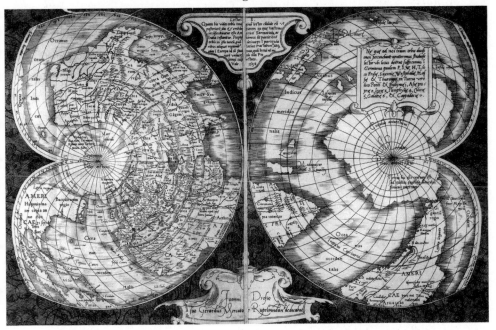

Fig. 12. Mercator's first world map, the heart-shaped 'Orbis imago' of 1538. (New York Public Library – Rare Books Division)

Gospel, just as Fine's projection was central to comprehending the true proportions of the planar world.

As Mercator would have known, Fine was no model of Catholic orthodoxy, spending – according to the university of Paris, who negotiated for his release – 'a long time'[16] in prison in the mid 1520s. The reason was alluded to in a letter of 1525: 'Wonderful it is', wrote Fine's friend Joannes Angelus, 'how in these days the theologians, who rave against every type of learning, would not be difficult to conquer if faith were strong and constant in those things in which it ought to be.'[17] The following year, the astrologer Henricus Agrippa von Nettesheim referred to Fine as 'a worthy mathematician and astrologer'[18] and implied that it was his notoriety in the predictive arts that had brought him to the attention of the theologians. Justifiable on the grounds of mathematical merit, Fine's hearts were touched with heresy.

And Mercator's choice of dedicatee reinforced the suggestion that his new map was intended to have a meaning beyond mathematics. Instead of Cranevelt, the map would be dedicated to one Johannes Drosius of Brussels, an ex-student of the university who had matriculated two years ahead of Mercator. A monk or a priest, Drosius had not been picked up by the Inquisition, but later events would reveal that he was one of many in Louvain whose inclinations were closer to Gospel study than to the rites of the orthodox Church.[19] Mercator had associated himself with someone who may have been under suspicion of heresy. He could be sure that his new world map – a revised geographical image of God's creation in the form of an opened heart – would settle eventually upon the desks of northern Europe's hardline theologians. Who happened to live a few minutes' walk away from Mercator's own door.

Mercator planned to use his new world map as a progress report on exploration. Coastlines which were known and charted would be indicated with a conventional inscribed line, hatched on the seaward side, while speculated, unexplored coasts would be depicted by hatching alone. In one glance, a viewer would be able to summarize the level of knowledge concerning the outlines of major landmasses. The map would also look stunning. Unlike Fine, whose map was a woodcut, Mercator would naturally engrave onto copper, lettering in his fine italic. To maximize sales and distribution, the map would be sized so that it could be bound – like Fine's – into books, or sold separately.[20]

Mercator's first exercise in converting the three-dimensional sphere into a two-dimensional map also provided the young mapmaker with an opportunity to correct fundamental errors he now recognized in the globe which he'd helped to engrave two years earlier.

Some parts of his new world were transposed more or less unchanged from the globe: Mercator repeated the coastlines of Antarctica, Africa and America, labelling the unexplored west coast 'Littora incognita', and fuzzing it with hatching south of the tenth parallel. (He did, however, introduce a river system to the east coast.)[21] The Mediterranean remained at its old Ptolemaic span of 62 degrees of longitude.

Elsewhere, there were significant changes. Up in the Arctic, Mercator no longer saw the Mare glaciale as an inland sea, and opened it into a gulf with seagoing access past the north-eastern corner of Asia. But this was a minor adjustment compared to the three regions of the world which he specifically drew to his viewer's attention, instructing them in the map's central cartouche to study the areas labelled 'India', 'Sarmatia' and 'America'. It was in these three regions that he had modified the geography that he'd engraved onto Gemma's globe a couple of years earlier.

'India', which on maps such as Apian's had occupied half of Asia, grew even larger under Mercator's burin as he pushed the continent's seaboard an astonishing 30 degrees further east, a decision which reduced the width of the Pacific to less than half that of the Mediterranean. It would now take a ship less time to sail from America to Cathay than from Seville to Antwerp. In accommodating this awesome tectonic shift, Mercator could not very well interfere with 'India intra Gangem' and 'India extra Gangem' – the regions each side of the Ganges – since their locations had already been established by the Portuguese. The part he stretched was lesser-known eastern Asia: 'Thebet', 'Cathai' and the stumpy peninsula labelled 'Lequii populi' by Gemma in 1536.[22] Under Mercator, the 'land of the Lequii' expanded to a great rectangular promontory four times the size of the Indian peninsula.

The second region Mercator wished his readers to examine was Sarmatia. This was the Sauromatae, or Sarmatae, of the Greeks and Romans, located according to Herodotus beyond the river Tanais, on the eastern boundary of Scythia, and occupied by people who were descended from Amazons and young Scythian men. Ptolemy had confused matters by stating that the border between Asia and Europe ran from 'the river Tanais and the meridian from this to the unknown land [to the north]',[23] yet at the same time, he described Sarmatia as an enormous region which occupied both northern Europe and northern Asia, with its own, northern, 'Sarmatian Ocean'. In another part of the world, this ambiguity would have mattered less, but the precise location and nature of Sarmatia now appeared fundamental to fixing the eastern boundary of Europe, the scale of which had recently become a subject of great fascination. The need to define exactly where Europe ended prompted Mercator to mark the only internal borders he was to include on his map. Having engraved the river Tanais, he pecked a line from its source north towards the Arctic Circle. Just over ten degrees to the east, he pecked out a parallel line running north from the Volga. The region between the two rows of dots now became 'Sarmatia Asie', a buffer zone between terrifying voids of Scythia and the lands which belonged to the Russians, Prussians and Poles: 'Sarmatia Europe'.

The third and final planetary modification turned a little-known island into a major sub-continent.

Ever since Waldseemüller had prised America from Asia, a residual northern landmass had floated off the island of Isabella. The German cosmographer had labelled it 'TERRA ULTERI INCOGNITA'. Not much bigger than Java, this landmass had been marginalized by map projections designed to illustrate Africa, Asia and Europe. 'Terra ulteri incognita' could usually be found high on the left edge of a world map, symmetrically opposing the equally inconspicuous sliver of Sipangi[24] on the right side.

But Mercator was also a globe-maker, and those who had worked on a sphere knew the truth of distance and relative area. By spinning the globe that he'd engraved, Mercator would have been able to place 'Terra ulteri incognita' in the centre of his field of view. The double cordiform projection offered a similar effect. The twin hearts placed 'Terra ulteri incognita' in the map's foreground, where it looked vast: bigger even than Europe. The continental newcomer was conspicuous, huge, unexplored and nameless.

While Mercator had no obligation on this map to provide detail, he could hardly allow such a large landmass to be seen without a label. By spinning the globe, by changing projection, by altering the perspective, the unknown had become nominable. And once named, it could claim a cartographic existence. But what to call it? Apian had followed Waldseemüller by labelling it 'Ulteri terra incognita'. Münster had used 'Terra de Cuba'. To Gemma – and Mercator – it had been 'Hispania major a Nuñno Gusmano'. As part of the Asian mainland, the regional labels applied to this part of the globe included 'Baccalearum Regio' and 'Terra Francisca' – the name Fine had used for the coast north of Florida.

Mercator's solution may not have been of his making. The one locative certainty concerning 'Terra ulteri incognita' was that it lay north of America, so there was a certain logic in naming it 'Americae pars Septentrionalis' – North America. But in naming the new continent after Spain's pilot major, could Mercator have been influenced by the proximity of the Spanish court, just down the road in Brussels? Since his only two previous works of note had been dedicated to Imperial councillors, he had opened himself to influence. And Antoine Perronet's father, the supremely influential first councillor Nicolas Perronet de Granvelle (who reported to Charles v on matters concerning the Low Countries) must by this time have been aware that the Habsburgs had a rising young mapmaker. Surely, it was these councillors who prompted Mercator to occupy 'North America' with the only historical caption of its size on the map, a caption which read 'Hispania maior capta anno 1530' – 'conquered by Spain 1530'.

Thus, the most modern world map in the Low Countries confirmed that America was not only separate from Asia, but that it had a northern relative who was larger even than Europe, and that both Americas belonged to Spain. With a single nomenclative flourish, Mercator had erected an Imperial barrier to the Indies that ran from pole to pole.

*   *   *

While Mercator had been mapping the world, Barbara gave birth again, this time to a girl.[25] The parents named their first daughter Emerentia, after Mercator's mother.

Short of tools, and without a workshop, Mercator struggled to reconcile his cosmographic ambition with the need to earn money. On completing the world map, his intention had been to produce the first of the regional maps: Europe. The legend on his world map informed readers that work had already begun and that Europe would soon be published, at a size no smaller than Ptolemy's 'universal' map.

This was something of an understatement. Mercator's readers – and Mercator himself – had no idea of the immense scale that this map would finally assume. Compiled from the latest coordinates and regional surveys, with reworked coasts, ranges and rivers, the new map would have a more complete plot of cities, towns and villages than had ever been achieved before. It was, he admitted later to Antoine Perronet, a 'huge' project.[26] Remarkably, the task was well advanced by the time Mercator was pulled away on an unexpected commission.

11

The Fall of Ghent

Mercator's adopted homeland was in trouble.

At the centre of the crisis was Ghent, the historic seat of the Flanders counts, the birthplace of the emperor, Charles v. The old cloth emporium whose wealth had once been so fabulous that it could field 20,000 armed citizens had always been a source of tension for the Habsburgs, but the decline of the textile industry had added terminal stresses. By the early 1500s, at least half of the population of Ghent had been reduced to poverty, while a residual core of aldermen, nobles and merchants (who still lived well from the grain staple on the Leie and the Schelde), continued to benefit from the city's ancient privileges. The city council and aldermen found themselves caught between the demands of Ghent's impoverished citizens for a return to a wealthy, independent past and Habsburg plans to forge a centralized empire.

Rebellion became apparent in April 1537, when Ghent's council defiantly refused to contribute money to the war chest which Queen Maria of Hungary – as regent of the Low Countries – was accumulating for her brother Charles' campaign against France. Suspecting that the council was colluding with the Imperial court rather than defending the reciprocal rights and privileges that existed between Ghent's citizens and her princes, the city's weavers and lesser guildsmen took desperate action. In August 1537, they armed themselves, seized Ghent's gates and appointed a committee of nine to administer the city. While former aldermen fled, the rebel Gentenaars set about restoring the city's image.

The senior of the city's four chambers of rhetoric – De Fonteine – conceived a drama and poetry festival whose purpose would be the promotion of 'trade and livelihood'[1] in Ghent. Through the art of rhetoric, Ghent's economy could be revived and the city's importance celebrated. Working to the same ends, a Gentenaar called Pierre Van der Beke produced a splendid new woodcut wall-map of Flanders which portrayed the county as a powerful entrepôt, strategically located between the Imperial court at Brussels and the sea lanes of Europe.[2] The map was printed in 1538 by another Gentenaar, Pierre de Keyzere.

Addressing his readers in French, Dutch and Latin (and thus assuring maximum penetration of Flemish culture), Van der Beke explained that the new map had been made necessary because 'until now no description has been accurately made, appropriate to the situation of the said country'.[3] Implied by these lines was a rejection of earlier maps which had failed to take 'appropriate' account of Flemish geography. Now, in 1538, the author 'judged it necessary and most useful to portray and to map again the said country'.[4] Below the three legends, Van der Beke's flag-decked Flanders had been geographically reorganized to reflect its princely honour and civic might. At the hub of a web of radiating rivers (specifically labelled as 'navigable' in the map's legends) rose the multiple towers, spires and turrets of Ghent, while a diminutive Antwerp clung to the edge of the map. Cogs, caravels and galleys flying the flags of the principal trading powers bore down on Ghent's multiple points of access. Echoing an earlier claim that the city could also speak for eastern Flanders in refusing to subsidize Queen Maria's war effort, Van der Beke had enclosed these areas in dotted lines and labelled them in capitals as lands 'ONDER GHENT'. Armorial shields and flags scattered across the map drew attention to the feudal privileges and independence of the county's towns and noble families.

The map – and the forthcoming festival of drama and poetry – demanded that men and women of Flanders stand by their birthright. In his address to the reader, Van der Beke pointedly introduced himself as a 'native of Ghent', and concluded the address by encouraging readers 'to learn about the places and regions of the said county of Flanders, of which the most powerful and most illustrious and excellent Emperor of the Romans, Charles v is a native'.[5] By implication, the emperor was indebted to Flanders. Conspicuously absent from the map was any dedication to Ghent's Imperial son.

By the end of 1538, Flanders was seething with revolt. The map was followed in 1539 by the anticipated festival. Blamed by many for triggering the cataclysm which followed, the festival opened innocently enough in April, with nineteen chambers of rhetoric (fifteen from Flanders and the remainder from Brabant and Hainaut) competing in the fields of didactic poetry, comic poetry and love poetry. By the time the festival climaxed in June with its morality plays, emboldened rhetoricians were inviting their audiences to consider the relationship between faith, good works and salvation. (The festival coincided with the circulation of a new Lutheran morality play called *The Tree of Scripture*, which mocked the sale of indulgences, simony, the veneration of images and the clergy's hunger for benefices.) A visiting Englishman, Richard Clough, was amazed: 'There was at thatt time syche plays played, that hath cost many a thowsanntt man's lyves; for in those plays was the worde of God fyrst opend in this countrey. Weche plays were and are forbeden, moche more strettly than any of the boks of Martyn Luter.'[6] To the chancellor of Brabant, Adolf van der Noot, the plays were 'full of evil and abusive doctrines and seductions, espousing Lutheranism'.[7] Later, Clough would write that the festival of 1539 'was one of the prynsypall occasyons of the dystrouccyon of the towne of Gantt'.[8]

The trouble erupted after the festival, as rumours circulated among Ghent's weavers and guildsmen that certain aldermen had removed from the city archives the document which awarded Gentenaars the right to withhold the payment of all taxes. At the end of August, Lieven Pyn, a 75-year-old former alderman who had been one of those involved in the tax negotiations of 1537, was ritually killed. After having his beard and head shaved so that there should be no hiding place for demons who might assist him during his torture, Pyn was hauled to the rack. Others followed. Ghent's nobles and patricians fled the city.

In distant Spain, Charles v heard that his Habsburg inheritance was on the verge of disintegration. An experienced reader – and critic – of cartography, Charles was 'most fond'[9] of maps and may well have seen a print of Van der Beke's vision of an independent Flanders. The Emperor reacted swiftly: in October he sent Adrien de Croy, the Lord of Beuraing and Count of Roeulx, to warn the Gentenaars that he was going to pay them a visit. One month later, Charles began his journey to the Low Countries.

This was probably the moment that 'certain merchants'[10] (*mercatores*) hatched a plan to avert the emperor's wrath by commissioning Mercator to correct Van der Beke's image of a defiantly independent Flanders.[11] The request from the 'merchants' was 'urgent'.[12]

Two of these influential merchants appear to have been the Louvain bookseller, Bartholomeus Gravius, and Pierre de Keyzere, the publisher of Van der Beke's treacherous map.[13] Presumably, de Keyzere had decided that his association with Van der Beke might be regarded as disloyal to his fast-approaching ruler, and that his mercantile prospects (and life) might stand a better chance of preservation should he promote – immediately – the printing of a new description which restored Flanders as a loyal, devout subject of a centralized Empire. These were Mercator's instincts too. An adoptive Fleming himself, he had no wish to see Flanders torn from the protective embrace of the Habsburgs. And neither, as a Fleming, would he wish to be associated with a county which threatened the Imperial dream. The rebel-busting 'merchants' were potential saviours of an increasingly bleak outlook for Mercator's home-land.

There was another in Ghent who may also have developed a sudden interest in cartography: in 1539 the town elected a new alderman called Karel Utenhove, a Louvain-educated Gentenaar whose admiration for Erasmus had carried him across the continent to visit the great humanist in Basel. A regular correspondent (Erasmus regarded the young man as well bred but ill read) Utenhove was an ardent reformer and one of the more influential rebels. Utenhove must have known of Mercator; Utenhove must have known that his own life was in peril; Utenhove may have been one of 'the merchants'.

Mercator responded to the request 'with enthusiasm'.[14] And so, it would seem, did the surveyor Jacob Van Deventer, who had already surveyed Flanders as part of his programme to create regional maps of the entire Low Countries.[15] As the leading practitioner of the method of triangulation described by Gemma in *Libellus de locorum*, Van Deventer had the mathematical raw material required to create the first truly accurate printed map in the Low Countries. Apart from the urgent need to create an Imperial map of Flanders, Mercator could look at this commission as an opportunity to be the first to engrave onto copper a triangulated regional survey.

The government in Brussels must have been behind the new map too. The four-year licence on Van der Beke's map had two or three years to run, and Mercator would have needed court approval to break its terms. One who may have shown a personal interest in erasing Van der Beke's map was Chancellor Granvelle. Not only had Granvelle been advising Charles on the crisis in Ghent, but his son Antoine Perronet would soon be consecrated Bishop of Arras, whose territory was uncomfortably pinched between this fractious version of Flanders and a perennially hostile France. Another who supported an imperial map of Flanders was Liéven Algoet. The occasional mapmaker and ex-courier for Erasmus had enjoyed a remarkable rise through court circles: having been appointed master of Queen Maria's school for pages in 1534, he had just been elevated once again, and was now clerk of the Imperial Chancery. Algoet had been born in Ghent, and the year that Van der Beke published his map, he had also been appointed herald of arms for the county of Flanders. As a Gentenaar, as a Fleming, and as 'official amanuensis and court scribe' whose 'regular work for the Emperor' was the composition of 'geographical works and maps',[16] Algoet had a greater motive than many for encouraging Mercator to disarm Flanders.

Mercator worked fast.[17] Van Deventer's survey probably took the form of a working map criss-crossed with sight lines. All Mercator had to do was to trace Van Deventer's geography and then help himself to descriptive (and decorative) pickings from Van der Beke.[18] Systematically, Mercator re-created Flanders along Imperial lines. The eighty-nine banners and armorial shields that had defiantly blazed beside Van der Beke's privileged towns, villages and strongholds were pulled down, then re-engraved as part of the map's border; in a neat stack down each side of the map, the feudal hierarchy of Flanders had been relegated to historic emblems. (Motivated presumably by sentimentality, Mercator added the arms of Rupelmonde; curiously this was the only one of the eighty armorial emblems not to be named.) The upper and lower borders of the map were occupied by a sequence of circular medallions which recorded the succession of Flanders' counts from Baudouin the Iron Arm to the current incumbent, Emperor Charles v. Inside each medallion, Mercator noted the count's name, his father, length of rule and date of death.[19]

Mercator also corrected Van der Beke's orientation. The Gentenaar had oriented his Flanders with south-east at the top. This could have been explained by a need to match the shape of Flanders to that of his four woodblocks, or a compositional desire to represent the sea as a symmetrical, parallel-sided channel along the foot of the map. But Van der Beke had used his quaint orientation to establish a false proximity between Flanders and its old ally England. Moving England eastwards, he was able to place Dover as close to the Flemish port of 'Grevelinghe' as Antwerp was to Mechelen. In reality England was nowhere near the location on Van der Beke's map, but it was impossible to show any part of England on a map which portrayed Flanders on a Ptolemaic, north–south orientation. England had always been closer to France than to Flanders. Mercator revolved Van der Beke's Flanders to a recognizable Ptolemaic orientation. Because the basic coordinates of Mercator's map had been compiled from angles and distances gathered by triangulated survey, his Flanders map was tilted nine degrees to the west. (The recurrence of this discrepancy between magnetic north and 'true' polar north was by now kindling Mercator's interest in the location of the magnetic pole.) Van Deventer's survey also enabled Mercator to achieve a scarcely believable accuracy in placing around 1,000 Flemish locations, 200 more than Van der Beke.[20] Correctly oriented and accurately located, Flanders lost all sight of England.

By repeatedly incorporating references to Van der Beke's flawed Flanders, Mercator reminded his readers that the new map was not an alternative edition, but a replacement. Van der Beke's letter codes, A and P (for abbeys and priories), and M and F (for communities occupied by men and women), reappeared in a few locations (such as Marchiennes AM; Loo PM; Dorceel AF; Berghen AM AF) on Mercator's map. The four rampant bears (representing the quartet of ancient Flemish lordships) that occupied each corner of Van der Beke's map were transposed to Mercator's. And so was the military vignette down on the southwestern border: where Van der Beke had marked 'Den nieuwen dyck' as a long severed worm with a trio of mute cannons aimed at France, Mercator turned the dyke into an earthen version of one of his Red Sea walls, with a single cannon and a lunging swordsman, sword-tip aimed at St Omer to symbolize a Flanders that would repel rather than invite French interference.

Mercator also corrected Van der Beke's spelling of place-names – presumably from information provided by Van Deventer. Occasional windmills, bridges and internal borders crept onto the map, probably from Van Deventer too. The two ships, and the pair of compasses whose points spread across the map's distance scale, were probably copied by Mercator from Olaus Magnus' map of Scandinavia, published in 1539. And Mercator was unable to resist improving upon Van der Beke's Rupelmonde: in place of the Gentenaar's crooked castle and pair of towers, Mercator marked two adjacent circles, one for the castle and one for the town, and then sketched in the towers and roofs of each with an attention to detail which he had not lavished on any other towns of similar size.

Copying Flanders took Mercator far less time than had been required for his own vision of the Holy Land. According to his friend Walter Ghim, Mercator 'planned, undertook and ... completed' the map 'in a short space of time.'[21] Engraved onto nine copper plates, Mercator's map was appreciably larger than Van der Beke's four-block woodcut.[22]

Not for the first time, Mercator provided his readers with a riddle. At the foot of the map, he inscribed his name in secretary hand onto a placard hung from a splintered tree. This too was a reference to Van der Beke, who had planted splintered trees across his map as hooks for his armorial shields. But Mercator may also have been splintering *The Tree of Scripture*, the Lutheran morality play which had helped to spark the conflagration in Ghent by mocking the Church and appealing to 'German doctors'.[23] And he may have been referring to St Paul's 'grafted tree', whose branches were broken off 'because of unbelief'.[24]

As the emperor bore down on hapless Ghent, one part of the new map remained blank. Mercator had engraved a square frame to take a panel of text but he – and presumably the mysterious 'merchants' – had been unable to commit themselves to cutting the words into the copper. Since the immediate future for Flanders was so unpredictable, they may have decided to wait until the map was printed before pasting typeset addresses into each engraved frame. As a precaution, it also had the merit that the wording of the panel could be modified to suit the outcome of the emperor's visit.[25]

To remove any doubt concerning the loyalty of those behind the new, Imperial image of Flanders, Mercator engraved an extravagantly flourished dedication to the emperor.

If Charles was presented with a copy of Mercator's map of Flanders, the gift had no appreciable effect on the fate he had in mind for Ghent.

The emperor strode into the city on 14 February 1540. With him was Queen Maria, the papal legate, a selection of princes, nobles, ambassadors and 5,000 German *Landsknechte*. The column took five hours to enter the gates. Thirteen of Ghent's ringleaders were beheaded. Justices, burghers, six representatives of every guild and fifty citizens were forced to walk from the court of justice to the castle, barefoot, in black, with halters hanging from their necks. At the castle, they fell to their knees and begged for pardon. Found guilty of treason, disobedience and revolt, Ghent had its rights and privileges revoked, its treasures confiscated, and its arms, artillery and ammunition removed. Henceforth, municipal posts would be occupied by appointed, rather than elected, officers. The great abbey of St Bavon – and its entire district – was razed to prepare ground for a new citadel. Finally, in a terminal act of spite, Charles removed time itself by taking away Ghent's great bell.

Already a shadow of its former mercantile glory, the turreted, flag-decked metropolis at the centre of Van der Beke's map had been reduced by a cataclysmic Imperial trashing to one of the many discrete urban nuclei depicted on Mercator's mathematical description.

With Flanders cartographically restored to the Imperial jigsaw, Mercator returned to his great work: the mapping of the world by regions. Despite the distractions, he was close to completing the preparatory outline of the map of Europe by the summer of 1540. Only Spain remained. The next stage would be to transfer the outline to copper plates, a task which Mercator estimated would take 'at least one year'.[26]

He probably knew, deep down, that 'one year' was ridiculous. Already, he was developing a character trait which would harry him to his death. The cobbler's son had a will to succeed which was shadowed by the need to please.

Nobody had ever mapped the enlarged, post-Ptolemaic world by regions. Mercator knew that he could do it, but the scale of the challenge was too huge to be contemplated in its entirety. Broken down into parts, it became plausible, but even the first region, Europe, had a habit of dissolving like a mirage on the horizon of possibility. Meanwhile, there were instruments to make and children to feed. Barbara had given birth again, to a girl they named Dorothée. Single-handed and short of money, Mercator strove to keep his improbable vision. He would always be short of time, always craving the patience of those he loved.

In August 1540,[27] Mercator finally finished working on an instrument for his old student acquaintance Antoine Perronet. In the letter which accompanied the new *calvariam*, Mercator admitted that the project had taken longer than he had expected. The 'unreasonably protracted' delay had been due to being 'forced to seek various people's workshops', being 'short of everything', and to the 'frequent occupations which have necessarily dragged me away'.[28]

Anticipating that the Europe map would take a lot more time, Mercator described to Perronet how he had become engaged in a more remunerative, associated project: 'I have concentrated my mind', he wrote, 'on the *spherica geographia* so as to have a shield for my domestic financial obligations.'[29] If these somewhat plaintive lines to the chancellor's son were a discreet invitation for financial succour or patronage, they were imparted with no sense of obligation or self-pity. 'I have decided', continued Mercator, 'to publish a terrestrial globe.'[30] The globe, he continued, would be more detailed than any other, more up to date and furthermore it already had support from high places. Perronet's old tutor, the Greek grammarian Adrien Amerot, was urging the construction of a new globe, and so was Gui Morillon, a Burgundian-born Imperial secretary who had recently returned from Spain and moved into a house on rue des Dominicans in Louvain.[31] An old supporter of Erasmus, Morillon had humanist connections which included the Portuguese geographer and part-time diplomat, Damião de Gois, and the printer Rutgerus Rescius. Mercator's globe had backers close to the emperor: 'Secretary Morillon and Adrianus Amerocius are extremely pleased,' he wrote, 'and encourage its being carried out'.[32]

The letter concluded with a wild underestimation of the time required to construct such a globe, and a gentle reminder to Perronet that the

Europe map had not been abandoned: 'I hope to finish this work easily within three months,' wrote Mercator. 'After that I will concentrate completely on Europe until that publication is finished.'[33]

12

Latin Letters

Work on the new globe was under way when Mercator was forced to undertake another short-term commercial project.[1]

This time it was a book, or rather a manual, and it would establish the cursive hand practised by Mercator as the standard form of lettering for maps.

Literarum latinarum, quas italicas, cursoriasque vocat, scribendarū ratio (How to write the Latin letters which they call italic or cursive), introduced the term '*italicas*' to the Low Countries.[2] To Erasmus, the sloping cursive Italian hand had been *literae latinae*, but Mercator's instinct for geographical concision settled on a term that had been mentioned nearly thirty years earlier by Sigismondo dei Fanti, who had referred to scripts associated with chancery hand as '*italicus*'.[3] Mercator was the first to use 'italic' for a book title, and in doing so, he confirmed its increasing significance in the world of cartography. The effect of this 'little manual'[4] would be disproportionate to its twenty-seven leaves.

With three financially impractical schemes in abeyance (the creation account, the grand scheme to map the world by regions and now the globe), it seems likely that the idea for the writing manual originated with a printer who thought that he had found a northern European author to rival the Italian masters of italic: Arrighi, Tagliente and da Carpi.

The perceptive printer who spotted the opportunity may well have been Bartholomeus Gravius, one of the three 'merchants' whose identities Mercator had carefully masked on his recent map of Flanders. In 1540, the year that Mercator's Flanders map appeared, Gravius had joined the infamous Rutgerus Rescius as a printing partner.

Rescius had achieved particular notoriety through his role in the Nesen/Mela prank and his subsequent arrest. One of Louvain's more colourful – and controversial – characters, Rescius had been born in the riverside town of Maaseik, just a stone's throw from Mercator's old home in Gangelt. After an education in Paris, Rescius matriculated in 1515 at the university of Louvain, where he had enjoyed a precarious double life, living and working with the printer Dirk Martens and taking the chair of Greek at the new Collegium Trilingue. Described by his friend Erasmus as 'more learned than he gives himself to be',[5] Rescius had wriggled out of his arrest and then – after marrying Anna Moons – found himself embroiled in an interminable dispute with the university, who suspected that his domestic concerns, his printing commitments and his precarious financial situation might perhaps be interfering with his obligations to the Trilingue. Rescius founded his own press in 1529, after Martens retired. To many in Louvain (though not to Erasmus, who defended him to the end) Rescius was a flaky, avaricious academic who once organized a series of lectures to promote the sales of his own book of Greek paraphrases.

By printing a manual which would encourage the use of cursive lettering, Rescius would alienate himself even further from Louvain's orthodox scholars, some of whom still regarded *literae latinae* as a defiant, Erasmian habit. But the printer's keen mercantile antennae had recognized in Mercator a practitioner who had proved the worth of cursive on two globes and three maps. Mercator was as good as – if not better than – the Italians. Rescius, the university outcast with little to lose, would facilitate the italic beachhead in northern Europe.

Mercator's motives were financial and humanistic. The market was eager, and this would be the first northern European manual on italic. As such, it had an Erasmian ring: 'Calligraphy like painting has a pleasure of its own,' the great humanist had written, 'and not only does the writer linger over its execution but also the reader over its appreciation.'[6] Fluidity and composition on the page were as salient as colour and perspective on a canvas. The lettered line connected the imagination of the reader with the intent of the author. While the value of text was derived from its meaning, its *effect* was measured by clarity and elegance. To an increasing number of humanists, italicized Latin was the language of knowledge.

Unlike Arrighi and Tagliente, who had written their manuals in the vernacular, Mercator would write in Latin, directly addressing the men of letters, and the mathematicians, the astronomers, the cartographers, botanists and anatomists whose investigations would be enhanced by the clarity, compression and decoration afforded by italic. His intention was not to convert every able-fingered scrivener to cursive; italic was ideal for formal and illustrative work; unnecessary for everyday, private writing. Mercator himself scrawled his notes and informal correspondence in secretary hand, while the exterior addresses on his letters were frequently written in formal italic.

For Rescius it would be a cut-price production. To save money, the pages would be printed from wood rather than copper, and Mercator would cut the blocks himself. The introduction and miscellaneous inserts which required smaller text would be printed using a new italic typeface which Rescius had recently received from Cologne.[7]

Literarum latinarum was published in Louvain in March 1541.

Sigismondo dei Fanti[8] had claimed in 1514 that it was impossible to engrave italics onto wood without blemishes; the great Arrighi had subcontracted his wood-engraving to others. Mercator's woodcut italic was blemish-free and beautifully executed. To allow for the large copy-book text, the blocks had been cut at only seventeen lines per page, with the lettering of each line centred so that each page of flowing script was symmetrical. Rescius was less conscientious, misaligning some of his inserts of italic type.

The manual began with an Erasmian defence of Latin letters. After presenting his compliments to 'the candid reader', Mercator opened his brief introduction with a subtle rebuff of cursive's 'carping critics', the old devotees of gothic and carolingian minuscule who regarded the new Italian script as 'simple' and 'unnecessary'.[9] They might condemn, suggested Mercator, but they were wrong to regard 'the art of writing italic script' as 'something of insignificant worth or something to be valued solely for its decorative quality'.[10] Mercator's placable character could be read between the measured lines. This was an author who fought his corner by subtle repudiation rather than blunt attack, the tactical wordplay revealing a sense of philosophical duty, exercised through gentleness of manner and a respect for his detractors.

This initial sally was followed by an invitation to 'reflect how dishonest it is for a king to take off his purple and to go about in beggar's clothing unworthy of his royal station, or again for a beggar to wear royal purple. This', continued Mercator, 'is just what happens when the Latin language abandons its proper characters and disguises itself in Greek or gothic lettering.'[11] Here, Mercator began to reveal one of the sources of his calligraphic inspiration, for it was Erasmus' *De recta pronuntiatione* which compared 'clumsy handwriting'[12] to a Ciceronian speech rendered in gothic letters. Where Erasmus had found it 'quite unbelievable how the contents of a piece of writing are improved by an elegant, clear, and readable script which presents Latin words in the Latin alphabet',[13] Mercator echoed that 'the Latin language ... has a script of its own, elegant, easy to write, and far more eligible than any other'.[14]

The introduction continued with a reminder that 'no craft is so simple that its correct execution does not require some prior study'.[15] In Vivian terms Mercator referred to his manual as 'a few brief rules' that had been presented in order that 'the student is not detained too long from other, more practical studies'.[16] In playing down his manual, Mercator was also reassuring his detractors that the manual was indeed the 'insignificant' trifle that they could expect from an enthusiast of connected lettering. Mercator concluded by making a distinction between his own humble guidelines and the intractable rules of the lawgivers: 'Should you ever want to leave my teaching for another's,' he offered, 'I shall be perfectly content, provided only that you can demonstrate its superior elegance and readability.'[17]

Literarum latinarum was only 2,900 words in length, but – in the manner of the best manuals – each sentence had been constructed with precision to fit a methodical step-by-step format. It was a work which had been subject to repeated drafts. Mercator had followed neither Arrighi nor Tagliente, but picked elements from both.

There were six chapters. One and two covered the trimming and manipulation of the quill; three dealt with the proportions of italic letters; four described how to create the letters; five dealt with the joining of letters and six with capitals and decorative flourishes. Each of the six chapters opened with a sentence which encouraged the reader onward and described the purpose of that chapter. He made it sound so simple: 'The tools which you require for writing are compasses, a straight edge, and a pen.'[18] And yet, of course, there was more to it than that. With the intuitive skill of a teacher who knew the nervousness of young minds, Mercator was careful not to intimidate: 'Choose a transparent medium-hard quill. I say "transparent" because, if it is clouded with white marks, it will be less easy to make the correct slit in it. It must be of medium hardness because, if it is too soft, it will supply too much ink, and, if it is too hard, it will usually supply too little . . .'[19]

Unlike Arrighi and Tagliente, who did not tackle the quill until the mid-stages of their books (and even then, they failed adequately to demonstrate how to hold it), Mercator's approach to his manual was that of the craftsman, establishing the role of tools, their preparation, their handling, before introducing the reader – after ten pages – to the abstractions of italic style.

Quill-equipped, Mercator's reader was led through slitting and trimming with an economy of words which reflected the author's ability to convey abstract notions within the limitations of grammar and vocabu-

ÆTATIS   *Mercator, tractusque nouos, terræque, marisque*   SVÆ. LXII.

Magna Pelusiacis debetur gratia chartis: *Magna tibi priscum tandem superasse laborem,*   *Mons Prasse, et magnum quod continet omnia calum &c.*   *I. Vhau.* ludd:

Polus magnus.

AMERICA

Peru

GERARDI MERCATORIS RVPELMVNDANI EFFIGIEM ANNOR·
DVORVM ET SEX — AGINTA, SVI ERGA IPSVM STVDII
CAVSA DEPINGI CVRABAT FRANC. HOG. CIƆ. IƆ. LXXIV.

Gerard Mercator aged sixty-two, engraved by Frans Hogenberg
in 1574. From *Atlas sive cosmographicæ meditationes de fabrica mundi
et fabricati figura*, 1595 (Royal Geographical Society 263.G.9)

Mercator's terrestrial globe of 1541
(National Maritime Museum,
Greenwich, GLB0096)

South America, on Mercator's terrestrial
globe of 1541. The opossum became a
representative beast of the South
American tropics following its first map
appearance on Martin Waldseemüller's
Carta marina of 1516. (National Maritime
Museum, Greenwich, GLB0096)

Mercator's celestial globe of 1551
(National Maritime Museum,
Greenwich, GLB0097)

The southern constellation
Centaurus, on Mercator's celestial globe
of 1551 (National Maritime Museum,
Greenwich, GLB0097)

A contemporary of Mercator, Pieter Bruegel (c.1525-69) made ample use of globes in his painting which illustrates over one hundred popular Flemish proverbs. On the inn wall, an upside down globe symbolises the 'topsy turvy' world of the sixteenth century. In a secondary allusion, a scornful fool shits on the world. On the right, a correctly oriented globe alludes to the man who can make the world dance on his thumb. Just above it, God's globe is clasped to the lap of Christ (who is being masked with a false beard by a hypocrite). At the foot of the painting, a fourth, larger globe is rolled along by the man who must writhe in order to make his way in the world. The inn's missing roof tiles are a reference to the Flemish proverb 'The roof has lathes' (Walls have ears). At the time, an indiscretion could lead to arrest and execution. Pieter Bruegel, *Flemish Proverbs*, 1559 (AKG London / Gemäldegalerie, Staatliche Museen zu Berlin - Preußischer Kulturbesitz)

Next page, Nova et aucta orbis terræ descriptio ad usum navigantium emendate accommodata, Duisburg, 1569. Mercator's 'New and more complete representation of the terrestrial globe properly adapted for use in navigation' took the form of a world map on a rectilinear grid of latitudes and longitudes, later known as 'Mercator's Projection' (Bibliothèque Nationale, Paris)

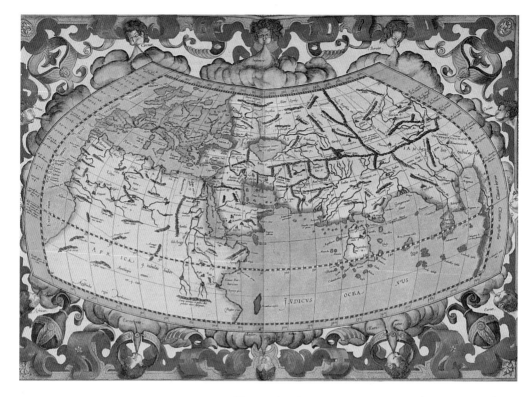

Universalis tabula iuxta Ptolemeum, Cologne, 1578. The
world map from Mercator's edition of Ptolemy's
Geography (Royal Geographical Society 265. G.12)

Asiae XII Tab, Cologne, 1578. The island of Taprobane, from Mercator's edition of Ptolemy's *Geography* (Royal Geographical Society 265.G.12). Note the map's engraver perched upon the cartouche.

Gerard Mercator and Jodocus Hondius, *Atlas sive cosmographicæ meditationes de fabrica mundi et fabricati figura*, 1619. When Hondius (1563-1611) acquired the copper plates for Mercator's Atlas in 1604, the Amsterdam cartographer sought to inherit Mercator's reputation by creating this twin portrait of the two men engaged in fruitful collaboration. Mercator had been dead since 1594. (AKG London/British Library)

lary. His procedure was so concise that the illustrations of a *mala fissio* (bad cut) and a *bona fissio* (good cut) were reassuring rather than essential.

In this ritualized tooling, Mercator was revealing the attitude he had towards his own mapmaking. Perfection depended upon preparation. And yet he wrote as one who knew the rigours of trial and error: 'When you have prepared your quill,' he instructed at the start of chapter two,

> hold it in two fingers – the index and the thumb – gently supported by the middle finger so that it is enclosed by them in a sort of triangle. Only the two former have the function of holding the pen; that of the middle finger is simply to stop it slipping away from the other two when they have to be pulled back in the act of writing. You will also find it a help if the two former fingers are stretched out straight in order that they can be drawn back a greater distance, since this will often be necessary, as you will see later.[20]

By chapter three, most readers would have sensed impending commitment: 'It will not be necessary to consider proportions here in minute detail …'[21] began the author, who then illustrated with the aid of a geometric drawing how to divide opposing sides of a square into twelve equal parts not dissimilar to a map's lines of longitude. Into the square, Mercator ruled the three principal strokes of the letter y, two strokes at 3.75 degrees from the vertical, and the diagonal formed from the hypotenuse of the inverted obtuse triangle. The principle – to create a parallel-sided script which sloped at just under 5 degrees – was straightforward and clearly illustrated in his diagram. But (excusably perhaps for a man whose first medium was the abbreviated image – although the fault may actually have lain with Rescius) Mercator made a small slip in his text, transposing two of his diagram's symbols in such a way that the reader's perfect y would be disfigured by a half-mast diagonal. The chapter ended with instructions on using the seven basic italic strokes required to write the alphabet.

It was in the book's penultimate chapter that subject and object became one. After the pages of deconstructed alphabet, the author was suddenly writing about italic, *in* italic. The reading effect was miraculous, for the eye now flowed along the lines as Mercator's own hand broke free of instruction in a celebration of cursive rhythm, pausing on '*immunis*' to illustrate how seven letters could be linked without a pen-lift.

Bringing the manual full circle, Mercator recalled the decorative value of italic, the increased writing speed it facilitated and the fact that it enabled the writer 'to avoid useless movements of the pen'.[22] Here, again,

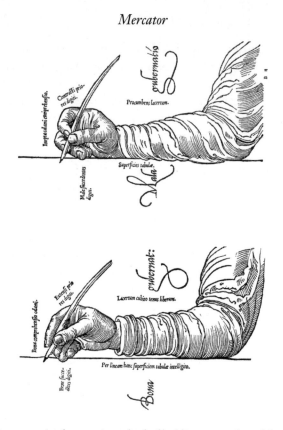

Fig. 13. The incorrect (and correct) method of holding a pen, from Mercator's manual of italic, *Literarum latinarum*. (Private collection)

was the essence of italic. Mercator distinguished two types of letter combinations, the 'impure', which occurred when part of a letter was 'hidden or distorted from its natural shape', and the 'pure', in which 'one letter is joined organically to another like a living creature'.[23]

It was no arbitrary whim which placed the ampersand at the end of this climactic chapter. Nearly three pages – more than for any of the alphabetical letters – were devoted to this symbol of inclusivity.

Recalling the structure of a Macropedius morality play, Mercator reserved the sixth and final chapter for the tidying of loose ends, in particular the taming of flourishes that could curl and loop from those letters 'that riot with greater exuberance than their fellows outside the body of the text'.[24]

13

A More Complete Globe

Only six or so years had passed since Mercator had picked up a burin and committed himself to describing the world.

In that time, he had engraved Gemma's terrestrial and celestial globes, produced a map of the Holy Land, a world map and a regional map, and constructed an unknown number of instruments. And he had published the first manual of italic to appear north of the Alps. In the same six years, Barbara had produced two sons, two daughters (the latest arrival had been named Bartholomeus), and held together a home while Mercator worked through every waking hour.

Mercator had every reason to assume that his fortunes – and his family – would continue to accumulate. His origins may have been humble, but his rise had been steady and he was a stranger to setback. Realizing 'that the products of his apprenticeship were being widely praised by experts, he quickly began another task'.[1]

That task was the globe that he had advertised to Antoine Perronet in August 1540. It would mark a peak of productivity for this measurer of Louvain. And it would be followed by disaster. Two disasters.

So fast was the world view changing that Gemma's four-year-old globe was already out of date. And so was Mercator's two-year-old map of the world. Gemma had recently published a new map of the world.[2] Now Mercator felt bound to correct the planet's geography. He was also eager to remove from the record his heart-shaped map, whose geography he now recognized as being embarrassingly obsolete.[3]

The decision to construct a new globe was a signal of supreme confidence. A globe required an investment of time and money which could only be justified by expectations of substantial sales. But Mercator had backing at court, and presumably from book merchants too. More than a map, or a book, or any other instrument, this globe could be seen as the sum of his many parts: it would be a work of exquisite craftsmanship, of mathematical ingenuity, of geographical research and of calligraphic excellence.

It would also be big. Mercator had given much consideration to the issue of size. On the one hand, he wanted the globe to 'include many more places than others' and be 'a more complete globe than has been done so far', and yet this had to be achieved 'without making it so large as to be difficult to use'.[4] In committing to an optimal scale, he was facing a quandary familiar to any globe-maker. Where a cartographer could separate world and regional geographies into small-scale and large-scale maps, a globe-maker had to compress all of his geography onto a one-off scale model of the planet. Initially, printed globes had been small spheres whose purpose corresponded with Ptolemy's parameters for a world map. Such a globe would show the 'broader, general' outline of the world 'as a single and continuous entity'.[5] But as more information became available, and as the expectations of globe-users increased, printed globes had expanded in diameter. Schöner's globe had been larger than Waldseemüller's, and Gemma's had been larger than Schöner's. There was, however, a limit to a globe's size. Notoriously fragile, a printed globe had to be sufficiently compact that it could be shipped without damage or undue cost. For convenience, a globe should also fit on a desk top. And since the size could not be altered once the copper plates were cut, one standard scale had to satisfy the full spectrum of potential customers, from professors to diplomats.

Fortunately, mathematics was on the side of the globe-makers, for it was a surprising but true fact that a relatively small increase in the diameter of a sphere would yield a considerable increase in surface area. Mercator decided to push the size of his new globe as far as he dare. Limited by the average size of a desk top, and by the need to construct a globe which was sufficiently compact and robust to transport, he settled on a diameter that would be just over one-eighth bigger than that of Gemma's globe. In doing so, he gained an additional surface area of nearly one-third.[6] This would be the largest and the most detailed printed globe ever constructed.

The enterprise marked Mercator's emergence as an independent earth-modeller. Since the printing of his world map two years earlier, he had been devoting his 'spare hours' to 'comparing the old geography with the new',[7] and had come to the view that currently accepted continental outlines of the world were deeply flawed. The problem was particularly acute along the southern reaches of Asia.

'The more carefully I examine,' he had admitted to Perronet, 'the more errors I find in which we are enmeshed.'[8] By 'we', Mercator was referring obliquely to Gemma and to himself, for the errors he now elaborated upon had been cut into copper on Gemma's globe, and then copied by Mercator on his map of 1538. 'It seems particularly erroneous', he continued, 'to take Malacca for the Aurea Chersonesus...'[9]

The problem, he explained, was one of peninsulas. There were not enough of them, and they were in the wrong place. Ptolemy had shown the southern Asian coast to possess two major distinguishing features: the huge island of Taprobana, and the bulb-ended peninsula of Aurea Chersonesus – the Golden Peninsula. Now Taprobana had been replaced by the triangular peninsula of India, and the Golden Peninsula had moved westwards so that it could correspond with recently discovered Malacca. Meanwhile, a third peninsula had been created by severing Ptolemy's far-eastern land bridge and turning its stump into the land of the 'Lequii'. Gemma's globe, and Mercator's world map, had shown these three prominent peninsulas.

But Mercator no longer believed that Ptolemy's Golden Peninsula should have been shifted west to accommodate Malacca. He was also unhappy that Gemma – and then he – had moored a shrunken Taprobana just off Malacca, despite the fact that Ptolemy had written that the island lay much further – 'about 30 degrees' – from the Golden Peninsula.

The problem had been caused by their attempts to reconcile disparate sources: 'When we, blinded in this way, attempted to harmonise the irresolvable difference between the old and the new, we denounced both the ancient and more recent descriptions; in addition, by means of small adjustments, we undermined the current proportions of the coasts as well as the findings the ancient geographers had achieved through great effort.'

Mercator's research had uncovered a source which had convinced him of their folly: 'How deceived we were in our conception of the Far East will be sufficiently clear to anyone who attentively reads Marco Polo the Venetian.'[10]

Marco Polo had been dead for two hundred years, but his book of travels, *Divisament dou Monde* (Description of the World) had circulated widely in manuscript form. Mercator had studied very carefully the Venetian merchant's itinerary. The distances and descriptions that could be extracted from the account of Polo's sea voyage from 'Mangi'[11] to India made it possible to reinstall the Golden Peninsula in its Ptolemaic location. The peninsula labelled by Gemma and Mercator as that of the Lequii, was clearly Polo's Mangi. To the west of the Golden Peninsula, lay the peninsula of Malacca, and west of that, the peninsula of India.

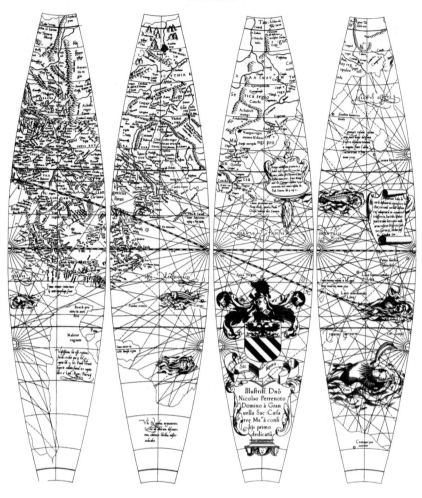

Fig. 14. A detail showing Asia, on the gores engraved by Mercator for his magnificent globe of 1541. (Bibliotheque Nationale Ier, Brussels)

Thus, there were four – not three – peninsulas along the coast of southern Asia.

There were other errors which needed addressing by the new globe. Mercator also intended to 'include in the edition corrections for ... Scythia, the two Sarmatias, Scondia, Scotland, Iceland and the islands in the Deucaledonius Sea, for Madagascar and for many surrounding islands, and finally for the lowest parts of Africa'.[12] That was not all: 'In addition, as far as America is concerned, although we will not change the coastlines from the way they have been presented up until now, we will give some more details of the lands in the interior.'[13]

Marco Polo proved a fruitful source. Mercator was able to use the Venetian's voyages to add detail to the vast Antarctic continent. The new globe showed a great polar promontory reaching almost as far north as 'Java maior'. Mercator tipped it with Marco Polo's province of 'Beach',[14] the 'wealthy province of the mainland' whose 'idolaters'[15] had elephants and so much gold 'that no one who did not see it could believe it'.[16] South of Beach, Mercator located Marco Polo's spice-drenched 'Maletur'.[17] Mercator provided further evidence of a luxuriant Antarctica by locating Pedro Alvares Cabral's 'Psitacorum regio' (Parrot land) west of Beach.

Mercator was impressed by Marco Polo. The Venetian had yet to be printed, and although others had used manuscript copies of his travels to modify their maps, Mercator latched onto Polo with the reverence of a new convert: in a decorated cartouche off Mangi, Mercator referred to Marco Polo's description of a disastrous attack on 'Zipangri'[18] by the Great Khan in the year 1268, and of the 7,448 islands that the Venetian recorded as lying between Mangi and Zipangri.

In the northern hemisphere, Mercator subjected Ptolemy to doubts that he'd felt unable to exercise on his heart-shaped world map. At last, he rejected Ptolemy's kinked Scotland, straightening Britain so that the meridian of 15 degrees east formed a central axis from the island of 'Vicht'[19] in the south to the 'Orcades'[20] in the north.[21] Mercator also corrected Ptolemy's Mediterranean, reducing its span of 62 degrees of longitude (which Mercator had repeated on his 1538 map of the world) to 58.5 degrees.

Up north, there was a more consequential modification: without providing his readers with an explanation, Mercator closed off the north-west passage and opened a north-east passage. In doing so, he tacitly instructed navigators that the route to Cathay would be found around the top of Scandinavia and not by way of the American strait, as he had indicated on his map of 1538.

There was much more to this globe than size and a revised geography.

The single most innovative feature arose from Mercator's developing preoccupation with magnetism. In the three years since he had seen Ziegler's woodcut illustration of compass declination, Mercator had become increasingly absorbed by the nature and location of the magnetic pole, and by the behaviour of the compass needle at various points on the earth's surface.

Others were thinking along the same lines. Just two years earlier, Olaus Magnus had published a huge woodcut wall-map of Scandinavia –

the *Carta marina*. Off the top of Scandinavia, the Swede had placed a small island labelled 'Insula magnetus', the island thought to draw compass needles towards a magnetic pole. To emphasize that the island should be considered as a second pole, Olaus Magnus had placed the geographical pole, 'Polus Articus', immediately to the north. Mercator copied the island and its location onto his globe and labelled it 'Magnetū insula'.

The other magnetic feature which appeared on the globe laid the foundations for Mercator's ultimate fame.

Hundreds of hours spent scrutinizing maps and charts had confirmed the incredible inconsistency of geographical data. Although he had never been to sea himself, Mercator believed that navigators were being led astray by their own compasses. The evidence could be demonstrated in two minutes to anyone with a globe and a compass: if a compass was laid on the surface of the globe, and a line of constant bearing – a rhumb line[22] – was plotted, that line would curve towards a pole; any rhumb crossed meridians at a constant angle, a characteristic which caused rhumbs to form a gentle, poleward spiral. Most navigators sailed under the false impression that a constant compass bearing took them along a straight course, and the inaccurate charts they made reflected that misconception.

The problem had recently become particularly acute for Portuguese navigators, whose immensely long voyages around the curved surface of the earth were proving increasingly difficult to record on a flat map. Returning to Europe from Brazil, some captains discovered that they were as much as 70 leagues further west of the Azores than their dead-reckoning suggested. In 1537, shortly before Mercator began considering his new globe, the Cosmographer Royal of Portugal – Pedro Nuñez – explained the phenomenon in his treatise *Tratado da Sphera*. (If Mercator saw this before completing his globe, he kept it quiet.)

In his treatise, Nuñez demonstrated the relationship between rhumb lines and meridians. The shortest possible distance between two locations on the earth's surface was a 'great circle', a straight course which required continuously changing compass bearings. The alternative was the constant 'rhumb line' bearing, which was far easier for the navigator, but involved a longer, curving passage.

With inspired prescience, Mercator decided to mark rhumb lines onto his new globe. Curving in spirals towards poles they could never reach, each rhumb line represented the hypothetical course a ship would take were it to steer on a constant compass bearing. Mercator was the first to engrave rhumbs onto a globe.

The next step would be to present rhumb lines on a flat map. But that would need a new projection.

Engraved with its spiralling rhumbs, radical new geography and multitudinous place-names, the new globe was an astonishing sight. Not only was it the largest globe ever to be printed, but Mercator's superlative italic lettering had been rendered at a barely believable level of precision and miniaturization. His lower-case letters were so small that some readers would require a magnifying glass for detailed inspection, and yet each was as perfectly formed as his own manual had prescribed. Where particular letters lent themselves to a flourish or three, Mercator had engraved minute embellishments with the needle-point of his burin. Rivers, lakes and mountain ranges swam into vision beneath the lens in a way which made the reader feel that they were engaged in an exploration of the earth's surface.

For all of its beauty, Mercator had designed his new globe as a utilitarian device. The sturdy horizon ring around the globe's circumference was supported on four legs of turned wood which were anchored in a solid cross-braced base. Thus protected, the globe could easily be carried about a house or court, and transported by ship or cart.

Mercator dedicated his new globe to Antoine Perronet's father, the most eminent of the emperor's ministers, Nicolas Perronet de Granvelle. In doing so, he confirmed his role as an Imperial asset; a loyal, extraordinarily productive subject who might match the cosmographical talents of Lutheran geographers like Sebastian Münster.

Although the Emperor had revised his vision of *dominium mundi*, his dependence upon accurate instruments, maps and globes had never been greater. Maritime navigation, artillery-ranging, location-finding and strategic planning required the latest data and equipment. A globe was a geopolitical tool; the only means an emperor had of visualizing his vast, geographically fragmented interests.

Granvelle had left the Low Countries before the latest globe was finished, although he may have received it as early as spring 1541,[23] while the emperor and his entourage were tied up in Germany with the Diet of Ratisbon. Among the entourage that spring was Mercator's other Imperial ally, Granvelle's son Antoine Perronet.

It was a frustrating period for Charles v, and his last serious attempt to compromise with the Lutherans. Assembled theologians from both sides of the schism had agreed on a set of articles but the Pope and Luther had condemned them, and they were rejected by the Diet. Unable to settle the issue by force because he needed the German princes to help repel the Turks, Charles decided to attack Algiers. In August, the emperor moved on to Italy, Spain and his disastrous African campaign, and by November 1541, he was back in Spain, where he suffered his ninth and worst attack of gout and learned that his American treasure receipts had begun to fall.

It was against this background of Imperial peril that Chancellor Granvelle recommended Mercator to the emperor. Mercator began constructing for Charles v 'numerous mathematical instruments',[24] among them celestial and terrestrial globes, a compass and quadrant, and an astronomical ring.[25] War had become a highly technical business. Set-piece collisions on open ground had been superseded by tactical clashes in which the advantage could be won in reconnaissance and analysis of terrain.

A bulk order of complicated instruments was a huge commission for one man to fulfil. But for Mercator it meant an end to borrowed tools. With a fifth child to feed (another boy, Rumold, having been born during 1541) he could at least depend on some measure of financial security. It did not last.

14

Enemy at the Ramparts

While Charles was in Spain, a firestorm swept the Low Countries.

Through the early months of 1542, the regent, Queen Maria of Hungary, watched the embers being fanned on all sides: 'Since the time of our grandfather,' she fretted, 'the Netherlands were never in such danger.'[1] The coastal towns were braced for an attack from the Danes. Artois and Flanders were being marched upon by the French. Luxembourg looked likely to fall to the Duke of Orléans. But the greatest threat of all was stirring beyond 's-Hertogenbosch, beyond the rivers that kept Brabant safe from flammable Gelderland, the province allied to the Turk, the source of French attacks on the Holy Roman Empire, the lair of Duke Karel's old commander, ferocious, bristle-moustached 'Black' Maarten van Rossem.

With the death of Gelderland's Duke Karel in 1538, the leading anti-Habsburgian trouble-maker in the north had been removed from the Low Countries. But Karel had died without an heir, and in 1539 Gelderland passed to the Duke of Cleves-Mark-Jülich-Berg, Wilhelm v, twenty-three and primed to ignite the Low Countries. Jülich, the duchy that had been Mercator's boyhood home, had become part of a strategic block which straddled the lower Rhine, under the rule of an Erasmian duke with Lutheran enthusiasms. Wilhelm v raised his anti-Habsburgian profile by making alliances with the Lutheran princes, and also with Denmark and France. It was no surprise to Queen Maria when Wilhelm's Gelderlanders burst across the Maas.

Led by Black Maarten, they came with Danes and Swedes and men from Jülich, and 600 horses under de Longueval, 16,000 in all, a ragtag army hellbent on easy plunder. It was a hot summer and the thatch flared fast. Past well-defended 's-Hertogenbosch they swept, torching northern Brabant, razing villages and burning windmills. 'Fire', exclaimed Black Maarten, 'is the *magnificat* of war.'[2]

In Louvain, it looked like the end: if Black Maarten smashed Antwerp and Ghent, he could link with the Duke of Vendôme's Frenchmen heading up through Flanders. The Habsburgian Low Countries would be split in two. Fearing that the invading force was swelling en route with Lutheran sympathizers from overrun towns, Queen Maria began arresting malcontents and suspected heretics for interrogation, torture and execution. Militias were called up and lords were instructed to recruit infantry and light cavalry.

As the invaders bore down on Antwerp's unfinished ramparts, the Prince of Orange rushed to the city's aid and a well-born Italian, Jean-Charles d'Affaytadi, called his fellow merchants to defend Europe's mercantile headquarters. Black Maarten breached dykes to cover his flank, but after one assault, he shied from mounting a siege and – swearing that his spurs would strike fire from the land – he wheeled south-east towards a softer target: Louvain.

But Louvain too would be saved by a foreigner. The hero here was the Portuguese poet-musician who had gone to print in defence of the Lapps and Ethiopians. Just as d'Affaytadi had rallied the merchants of Antwerp, Damião de Gois rallied the students of Louvain. The city had cannon, but no commanders, and the council seemed incapable of organizing a coherent defence.

On 21 July, the rector ordered that students and staff who had a house or dwelling in the city should stand watch on the walls and the following day, they began clearing the moats and repairing the battlements. Among the students was Mercator's mentor; the thin physician Gemma Frisius spent four days on the ramparts watching for fires that signalled Black Maarten's approach.

As the onrushing horde spread panic in Louvain's streets, scarcely a day passed without the town council issuing another nervous order. Queen Maria's relief force failed to reach Louvain, and in a measure which betrayed the regent's distrust of the student population, she ordered Louvain's council to inform all *scholares* over sixteen from Cleves, Berg, Mark and Jülich that they should leave Imperial territory within two days. Mercator had lived in Jülich for his first six years, and for him, Maria's edict was reaffirmation of Imperial prejudices. Like others who had lived beyond the Maas, he was tainted with foreign, Lutheran habits.

On Friday 28 July, the university authorities debated whether to hide their treasured sceptre. A solemn Mass was heard in St Pierre's, and prayers said for the city. By now, the rector and his deputies had lost faith in Maria's promise of troops, and in the town council's reassurances that the defence was under control. Students whose parents had requested their evacuation of Louvain were allowed to leave, and so were those who had no permanent home in the city.

The students who remained were those compelled by domicile, and volunteers. They became the thin courageous line between the approaching barbarians and the Habsburg seat of learning. On Sunday 30 July, the students appointed their commander: *capitaneus* Gois.

Was Mercator up on the walls with Gemma? Mercator came from migrant stock whose instinct when threatened was to move to safer ground. Plagues, famines and wars were all the same: migrants moved ahead of them. But Barbara was expecting (or had recently given birth to) their sixth child, Katherina, and Mercator was encumbered by his invaluable collection of tools, copper plates, books, maps and associated cartographic paraphernalia. He knew better than anyone how to assess a map for suitable sanctuaries, and it must have been clear that the open road was no safer than Louvain.

Three days after Gois was appointed student commander, the enemy reached Louvain and pitched their camp at Ter Banck. At the sight of Black Maarten's marauders, the bailiff of Brabant and the captains of the recently arrived auxiliary force decided that defence was impossible and – without informing Gois and his students – offered the enemy money to leave. An exchange of heralds formalized the decision to parley and Gois was deputized with three others to negotiate a surrender with Black Maarten and de Longueval.

By the time Gois returned from the enemy camp with the besiegers' conditions, the French herald had informed de Longueval that Louvain's authorities had suffered a collapse of morale and that the city was helpless. Returning to de Longueval, the four Louvain negotiators were told that the sum required to spare the city had been increased. This time, only three negotiators returned to Louvain. Gois remained in the enemy camp. (Later, he would relate that he had stayed in order to restrain the northerners from attacking the city.)

Meanwhile, the motley army of besiegers had begun milling about the moat and walls of the Brussels Gate. Above them, the students on the ramparts suddenly opened fire. Men fell and withdrew in confusion. Pandemonium erupted on the walls. Appalled by the consequences of breaking the truce, Louvain's mayor, the Lord of Blehen, tried to halt the firing by riding from the city gates. Before the walls, he ordered the students to cease their firing. When they refused, he galloped up Ter Banck to find that Gois had already convinced de Longueval that the source of the firing had been the sudden arrival of Queen Maria's long overdue relief force.

Persuaded that the siege had been relieved, de Longueval proposed to his commanders that they leave Louvain and press on south to meet the Duke of Orléans' force. (Later, Black Maarten and de Longueval would admit that they had another reason for lifting the siege: the fear that their men would become so satiated on the city's stupendous riches that they would have no motive for following their commanders to France.) Leaving a trail of smouldering buildings, the invaders melted into the low hills beyond the woods of Heverlee, bound for Yvoy and the long-anticipated link-up with the Duke of Orléans.

For three weeks after the siege was lifted, students were kept in arms, and 24-hour watches mounted on the walls. On 20 August, at eight in the morning, a thanksgiving procession moved through the streets, with torches and candles held aloft by students and teachers. Missing from the procession were the siege's two notables, Danião de Gois and the mayor, Adrian de Blehen. Both had been carried off by Black Maarten and de Longueval. A few days after their abduction, letters were received in Louvain from the two hostages. The city could reclaim its student leader and mayor on delivery to Black Maarten of 70,000 gold crowns.

Mercator might have been back at his maps and instruments by September. But Louvain was a city in fear. Luxembourg had fallen, and although the French appeared to have run out of fight, nobody could be sure when Black Maarten would return to avenge his thwarted siege. As impecunious as they were spineless, the city's authorities began quibbling over the financial costs of the siege. In October 1542, the university reluctantly bowed before pressure from the emperor, and began contributing funds to reconstruct the city's ramparts. To the outrage of Gois' friends – and distress of his wife – the city council and the university refused to pay the ransom demanded for the return of the two hostages. Claiming a shortage of funds, they held that Gois and de Blehem were not heroes but victims of their own improvidence. They should never have ventured outside the walls.

Mercator would have done well to note the failure of Louvain's authorities to rally to a lost alumnus.

By the autumn, Queen Maria's commanders were pressing her for a punitive attack on Duke Wilhelm of Cleves. In October, Imperial troops fought through to Jülich, and the Prince of Orange took Sittard, the neighbouring town to the Kremers' old home in Gangelt. But Sittard was quickly taken back by the Duke of Cleves, and Düren too. Misery amid the cinders of Brabant was compounded by snow storms and gales during a wild, wet winter. Through war and weather, the road to Cologne remained cut for a year. Louvain bought back its mayor for 2,000 crowns, but Gois was carted off to Picardy.

Then Black Maarten reappeared. Supported by the headstrong Duke Wilhelm, the terror of Louvain exploded again from the north at the head of a huge, French-funded force. In deep snow, Heinsberg was besieged, and yet another battle raged about Sittard. Queen Maria clung to her fortresses and her cannon, waiting for her brother.

Charles embarked at Palamós in Spain, on 1 May. The time was ripe for Gelderland to learn the Imperial way. Gelderland had become the last piece of the Low Countries jigsaw to lie outside Imperial control. Eradicate the errant Lutheran Duke Wilhelm, reasoned Charles, and a ripple of resolutions would return peace, prosperity and Catholicism to the truly united Low Countries. Gelderland symbolized the emperor's universal struggle against the enemies of Christendom.

From Spain, the emperor sailed to Italy, meeting with Granvelle in Pavia before crossing the Alps and assembling an army of 36,000 Germans, Italians and Spaniards. Certain of their own annihilation, Duke Wilhelm's allies evaporated. Jülich, Düren and Roermond capitulated. In September, the petrified duke went on foot to the emperor's camp outside Venloo and renounced both Gelderland and Lutheranism. In return, he was allowed to keep his original German duchies.

The residual duchy of Cleves-Jülich-Berg-Mark would form the stage for the second half of Mercator's life.

In Louvain, the news of Duke Wilhelm's humiliation was greeted with relief and fear. No sooner had the enemy at the gates been vanquished than the authorities turned to the enemy within: heresy.

The Low Countries were saturated with Lutheran print. Twenty-six years had passed since the Kremer family had migrated from Germany to the Low Countries, and in that time the number of printings and reprintings of Luther's works – *excluding* Bible translations – already exceeded 2,500; over the same period, Catholic printings had barely reached 500.[3] And for every five treatises by Luther, there were another four by other evangelicals; by the mid 1540s, nearly 6 million evangelical tracts had been printed, or one tract for every two people in the Holy Roman Empire.[4]

Sects had proliferated, though few had been as severely persecuted as those who rebaptized adults into their Godly community. The 'Anabaptists' (from the Greek 'ana' for again) were alone in rejecting Christ's affirmation to 'render unto Caesar the things which are Caesar's'.[5] In refusing to pay taxes and to serve in the military, they had invited a wrath reserved for traitors to the Empire, as well as alienating themselves from moderate sects who pursued a more cooperative relationship with the authorities.

Among the sects local to Mercator was a shadowy group centred on the divinely appointed David Joris – who had seen visions in which the world lay at his feet. Charismatic, yet professing a spirituality which left plenty of room for interpretation (a precaution which encouraged recruits) Joris and his followers were being hounded by the Inquisition. In 1538, his mother had been beheaded in Delft, his wife and daughter banished, and a price put on his head. The same year, twenty-seven of his less fortunate followers had faced the executioner. For the last few years, Joris had been hiding outside Antwerp, where he had written *TWonder-boeck*, the Jorisian 'Bible' – printed in Deventer in 1542.

Conciliatory by nature, Mercator was not the kind of individual to be drawn to divisive sects, but their presence in and around Louvain fuelled the paranoia that would soon engulf him.

To the established sects could be added any number of individuals and non-affiliated cells ranging from the curious to the openly sceptical. Across Louvain, in back rooms and bedrooms, secretive knots of evangelists had been gathering for years to explore their faith. Believing themselves to be beyond the reach of the authorities, they held readings from the New Testament and from printed sermons (Postills), and participated in controversial discussions about baptism, purgatory, confession and Communion. Some cells were more active than others. Descended nobly from a civic magistrate, widowed Antoinette Van Roesmaels had hosted several meetings, and evangelicals had gathered under the sign of the *Palmier* to be fed by its innkeeper, Jean Bosschverckere. Among those who had regularly slipped through the dusk were Josse Van Ousberghen, a furrier's assistant, Calleken (or Catherine) Sclerckx and her widowed sister Betteken (Elizabeth), the tanner Chrétien Broyaerts (a nephew of Antoinette Van Roesmaels), Jacques Ghysels the potmaker on Cortestrate, Laurent the tailor on Bieststrate, Thierri Gheylaerts the hosier (whose habit it was to sing in inns 'beware the leaven of the pharisees'),[6] Jean Vicart the haberdasher on Scipstrate and his neighbour, Jérôme Cloet the bookseller. Many of them were civic figures of some prominence; people Mercator must have known. The Lord of Grembergen's son, Paul Roels, was known for mixing with Lutherans, and most conspicuously with the reformers Jean de Lasco, Albert Hardenberg and Jacques Enzinas from Spain, whom he lodged for a month in 1540. Jacques' brother Francisco[7] (a graduate of Wittenburg and an ardent Melanchthonian) had just been arrested in Brussels for translating the New Testament from Greek to Spanish without a licence. Roels had been granted his licentiate degree in medicine four years ahead of Gemma and been appointed university rector in 1532, the year Mercator graduated. Equally prominent in Louvain was Jean Utenhove, whose father had been president of the council of Flanders, and whose half-brother was the Erasmian humanist, poet and one-time rebel burgomaster of Ghent, Karel Utenhove. Catherine Metsys was also well known, as the daughter of Josse Metsys, the clock-maker, sculptor and architect of the incredible plans that were intended to make the unfinished tower of St Pierre's the tallest in Europe. Catherine's evangelical husband was the sculptor Jean Beyaerts (a grandson of the town secretary and nephew of the town sculptor), while her uncle was the artist Quentin Metsys, famous among humanists for painting the first portrait of Erasmus. Catherine's cousin Jean Metsys (Quinten's eldest son) was also involved with the evangelists. And priests were implicated: Paul de Roevere, Pierre Rythove and Matthieu op de Brug van Rillar would all find themselves implicated in heretical cells. The risks were considerable: in July 1543, the Spanish reformer Francisco Enzinas witnessed the execution in Louvain of two elderly women accused of Lutheranism. They were buried alive. The lesser crime of blasphemy was punishable by a hot iron through the tongue.

This diverse, secretive community included Johannes Drosius, to whom Mercator had dedicated his map of the world shaped like a heart.

Mercator and his friends dwelled in a *terra incognita* between the Anabaptist and Catholic polarities. Mercator was one of many Erasmians who quietly shared the view that faith in Christ was more important than ritual and ceremony; that the Gospels could be read as a communal source of spiritual guidance; that through dialogue and worship some kind of reconciliation could be achieved between the opposing religious factions. He was, like Louvain's saviour Damião de Gois (who had published his own plea for religious tolerance)[8] a seeker of religious harmony.

Those who had spun a globe, who had laboured to create a working model of the world from seemingly irreconcilable raw material, were familiar with diversity. Mercator's journey towards a unified cosmography was a mirror of the syncretic ideal, his reconciliation of conflicting geographical data a reflection of the compromises required to harmonize Christianity. But the middle ground could be more dangerous than the extremities: caught in the theological crossfire, despised by Catholics and Lutherans alike, the eirenic idealists of Louvain were no less vulnerable than the Anabaptists.

That Mercator had formed his own views there is no doubt. But it would be many years before he would commit those views to paper. In the mean time, there were discreet conversations, prayers, contemplation, letters perhaps, delivered by trusted friends.

Two years of war had left Mercator with little to show since producing his globe. His Imperial commission for instruments promised some security, but the siege and its aftermath had cost Mercator his creative momentum. And his income.

Regrettably, he had relinquished his financial stake in his manual of italic. Following the success of the first edition, Mercator and the avaricious Rescius had sold the woodblocks for *Literarum latinarum* to an Antwerp book-printer, Jean Richard, who had produced a shoddy second edition scattered with printing errors and missing its last page.[9]

The manual, however, had indelibly associated Mercator with italic script and Italian humanism. Gemma was one of many in the Low Countries whose recent editions had been enhanced by italic, and Mercator was admired by printers for his lettering and illustrative skills. At the Golden Sun printing house, Rescius' partner Bartholomeus Gravius commissioned Mercator to create a personalized printer's device which he could use on his title pages. Mercator devised a spiky sun whose thirty-two principal 'flames' were so precise that he may have lifted them from the 32-point compass he had placed at the top of the map of Flanders with which Gravius had been associated. In place of a central compass dial, he engraved the Christ Child bearing a cross. The three lines of italic were characteristically clear and elegant.

At about the same time, the Antwerp printer Matthew Crom also commissioned Mercator to cut an italicized block which he could use on his title pages. Crom had published around forty books and pamphlets since 1537 and Mercator must have known that these included Lutheran editions of the New Testament in English, and at least one Flemish work which would soon attract the ire of Louvain's theologians.

But these were minor commissions undertaken by a troubled mind. In the aftermath of the war, Mercator even found himself surveying land in Ghent, where the prior of St Bavon was in dispute with the abbot of St Pierre. Ghent was still recovering from the emperor's punitive visitation and Mercator may have been assisting with the construction of the new citadel on the razed lands of St Bavon's abbey.

In November 1543, Mercator was summoned to Brussels.[10] There he

Fig. 15. Mercator, as he would have appeared at around the time that he left Louvain for Duisburg, engraved by Nicolas II de Larmessin (1654–94). (Private collection)

met his old university acquaintance Antoine Perronet (who had been consecrated Bishop of Arras in Valladolid that May) and the Archbishop of Valenciennes, both of whose lands lay on the front line with France.

The purpose of their meeting can only have been the resolving of border geographies. Towns of over 10,000, Arras and Valenciennes were the two main centres of population on the southern borders of the Habsburgian Low Countries, and they lay each side of the French salient of Cambrai. Back in Brussels, the emperor had reason to focus on this section of his border. At the beginning of the month, he'd faced off Francis and the French army just outside Valenciennes, then marched into Cambrai and taken it from the French. After five days in Cambrai, Charles had spent four days in Valenciennes. The Valenciennes–Cambrai–Arras front would be the start-line for the next French incursion and as such, was of significant strategic importance.

That Charles, the two Granvelles and the Archbishop of Valenciennes were poring over Mercator's map of Flanders there is little doubt. The framing of the map had prevented Mercator from marking Arras and Cambrai, while Valenciennes clung to the edge. Charles would have had urgent cause to establish how much Mercator knew of the land beyond the edge of his map. The lone swordsman engraved onto Mercator's map can only have reminded Charles of his current vulnerability.

Mercator was vulnerable too, although he may not have known it. Below the clear-cut world of his cartography lurked a murky netherworld of whispers, intrigue and betrayal. Beholden to the court for his privileges and patrons, dependent upon liberated humanists for his geography, and upon the Gospels for his faith, Mercator had his own enemies.

15

The Most Unjust Persecution

Mercator's long winter began at the end of 1543.

The wind from the east began to blow at the start of December, a bitter wind that turned the soil to stone. Across the lowlands of Flanders and Brabant the wind cut unchecked. Frost turned the polders to blinding panes. Meers froze. Food ran short. The price of wood soared. At the beginning of January, the emperor left Brussels for one night in Louvain, then departed again for Germany – equipped perhaps with some of the instruments which had been commissioned from Mercator through Granvelle. By the beginning of February, the Schelde at Rupelmonde had frozen over. One of the many who died that long, cold, winter was Mercator's benefactor and paterfamilias, his uncle Gisbert Kremer. Following the troubles of war and interrupted work, the death of Gisbert struck an unnecessarily cruel blow. There was worse to come.

While the east wind had been icing the Low Countries, Queen Maria of Hungary had been preparing to purge her regency of heresy.

In one swoop, Louvain would be cleared of Lutherans. A list of suspects was compiled. The fact that the procurer-general of Brabant was still Pierre Dufief – the man who had given the nod to Tyndale's strangler – did not bode well for individuals whose names were on the list.

By February, Dufief had forty-three names from Louvain, five from Brussels, one from Antwerp and one each from Bethléem, Groenendael and Engien.

Among those on the list were the widow Antoinette Van Roesmaels and her nephew Chrétien Broyaerts, the architect's daughter Catherine Metsys and her sculptor husband Jean Beyaerts, the ex-Rector of the university Paul Roels, the three priests Paul de Roevere, Pierre Rythove and Matthieu op de Brug van Riller, the furrier's assistant Van Ousberghen – and his boss, Louis Van Malcote, Cloet the bookseller, Ghysels the pot-maker, Gauthier the glazier, Laurent the tailor, Vicart the haberdasher, Gheylaerts the hosier (and his brother Baudouin, and his wife, Marie, sister of the arrested furrier Malcote), Bosschverckere the innkeeper, the sisters Calleken and Betteken, Paul and Jean Hersthals, brothers, André the town flautist, Pierre the painter, a monk, a midwife, a cobbler, the guardian of the fountains at Groenendael, Gaspar Vander Heyden – related perhaps to Mercator's fellow engraver on Gemma's globe. Also on the list was Mercator's friend Drosius.

Mercator appeared as 'Meester Gheert Scellekens'.[1] The use of Barbara's maiden name was odd, but unambiguous. Queen Maria wanted Mercator. After the name Scellekens was the accusatory note: *'woenende achter den Augustynen. Minores Mechlinienses habent litteres suspectes'* ('resident behind the Augustins. The Brothers Minor of Mechelen have suspect letters of his').[2] Mercator was to be investigated for writing 'suspicious letters' to Minorite friars in Mechelen. He was to be charged with *'lutherye'*. Could that Mechelen friar be Franciscus Monachus, the first man in the Low Countries to construct a printed globe?

Dufief's men came to Mercator's door early in February. But their quarry had gone. It was presumably an anxious Barbara who attempted to explain to her unwelcome visitors that Mercator had been obliged to travel that day to Rupelmonde, on the Schelde, in order to see to the small inheritance left by his recently deceased uncle. At best, Barbara's tale was suspicious; at worst, implausible.[3]

Mercator was declared a 'fugitive', and the bailiff of the Pays de Waas, Louis Van Steelandt, was ordered to seek his arrest. The fugitive was apprehended in Rupelmonde and taken through the streets of his boyhood, down the rutted slope towards the frozen Schelde and the shadowed cliff of the castle.

From the shore, a timber bridge supported on tall stilts sloped upward to a narrow aperture in the castle's landward tower. This was the only way in, and out. Ringed by its decrepit towers and pierced by innumerable arrow slits, Rupelmonde's castle was vast, lightless, damp and very cold. Inside, narrow bricked passageways twisted and turned past low doorways. Mercator had grown up with the castle and its brutal past: the sieges, the tales of treachery and execution, and of Zeger De Kortrijkzaan, the Ghent militia commander decapitated in one of the castle's beds.

Isolated in freezing darkness, Mercator must have wondered who had betrayed him.

For a decade he had been walking on ice. Perhaps the theologians of

Fig. 16. The castle at Rupelmonde. Mercator grew up in its shadow and was imprisoned within its walls. From Sanderus, *Flandria illustrata*, 1641–4. (British Library 177h.11)

Louvain had at last seen through the opaque film that Mercator had so carefully laid across his engravings. There was little doubt that the suspicions of Tapper and Latomus would have been raised by the Holy Land map. And the map shaped like a Melanchthonian heart might have confirmed those suspicions. They would also have recognized Erasmus in the introduction of *Literarum latinarum*, itself a symbol of humanist frivolity and worse, printed by one of the university's oldest antagonists.

Inside the castle, Mercator can have known little of the attempts being made to secure his release.

Immediately she heard of her husband's arrest, Barbara turned to the one man who might be able to negotiate a release. As a past rector of the university, and as priest of St Pierre's, Pierre de Corte was not without influence.

De Corte immediately wrote a testimony to Mercator, vouching for his sound character and beliefs. The testimony was sent to Van Steelandt, who sent it on to Queen Maria. De Corte also enlisted the support of Pierre Was, the abbot of St Gertrude in Louvain. Mercator had taken the precaution of remaining on the university roll, and the abbot had a statutory responsibility as guardian of university privileges to protect its alumni. Taking de Corte's lead, Was protested to the bailiff Van Steelandt that Mercator was innocent, threatening to pursue the bailiff if he continued to infringe Mercator's privileges.

On 19 February, Maria of Hungary wrote to de Corte, demanding that the priest explain why he was defending a fugitive, and how he knew that Mercator was not guilty of heresy. Twice in the letter, Maria insisted that Mercator's absence from Louvain was evidence that he had been attempting to evade the law. The same day, Maria also wrote to Pierre Was informing the abbot that Mercator had forfeited the privilege of university protection because of his attempt to avoid arrest. Not only did the regent order the abbot to refrain from lobbying Van Steelandt, but she threatened him with judicial investigation.

Somebody, or some persons, were determined that Mercator should burn.

The horror of his situation would have been all too apparent. Dufief was not a man to be denied a victim.

Alone in his cell, Mercator had time to recall the fate of Tyndale and the various other victims of the procurer-general. The details of these executions were well known.

Tyndale's cell door had been opened early on that autumn morning. He had been led down the cold stone stairs to the castle gate where the crowds had gathered, many of them citizens of Louvain. A great timber cross had been erected in a space cleared of spectators by barricades. Within the circle sat the procurer-general and his commissioners. Tyndale prayed briefly and was chained by his feet and hands to the cross. Straw, brushwood and logs were piled against his body, then sprinkled with gunpowder.

The Englishman would have felt the hands of the executioner drag the hemp rope around his neck from behind. He may have seen the sign given by the procurer-general, the sign which ordered the executioner to tighten the hemp. Tyndale died by strangulation. When the last breath had been crushed from his shattered windpipe, the procurer-general handed the executioner a wax torch.

Much later, one of Mercator's closest friends would remark that the cartographer's response to adversity was one of 'great patience'.[4]

Outside the walls of Rupelmonde's castle, the efforts to release Mercator were going from bad to worse. Four days after Maria of Hungary had written to de Corte, the priest picked up his pen and attempted to convince the regent that Mercator had not run from the law but had been away on legitimate business.

'Gerardi Mercatoris', wrote de Corte, 'did not quit the town out of fear'.[5] The cartographer, explained the priest, was frequently called away, most recently to Flanders concerning an inheritance and before that, on a surveying trip 'to settle a dispute between the abbot of St Pierre and the provost of St Bavon in Ghent'.[6] On another occasion he had been called to Brussels to see the Archbishop of Valenciennes and the Bishop of Arras – Antoine Perronet. At the time of his attempted arrest, Mercator was – insisted de Corte – seeing to his uncle's affairs. Finally, with God as his witness, de Corte confirmed that he would never have attempted to exonerate Mercator if he had believed him to be a heretic.

Three weeks after the arrests, word was spreading through Louvain of the victims' identities. Old friends like Gemma and Van der Heyden must surely have been concerned. On 8 March, the university itself raised a response, sending a letter and other documentary support to Van Steelandt, with a request that he forward it to Dufief with the intention of establishing whether or not the procurer-general had learned that Mercator had fled out of fear, or because he had been accused of heresy. Van Steelandt was asked to convey the response to Maria of Hungary so that she could respond to the university authorities.

Mercator now had two charges to defend: that of *lutherye* and that of avoiding arrest, with the latter charge confirming the former. His situation could not have been much worse. Neither were the predicaments of the other 'heretics' much better. Dufief was sure of several executions. With every week that passed, Mercator must have known that his chances of release were being reduced.

The initial terror of his arrest gradually subsided into fear, listlessness and anger at the injustice of his incarceration. How often he must have turned to St Paul: 'For to me, living is Christ and dying is gain. If I am to live in the flesh, that means fruitful labour for me; and I do not know which I prefer. I am hard pressed between the two: my desire is to depart and be with Christ, for that is far better; but to remain in the flesh is more necessary for you.'[7]

Among Mercator's reasons for remaining in the flesh were Barbara, his children and his unfinished geography.

March slipped into April. Through a barred window he detected the beginnings of spring; a return of bird-life; the calls of the boatmen he had known as a boy.

April slipped into May.

The epistolary wrangling caused by Mercator's absence from Louvain when Dufief's men called had delayed an examination of his original charge. He had been locked away for over three months before Maria of Hungary instructed Van Steelandt to travel to the monastery of the Friars Minor of Mechelen. There, he was to engage the guardian to secretly procure correspondence that Mercator had sent to one of the monks. The letters were to be handed to the bailiff.

The search produced no letters.

Neither did the interrogation of the other heretics produce any incriminating connections with Mercator. Begging not to be tortured, Catherine Metsys told her interrogators that she had been to Antoinette Van Roesmaels' house 'on a few occasions',[8] and named those she'd seen there. Sometimes there had been four, sometimes five, and there had been 'frequent readings ... sometimes from the Bible, sometimes from the Postilles'.[9] She named the readers, and confessed that one of them had been herself. She denied that baptism had been discussed at the house, but Communion had been a topic. She admitted that she was of the opinion that the body of Christ was not present in the Host, and that the sacrament was nothing more than a commemoration. She did not believe that there was a purgatory, and described the 'great villainy' that had occurred 'beneath the bells of St Pierre', where her husband had removed 'a small panel representing Purgatory'[10] from the shrine of St Jacques. She admitted to breaking up the panel and burning it. Catherine's husband Jean Beyaerts testified against Paul de Roevere, declaring that the priest regarded the Holy Sacrament as a symbolic token. And a search of de Roevere's library revealed the writings of the Swiss reformer Heinrich Bullinger.

Mercator's home must have been searched too, and his possessions confiscated. Nothing incriminating was found.[11] As the interrogations progressed, Antoinette Van Roesmaels emerged as a main defendant. She confessed to the gatherings, the readings, and provided names (the sisters Calleken and Betteken, Chrétien Broyaerts, her niece Marie, the priest de Roevere, Gilles the cutler from Brussels). She named the furrier's assistant Josse Van Ousberghen as another who did not believe in purgatory. Condemning Josse, she recalled that he did not go to confession often, and that he believed in addressing God alone and not the saints. In a futile attempt to exonerate herself, Antoinette claimed not to share Josse's opinions. On the contrary, one should 'follow the ordinances of the Church'.[12]

Fortunately for Mercator, 'Master Geert who married the daughter of Scellekens'[13] was one of those listed by Antoinette as never having visited her house.

On 12 June, Van Roesmaels was sentenced to death. Three days later, she was buried alive, to die slowly by suffocation. Over the following days, sentences were passed on several others from the group of 'heretics' arrested with Mercator. There can be little doubt that Dufief did not fail to transmit their fate to his prisoner in Rupelmonde.

The man whose imagination had grown used to roving continents was confined to a cold, damp cell while, one by one, his co-accused were executed. Van Roesmaels' live burial was followed by that of Catherine Metsys, the daughter of Louvain's tower-builder. Catherine's husband, the sculptor Jean Beyaerts, was burned at the stake. So was Jean Vicart, the haberdasher from the house called Porte d'or. Josse the furrier's assistant was beheaded. Chrétien Broyaerts escaped with banishment 'in perpetuity, on pain of death',[14] and the confiscation of his property.

Spring passed to summer.

16

The Slight Youth from the North

He referred to it as 'the most unjust persecution'.[1]

In September, the heavy door to Mercator's cell swung open and he was escorted through the narrow, stepped, stone tunnels to the castle's main gate, and the timber causeway that led back to the green fields of Flanders. The ice had disappeared during his incarceration. The land was heavy with late summer foliage. He had been locked away in his island prison for seven months.

At the landward end of the causeway rose the familiar mill of his child-hood, the tidal leat, the wharves and the moored barges. The rutted climb to the square and the church. Unused to walking, and blinded by light, Mercator walked with the tentative steps of an uncertain child. But this was his village; he *was* Gerardus of Rupelmonde. There must have been hands outstretched, from one of his five brothers, or his sister Barbara (now married to Michiel, with two children). He may have been greeted – discreetly – by old neighbours and friends of his great-uncle Gisbert.

Mercator made his way back to Louvain, to Barbara and his children. The family finances can only have been desperate. More than two years had passed since his extraordinary run of productivity had been ter-minated by the siege of Louvain. He had six children to feed and may have had to pay – like Tyndale – for the cost of his own imprisonment. Some of his possessions may also have been confiscated by Dufief, perhaps even his tools and instruments, books and manuscript sketches.[2] His letters would have been destroyed by Barbara to prevent them betraying correspondents. Except for the work that had been printed, and therefore preserved through multiplicity, Mercator had lost his past.

And yet the double catastrophe of the siege and his arrest would eventually give Mercator the opportunity to begin again, with redis-covered vigour and originality. His was to be a life in two parts, and it would be the older, wiser Mercator who produced the great works.

*   *   *

In the mean time, life in Louvain for an inquiring humanist had become no safer.

Shortly after Mercator's release, the printer Jacob Liesvelt was led from a prison cell to face the fate that Mercator had evaded. The Antwerp printer whose illustrated Bible had introduced the concept of the map to many of Mercator's generation had for years been provoking Louvain's theologians with increasingly risky editions. Twice he had been accused of printing heretical works, and twice he had won acquittals. The third accusation stuck and he was beheaded. In the thirty years since, Liesvelt had printed the first Dutch edition of the Lutheran Bible, with its revolutionary map, he had printed at least two other maps, one describing the county of Champagne, and the other showing Zeeland. The association between cartography and heresy had never been so explicit.

The Spanish reformer Enzinas had also been caught in the round-up. But in a remarkable feat of escapology, Francisco walked through a series of unlocked doors in his Brussels prison, scaled the town walls and then walked through the night to Mechelen, where he found a horse and rode in two hours to Antwerp. One month later he reached the geographical heart of the Lutheran world, Wittenberg. 'Our Spaniard Francisco is back,' wrote Melanchthon to his friend Joachim Camerarius in Leipzig, 'saved by providence'.[3]

Latomus had died during Mercator's imprisonment, but Tapper was still at large, elected once again to the rectorship of the University. From 24 October 1545, matriculating students were sworn to reject 'the doctrines of Martin Luther and all heretics, in so far as they clash with the teachings of the ancient Roman Catholic Church'; at the same time they had to swear 'to live according to the Church's precepts and under the guidance of its supreme guardian, the Roman Pontiff'.[4] The Theology Faculty now ran regular checks on the suitability of books, censoring Bible translations and any books thought to refer to Lutheranism. With this in mind, the faculty compiled an index of prohibited works, the first list of its kind. By Imperial decree, this was published on 31 June 1546. Henceforth, printers had to apply to the central government for a permit to work.

The safety of Mercator and his family now depended upon continual vigilance. Letters and conversations would have to be guarded, friendships protected. As long as he lived in Louvain, nothing controversial could be committed to paper. Heeding the old Flemish proverb, that an accused man should not draw attention to himself by 'playing on the pillory',[5] Mercator kept a low profile.

Picking up the pieces of his life, he struggled to fulfil commitments that he had made before his arrest. Among these was an astronomical ring which he had promised Antoine Perronet. In connection with the ring, Mercator was also advised by Gui Morillon – the retired Imperial secretary who had supported his ambitions to construct a new terrestrial globe – to provide Perronet with a written description of the instrument. Mercator did so immediately he was released, writing to Perronet early in October 1544: 'I would even have explained the advantages of this instrument in greater detail in an adequate pamphlet,' wrote Mercator to the bishop, 'had I not been weighed down by a whole host of commitments, and had I not been tormented of late by a lack of time since leaving that most iniquitous of jails.'[6] The uses of the instrument would have been described earlier, added Mercator, 'if Morillon had not delayed in entrusting them for transcription'.[7] Morillon may have been nervous of his association with a suspected heretic, or anxious that Mercator might be betrayed or misinterpreted again. Apart from a lack of time, Mercator may also have been prevented from issuing his 'adequate pamphlet' through lack of a printing permit from the chancellor of Brabant. (Mercator was eventually issued with such a permit – both to print and to sell books – on the eve of his departure from Louvain.)

Instead of the pamphlet, Mercator provided the bishop with an epistolary instruction manual, written, excused Mercator, 'at a time when I was so greatly disturbed'.[8] A model of technical clarity, the letter was not only worthy of printing but sufficiently clear that a schoolboy could have mastered the astronomical ring in a few minutes. The sphere, explained the mathematician to the bishop, was 'the perfect representation of the celestial circles' and as a working model, was 'capable of reproducing any given situation in the sky'.[9] The bishop learned that there were two longitudinal movements in the heavens, 'that of the first moving part',[10] which turned about the poles of the earth, and that of the planets and the stars, which turned around the poles of the zodiac. The former turned from east to west, with the equinoctial line as its belt, and completed one revolution every twenty-four hours; the latter, which turned more slowly, turned from west to east following the signs of the zodiac, with the ecliptic as its belt.

Mercator concluded his letter by explaining that the astronomical ring itself was unfinished due to his imprisonment, and to 'a lack of materials'.[11] He added that he was 'now at work on it' and would 'send it as soon as it is completed'.[12]

Early in 1545, he was writing to Perronet again, this time describing how to use the 'threefold ring'. The ring, explained Mercator, was basically the same as the model produced by Gemma, the differences being found in the markings. With a hint of pride, Mercator added that he had calibrated the device for additional stars so that 'no part of the night, provided that it is calm, can escape mathematical calculations'.[13]

By the end of 1454, Mercator must have delivered the 'numerous mathematical instruments'[14] required by the emperor. Collectively, they represented hundreds of hours of painstaking labour. An astrolabe made by Mercator at this time bore the telltale signs of stress and exhaustion: an otherwise fine instrument was marred by two mispositioned star pointers, and an omitted star-name. The instrument also incorporated an engraved diagram whose inherent inaccuracy led Mercator to drop it from subsequent astrolabes. The deficient astrolabe was kept in the workshop and engraved with a 'GMR' monogram – for Gerardus Mercator Rupelmundanus.[15]

In December, the emperor left Brussels on a journey which would culminate in a series of fiery showdowns with the Lutheran princes. And the destruction of Mercator's instruments.

As the sun waned on Mercator's Louvain, a young irenicist came from the north. It was June 1547.

A 'tall, slighte youthe, lookyinge wise beyonde his yeares, with fair skin, good lookes and a brighte colour',[16] John Dee was ambitious, unpublished and inflamed with the quadrivial arts. Inspired at Cambridge by Sir John Cheke and his programme of Erasmian 'new learning', Dee was well connected and already notorious for his student production of Aristophanes' *Peace*, a play which required its principal character Trygaeus to fly to the palace of Zeus. Dee had constructed a giant dung beetle which appeared to leave the stage, its pneumatics, springs and mirrors unseen by the incredulous audience in Trinity College's main hall.

Still nineteen, John Dee had left the Fens 'for the purpose of investigating those sources from which, in our age, many channels of the best of those arts have been led to us, and of living on familiar terms with men whose lightest single day of writing would have furnished matter enough to require the labour of a full year for comprehension while I formerly sat at home'.[17] In Louvain, he intended 'to speak and confer with some learned men, and chiefly Mathematicians'.[18] The particular mathematicians he sought were 'Gemma Phrysius, Gerardus Mercator, Gaspar à Mirica, Antonius Gogava'.[19] But of the four, it was Mercator whom Dee fell 'first upon'.[20] Communicating in the universal media of Latin and mathematics, the English pupil and Flemish master quickly established a fruitful rapport, and after 'some months'[21] in the Low

Countries, Dee returned to Cambridge with his trophies: 'the first Astronomers staff in brass, that was made of Gemma Frisius devising, the two great globes of Gerardus Mercator's making, and the Astronomer's ring of brass, as Gemma Frisius had newly framed it'.[22]

The following summer, Dee was back in Louvain. But by now Gemma was in continual pain from kidney stones. Mercator found himself Dee's prime attraction. Later, the young Englishman recalled that 'it was the custom of our mutual friendship and intimacy that, during three whole years, neither of us willingly lacked the other's presence for as much as three whole days; and such was the eagerness of both for learning and philosophizing that, after we had come together, we scarcely left off the investigation of difficult and useful problems for three minutes of an hour.'[23]

Neither men (one an ex-prisoner, the other a future prisoner) left a record of their seemingly continuous dialogue, although it is fair to assume that they discussed religion, astronomy, astrology, geography, instrument-making and the curious properties of rhumb lines – Dee and Mercator sharing an admiration for the work of the Portuguese mathematician and cosmographer, Pedro Nuñez.

So Dee would have been more than interested to learn that Mercator had been busy tracing the source of the Portuguese cosmographer's magnetically derived rhumbs. Although Magnus (and Mercator) had placed magnetic islands on their depictions of the Arctic, the general view held that the magnetic needle was drawn to a point in the heavens, as Petrus Peregrinus[24] had stated so convincingly in his *Epistola de magnete*. Mercator knew otherwise. In one of the letters which he had written to Perronet after his release, he declared that he had located the magnetic pole.[25] 'In what place', wrote Mercator, 'that point lies, which the magnet so greatly seeks, I shall explain, in general, as far as is now possible, to your Reverence.'[26] Mercator then described to the bishop how he had taken the known declinations of compass needles at the island of Walcheren and at Danzig, and then followed the axes of both needles until they intersected 'at about 168 degrees longitude and 79 degrees latitude'. And that, added Mercator, is the 'place the magnetic pole must be'.[27] No wonder Dee found Louvain stimulating.

Heavenly bodies were also a particular interest for both mathematicians. During Dee's stay in Louvain, Mercator began working on a celestial globe to accompany the terrestrial globe which he had completed ten years earlier. Dee meanwhile wrote two astronomical works. Their titles (*Of the Great Conveniences of the Celestial Globe* and *Concerning the Distances of Planets, Fixed Stars, and Clouds from the Centre of the Earth, and Concerning the Discovery of the True Magnitudes of all the Stars*)[28] suggested a fruitful collaboration. Dee may have been able to return the favour, for one of Mercator's other works-in-progress was his long-promised map of Europe. As it happened, the Englishman was well-placed to provide geographical titbits on the British Isles: in December 1549, Dee began making trips to Brussels 'to eat at the house'[29] of Sir William Pickering, the respected and highly influential English ambassador to the court of Charles v. Dee was soon tutoring Pickering in mathematics and the use of various instruments, including globes, the astrolabe and the astronomer's ring. In return Pickering began to send Dee books. With little trouble, Dee could have asked the ambassador for maps and books describing regions of Britain. Dee may also have been able to help Mercator obtain maps through his old tutor Sir John Cheke (now an aide to the boy king Edward vi), and through Cheke's son-in-law, William Cecil.

The affairs of the English would periodically preoccupy Mercator for the rest of his life.

Apparently inseparable, Dee and Mercator were both in Louvain when the town received the emperor on 15 September 1548.

Weary and gout-ridden, Charles v had returned from Germany, over the hills to Jülich and across the Maas to Brabant. The Lutheran princes had been beaten in battle, but the war against the Church's enemies was far from won. For the first time, Germans had witnessed a European war on their own soil, a war in which both sides considered themselves the defenders of German unity; it was a conflict between Catholic and Lutheran, between emperor and prince, between a universal assumption and a national principle.

After two nights in Louvain, the emperor travelled on to Brussels, where he stayed until the following June.

Mercator was one of many who were summoned to Brussels during those months of recuperation. There, he learned that the instruments he'd built for the emperor had been destroyed. It had happened at the end of August, during one of the early exchanges with the Schmalkaldic League. The two forces had confronted each other outside the walls of Ingolstadt. (Mercator would have been fascinated to learn that one of those present had been the court mathematician and resident of Ingolstadt, Peter Apian, who had been summoned to amuse the emperor with a mathematics lecture; while an attentive Charles sat and listened, firing from a nearby battery had caused the author of *Astronomicum Caesareum* to visibly tremble with nerves.) For four days the two sides had skirmished and exchanged artillery rounds. Mercator's instruments had been 'melted and destroyed in a fire that had been secretly started by the enemy in a barn not far from Ingolstadt'.[30] The emperor 'commissioned Mercator to make him a new set'.[31]

Mercator knew by now that his Imperial patronage could not last. Forty-eight and wracked with gout, the man who had been monarch of the Low Countries since Mercator's schooldays was preparing to abdicate, an event which was already seeding dread among his Habsburgian subjects. Monarchial handovers ranked with plagues, floods and wars as sources of human disaster. But this one was especially ominous: the world was well advanced into the fourth and last of the successive monarchies predicted in the Book of Daniel and the passing of Charles brought the end even nearer. More immediately, and locally, the prospects for the Low Countries looked poor.

By the time Mercator received his commission to construct the replacement instruments, the emperor had already browbeaten the Imperial Diet into agreeing that the Low Countries should be prised away from the Holy Roman Empire. A virtually autonomous political unit, the seventeen provinces would pass to Charles' son Philip rather than to his brother Ferdinand. Philip would soon be King of Spain; Ferdinand would soon be Holy Roman Emperor. Once Charles abdicated, the Low Countries would be bonded to Spain, the seat of the Inquisition, rather than to Germany, the seat of reform.

Early in 1549, Crown Prince Philip left Spain to undertake his first, unwilling, visit to the distant wetlands of northern Europe. From April until September, the 22-year-old Spaniard undertook a grand tour of towns who welcomed their heir apparent with triumphal arches, processions, theatrical dramas and lion fights.

At the beginning of July 1549, the 'Spanish Prince'[32] came to Louvain to be invested as Duke of Brabant, an event which made its way into Dee's diary of noteworthy events. The following month, Antwerp blew 250,000 florins greeting the heir apparent and the emperor. Obscured by the pageantry and images of concord was an increasingly ungovernable collection of provinces characterized by differences in tradition, language, tax and judicial systems, riven with religious dissent, terrorized by persecution and fearful of life beyond Charles. An abstemious, grave workaholic with no empathy for the beer-swilling lowlanders, and no languages other than Castilian, Philip was ill-equipped for his role.

The fate of Mercator's most influential court ally, Antoine Perronet, was tied to that of the outgoing monarch. In recent years, Perronet had become indispensable to Charles, addressing the Council of Trent on the emperor's behalf, and playing crucial roles in the twin offensives against the French and the Lutherans. Anticipating his son's unpopularity in the Low Countries, Charles intended Perronet to help govern the ungovernable. Perronet would become a Spanish instrument.

To the 'slighte youthe' Dee, these momentous changes were a source of excitement rather than anxiety. He spent his last months in the Low Countries dashing between Louvain, Brussels and Antwerp. In May 1550, he spent over a fortnight in Antwerp, and at the end of the month, Mercator asked him to visit Brussels – perhaps to convey an important letter – or a new instrument – to the Imperial court. In the houses of these towns, Dee would have seen how cartographic works had become as collectable as portraits, reliefs, statues and carpets. Maps and globes reflected their owner's ease with novelty, and as subject matter they symbolized a universal awareness. In Antwerp, Michiel Van der Heyden's gaming house on the edge of the city had become a veritable palace of contemporary achievement. Among the mirrors, tapestries and biblical oils were pictures by Hieronymus Bosch and 'Two paper maps, framed – one of Brabant, the other of Flanders', 'A large map in a frame, made of landscapes' and 'A framed map of High Germany'.[33]

John Dee stayed on in Louvain until the summer of 1550, then departed for the five-day journey to Paris, where he 'did undertake to read freely and publicly Euclid's Elements Geometrical ... a thing never done publicly in any University of Christendom'.[34] Dee's departure for mathematically liberal Paris can only have reminded Mercator that Louvain was no place for exploring the boundaries of cosmography.

In the months following Dee's departure, Mercator completed his celestial globe. The same size as the terrestrial globe of 1541, it would complete the pair.

This was the first astronomical work that Mercator had attempted, and while it could have been criticized for its lack of new knowledge, it was exemplary in its precision and artistic elegance.[35] Mercator could not claim any Copernican breakthroughs, but he did tidy the heavens.

The most visible modification he made was to rationalize the orientation of the globe's gores. The normal practice (promoted by Gemma and Schöner) had been to arrange the twelve celestial gores so that each gore contained one zodiacal sign. This eased the task of the engraver or woodcutter, but it had the disadvantage that the axis of the globe was not the same as the point of convergence of the gores. Once mounted, globes constructed in this manner appeared to have two axes. Mercator reoriented his gores so that their tips converged at the equatorial rather than at the ecliptical poles. The axis of the globe now ran through the point of convergence of the gores – further tidied by Mercator's use of a separate pair of small circular gores covering the two poles, a device he had employed on his terrestrial globe. The overall effect was to create a neater, more aesthetic globe. In astronomical terms, it was a superficial development, but it demonstrated Mercator's confidence as an engraver and his ability to think in different dimensions.[36]

But the new globe was not entirely without astronomical innovation: to the forty-eight Ptolemaic constellations, Mercator added two new constellations – Cincinnis[37] and Antinous – which he had probably discovered on a globe printed in Cologne by Kaspar Vopel fifteen years earlier. Although they had been known since antiquity, the two constellations had never appeared as figures on a celestial globe. He also labelled his constellations in Latin and Greek, and in some cases in Arabic transliteration too.

Purchasers could be certain that they were buying a globe which displayed the heavens in their latest configuration, the star positions having been calculated for the 1550 epoch – a novelty recently made available to Mercator by Copernicus' revolutionary work on cosmic motion. In doing so, Mercator became the first globe-maker to use the theory to fix the true location of stars. On the globe's paper horizon ring, Mercator had printed scales to show the signs of the zodiac, the days of the Julian calendar, the twelve wind directions, various humours and astrological information that would catch the eye of physicians and those of superstitious bent.

The new globe also provided a platform for Mercator's developing artistic skills. Unlike the figures on Gemma's celestial globe, Mercator had dressed some of his in Roman clothing. A reflection of Mercator's aversion to crudity and his pleasure in humanist allusion, the new figures revealed an eye for shading and texture; among the many improvements was Andromeda; a naked dumpling on Gemma's globe, she was now elegantly gowned. Through defter use of hatching, muscles (like those on Centaurus' back) rippled to life and fabric appeared to fold and flow.

The engraved lettering was the most accomplished Mercator had displayed; one decade after publishing his manual on italic, the author was writing with the elegant, legible ease that he had urged upon his readers.[38] Like its partner, the celestial globe incorporated the ingenious spoon-shaped stabilizer, and the robust, portable four-legged stand Mercator had designed over a decade earlier.[39]

The last dedication which Mercator engraved in Louvain was addressed to the bishop under whose ecclesiastical authority he had lived since his imprisonment. George of Austria had been appointed Prince-Bishop of Liège in 1544, after a spell in prison himself. Captured by the French before the siege of Louvain, George had been released after the payment of a ransom. As the (illegitimate) son of Emperor Maximilian I, he had been educated alongside the future Charles v and was respected as a sound administrator, and had helped to secure for Erasmus the provostship of Deventer. Mercator may have known George since the 1530s, when he was attached to the court of Queen Maria. Mercator's dedication reflected gratitude and admiration for a figure of ecclesiastical authority who had honoured the cause of humanism: 'To the Magnificent Protector and Prince,' read the inscription, 'the very distinguished George of Austria, by the Grace of God, Bishop of Liège, Duke of Bouillon, Marquis of Francimontensi, Count of Lossensi, the very splendid patron of arts and science.'[40] To protect his work, Mercator obtained a privilege forbidding anyone 'under the penalties and fines prescribed' from reproducing or selling copies of the globe 'within the Empire or the Low Countries of His Imperial Majesty until after ten years'.[41]

The globe was completed in April 1551. In the same month, Mercator was granted a permit by the Chancellor of Brabant to print and sell books. But the moment had passed: after twenty-two years in Louvain, Mercator was preparing to return to the land of his ancestors.

Like his parents and grandparents, Mercator was fated to migrate. His youngest daughter Katharina had reached the age that Mercator had been when his own parents had loaded a cart in Gangelt to leave for a better life in the Low Countries.

There were many reasons for leaving Louvain. Having been arrested once, he would always be a suspected heretic. Like Liesvelt, he could expect future troubles. In the Low Countries, Philip had brought great uncertainties, and Calvin's calls to destroy Catholic images carried apocalyptic overtones. In September 1551, war had resumed between the emperor and France. And Chancellor Granvelle's death in 1550 had removed from the court Mercator's most influential ally.

A Fleming with German roots, living in Brabant, Mercator would never be accepted as a complete citizen of Louvain. Not only was he stigmatized by arrest and imprisonment, but he was prevented – as a non-Brabander – from holding public office. Despite his many achievements in the fields of mathematics and cartography, he had failed to win lasting recognition and support from court. He may have blamed his association with heresy, or his humble background. Mercator only had to look at Peter Apian to see what advantages were bestowed upon a favoured mathematician. Not only had the printing of Apian's *Astron-omicum Caesareum* been paid for by the emperor, but the author had apparently received a gift of 3,000 gold pieces. The Ingolstadt scholar had also been appointed court mathematician and a Knight of the Holy Roman Empire. A cardinal had granted him an ecclesiastical title.

Mercator's reasons for leaving were compounded by years of stalled productivity. By 1552, twelve years had passed since he had produced a map.

But a reason to leave is not enough to put a migrant on the road. There also has to be a geographical promise.

In the forty years since Mercator had been born, the Kremer homeland of Jülich had become a haven for liberal refugees; a working model of Erasmian government.

The Kremers had already moved to Rupelmonde when Johann III's father died in 1521 and the young duke was able to unite his own duchy of Cleves and Mark with that of Jülich-Berg. A gifted mediator and administrator, Johann III was also a great admirer of Erasmus, who had advised the duke on the drafting of the radical church ordinance of 1532 which proclaimed that all preaching should be drawn from Scripture and the early fathers, and that priests should be educated and free from polemic. Erasmus was granted an annuity by the duke and repeatedly urged to move to Cleves.

Significantly for Mercator, the Erasmian mantle passed from father to son; from Johann to Wilhelm V, who inherited the duchies in 1539. Following the Gelderland débâcle and his humiliating confrontation with the emperor at Venlo, Duke Wilhelm V had quietly continued the work of his father, upholding the Erasmian church reforms and allowing the spread of Lutheranism. In Louvain, it was also well known that Wilhelm was dedicated to improving the education of his subjects. Educated himself by a leading humanist, Wilhelm had founded a *Gym-nasium* in Düsseldorf in 1545, and as recently as 1551 had charged Mercator's old university acquaintance Andreas Masius with the task of soliciting the Pope for 'certain prebends or other ecclesiastical benefits in favour of the schools'.[42]

At around this time there was also talk of a new university,[43] to be founded in the small town of Duisburg in the centre of the duke's territories. Peaceful, tolerant and suggestive of an education boom which would require teachers and teaching aids (such as instruments and books), the Rhine duchies – and Duisburg in particular – exerted an irresistible pull. Duke Johann had been known as 'the Peaceful'; his son Wilhelm had become 'the Rich'.

For a suspected heretic of forty with great works to be done, the Rhine was as benign and as fertile as the Jordan.

This was the longest journey Mercator had made. Longer even than the treks between Gangelt and Rupelmonde. Moving his family and household effects, his tools, instruments, globes, library, the 'three or four'[44] fragile copper plates so far engraved for his unfinished map of Europe, must have required at least one waggon. This would be the final migration.

The family's destination was Duisburg. Not Wesel, the mercantile hub, nor the courtly towns of Cleves, Düsseldorf or Jülich, but Duisburg, a place which Albrecht Dürer had once dismissed as 'another little town'.[45]

Having spent his adult life within the extraordinary orbits of Antwerp, Mechelen, Louvain and Brussels, Mercator had decided to remove himself to the outer reaches of the commercial, theological and humanist firmament. The reward would be peace. Tolerant, small and yet well connected by river and road, Duisburg would be the perfect sanctuary for a man requiring space.

17

Somewhere Worthy of the Muses

Thirty years after leaving Louvain, Mercator would write of returning to the lands of his 'natural lords',[1] the dukes of Jülich, Cleves and Berg.

Mercator was back in the fold, behind Rhineland walls: 'I was conceived under your protection,' he would remind his new duke, 'in the territory of Jülich by parents from Jülich, and educated there from my earliest years (notwithstanding that I was born in Flanders)'.[2] Mercator left the duke in no doubt that he was a man of Jülich.

The Germany he returned to was quite unlike the Germany he had left as a boy. And even less like Louvain. As long ago as 1543, evangelical preachers had been admitted to Duisburg's two parish churches to give sermons in Mass, and the following year, both burgomasters attended Holy Communion at which Catholic and evangelical rites were jointly observed. Such freedoms would have been unthinkable in Louvain.

Duisburg offered Mercator a second chance. 'He was so fond of peace and tranquillity in both public and private affairs,' observed one of his neighbours, 'that, during the forty-two years that he resided here in Duisburg with his family, he never exchanged a harsh word with any of his fellow citizens; he neither entered into any dispute with any man nor was himself summoned to law by anyone'.[3] The incomer was a model of equability: 'He paid the magistrates the honour and respect that was due to them. Wherever he lived, he always got on well with his neighbours; he crossed nobody's path; had proper regard to the interests of others; and did not put himself over anyone else.'[4]

At forty, Mercator had almost reached the limit of his father's life span. There could be no more migrations, and neither should there be the need: nowhere else within the geography of Mercator's experience could match the duchy's political stability and religious tolerance. Duisburg was 'the sort of place where one can lead a good life in peace and quiet',[5] a place 'worthy of the Muses'.[6]

*   *   *

Duisburg had been greater, but the Rhine had changed course one year and the town had found itself separated from the river by a new flood-plain. The forest of beech and oak which rose behind the roofs added to the sense of seclusion.

With a population of less than three thousand, and civic monuments which amounted to little more than town walls and the solitary tower of Salvatorkirche, Duisburg had more in common with rustic Gangelt than it did with the metropolises of Louvain, Antwerp, Mechelen and even 's-Hertogenbosch. Within the walls were about five hundred houses, a Latin school, law courts, a poorhouse, a covered meat market (the 'Scharn'), a weighhouse and a horse-powered corn mill. In the ground floor of the *Rathaus* (town hall) was a school of wine-making which was used as a prison during times of war. The streets were cobbled with river pebbles and wooden blocks, with centre gutters to wash away rubbish and animal slurry. Each household was required by law to sweep its street frontage once a week, but enforcement was lax and the stench commensurate. Most of the houses were backed by orchards and gardens for growing vegetables and fruit, and for rearing pigs and cattle, some of which would change hands at the weekly market below the castle rock. Outside the walls, at the foot of the Duisburg forest, was the place of execution: a gallows and trio of punishment wheels.

First impressions suggested that the chances of a university rising over Duisburg's roofs were slight. Indeed, the town was still recovering from a fire that had incinerated many of its houses fifty years earlier. Many plots still lay unoccupied. The year Mercator arrived, the town council had finally decided to ban the use of thatch on roofs, handing out pantiles to those too poor to buy their own.

This was the place Mercator had picked from the map. Sited at the confluence of the Rhine and Roer, downstream of Cologne, the little town lay on one of Europe's busiest waterways; here, the route from Italy to the Low Countries crossed the route from Saxony to Paris. Duisburg was also roughly equidistant between Antwerp and Frankfurt, the venue for the world's most important book fair. Just a short journey upstream, Frankfurt was fundamental to Mercator's calculations. The mapmaker was not the only bibliophile to have examined Frankfurt's location with some care. The geographical reflections of the French scholar Henri Estienne spoke for all who made the annual pilgrimage to the Main: 'For if we measure its distance from certain remote towns from which, at the time of the Fair, it is want to call such a multitude of men to itself, we shall find that it is situated, as it were, at the very centre of a circle.'[7] Duisburg lay near the centre of that circle. Mercator's new home town may have been humble, but it was incomparably well connected.

Despite his sense of homecoming, Mercator would never 'belong' to Duisburg any more than he had 'belonged' to Rupelmonde, 's-Hertogenbosch or Louvain. In Duisburg, he could have obtained a sense of belonging by becoming a burgher. For a man of his reputation, this would have been a mere formality; those whose fathers were not burghers simply had to swear an oath and pay a registration fee – or establish their civic worth by performing a service before the town council. (The only residents likely to experience difficulties were servants, serfs and Anabaptists, who had to have lived in the town for ten years before qualifying.) Once a burgher, the benefits included the right to vote and to hold office, and also customs and trading concessions. There were, however, obligations, principally military. Burghers were required to keep their own arms and armour, to stand watch on the walls and to defend the town.

Not only was Mercator indisposed to bearing arms, but by becoming a burgher, he would have to ally himself to one of two factions within Duisburg: the council, led by two burgomasters and elected by the town's gentry; and the committee of 'the Sixteen' which represented the artisans. Comprised of four delegates from each of the town's four quarters, 'the Sixteen' were forbidden to attend meetings of the council, who held the political power. Twice in the last forty years, the town's burghers had turned against the town council, forcing interventions by the duke. As Mercator moved to Duisburg, another revolt was imminent.

But the boy who had been an accidental outsider had grown into a practitioner of conscious non-affiliation. Mercator was no more willing to associate himself with political factions than he was with religious movements. Just as his spirituality was drawn directly from God, his political allegiance lay directly with the duke, the ultimate arbiter of Duisburg's internal difference.

(Mercator's civic neutrality was well judged: three years after he settled in Duisburg, riots prompted the duke to revoke the town's privileges and to restrict the election of town councillors.)

Duisburg was as ordinary as its most recent incomer was exceptional.

Celebrated as 'the most skillful man of our time in the making of [astronomical] instruments',[8] Mercator brought more than workshop skills to Duisburg. His maps, engraving skills and his personal connections with the emperor's court contributed to a stature made all the more authoritative by his scholarly interests and religious conviction.

Few of Duisburg's burghers were as glad to welcome Mercator as

Walter Ghim. Born to one of the town's more distinguished families, Ghim had served both as a bailiff and as burgomaster. A well-read Latinist, Ghim was part of the tiny core of Duisburg's humanists. He found Mercator 'a man of calm temperament and of exceptional candour and sincerity'.⁹ In Ghim, Mercator found his own biographer. And in Mercator, the mayor had found the man who could put Duisburg on the map.

To Ghim's admiration, the 'indefatigable'¹⁰ Mercator wasted little time in resuming his cosmographical works: 'Shortly after he had made his home among us here,' wrote Ghim with undisguised pride,

> he constructed by order of the Emperor two small globes, one of purest blown crystal and one of wood. On the former, the planets and the more important constellations were engraved with a diamond and inlaid with shining gold; the latter, which was no bigger than the little ball with which boys play in a circle, depicted the world, in so far as its small size permitted, in exact detail.¹¹

These two exquisite globes were presented 'with other scientific instruments', to the emperor 'at Brussels'.¹²

At about the same time, Mercator wrote a manuscript treatise for the emperor. In fourteen pages, *Declaratio insigniorum utilitatum quae sunt in globo terrestri, coelesti, et annulo astronomico* described the 'particular advantages'¹³ of terrestrial and celestial globes, the use of the astronomical ring and the whereabouts of the magnetic pole.

In his description of the instruments which he had made to Imperial commission, Mercator was relieving himself of his old obligations. As he was no longer a neighbour to the court at Brussels, Imperial queries could not be serviced by Mercator in person. There was more than a whiff of valediction in this settling of outstanding obligations. Mercator had moved on and moved away. Henceforth, his political allegiance would be ducal rather than Imperial. Now that he was in Duisburg, Mercator's cartouches would carry the Erasmian imprint of Duke William v.

There was, however, one final item of unfinished Imperial business, a map which had been in intermittent production for over a decade. The map of Europe which Mercator had promised to Antoine Perronet back in 1540 would be the first – and last – work Mercator would dedicate to his old Castle ally.

Through sheer inscrutability, this peerless civil servant had made himself as indispensable to Philip ii as he had been to Charles v. In 1550,

Perronet had succeeded his father as keeper of the emperor's seals. 'My Lord of Arras', a Venetian ambassador had reported in 1551,

> has an income from the Church of four thousand crowns a year, besides some extraordinary donations which he, too, obtains from the emperor, so that, without anything else, these two [i.e. Granvelle and his father] have more than fifty thousand crowns a year. They have so enriched their house in a few years that at present their capital reaches millions.[14]

A fragment of the Perronet fortune was on the way to the Mercator household.

The map of Europe was eventually completed two years after Mercator settled in Duisburg.

It was a truly remarkable sight. When the three rows of five prints were pasted together, the map was as wide as a man was tall, and nearly as deep.[15] Beasts and satyrs glared and cavorted around the four decorative borders, each of them inset with a cardinal direction, engraved in Latin, Dutch and Italian.

But the wonder lay within, for here was a new Europe. The reason for the map's prolonged gestation was given in the largest legend, a block of italic text anchored in the Atlantic. Explaining to the reader that his purpose had been to reduce the surface of the globe to a plane with a minimum of distortion, Mercator described the projection[16] he had used, and the method he'd employed for the subsequent plotting. Having drawn in his central meridian, and the parallels, he had marked off westwards and eastwards the intersection points of the parallels with the other meridians. Maintaining faith with Ptolemy, Mercator's 'zero meridian', o degrees of longitude, was the meridian which ran through the Canary Islands.

Onto this grid of latitudes and longitudes Mercator had plotted the coordinates of places worthy of inclusion on the new map – beginning with Ptolemy's city of Alexandria.

Mercator had accumulated the map's geography from an extraordinarily diverse collection of sources. The British Isles stood as an example. Although Mercator had transcribed much of the British outline from his 1541 globe, he had also consulted printed and manuscript maps, portolan charts, coastal rutters and regional maps. Against these maps, he compared older sources ranging from Ptolemy and Bede to Gervase of Tilbury, Geoffrey of Monmouth and Giraldus Cambrensis.[17] And he also used the latest printed works on the British Isles, such as John

Major's *Historia Magnæ Britanniæ* of 1521 and Hector Boethius' *Historia Scotica* of 1526. Münster's map of Britain in his recent *Cosmographia* was helpful too. This diversity – and perhaps some of the maps and books themselves – must have been a part of the legacy left to Mercator by John Dee during his recent sojourn in Louvain. Thus did Mercator pick and choose his revisions: Wales benefitted from the map published by Lily in 1546; in a legend moored off the Norwegian coast, Mercator ascribed his new depiction of the Orkneys to information he'd obtained from a ship's captain; and from an unidentified source – or sources – Mercator managed to populate Ireland with no less than ninety-four place-names (Lily had managed twenty-four). Although it was a recent print, Mercator was not entirely convinced by the veracity of Lily's map; he followed Lily in modifying his 1541 outline of Wales, but decided that the Englishman had been incorrect in marking a single definitive mountain range in north Wales. Instead, Mercator depicted several long crooked ranges radiating from central Wales like the broken spokes of a cart wheel.

For every region in Europe, Mercator had repeated this laborious process, sifting, comparing, rejecting until he had reached a consensual geography that he could commit to copper. For northern Europe, Mercator had turned again to Ziegler's *Schondia* map, supported by Ruysch's map of 1508 and Olaus Magnus's *Carta Marina* of 1539. For France, he returned to the mathematician he had studied as a youth. Oronce Fine's map of 1525[18] was a rare example of the kind of modern material Mercator could trust, for the Frenchman's map – the first map of France made by a Frenchman and printed in France – carried 124 cities, towns and villages whose locations had been fixed by a careful determination of latitude and longitude. For Switzerland, Mercator had to use the map that Münster had copied without permission from Aegidius Tschudi, the remarkable ex-pupil of Zwingli who – aged nineteen –

> had hiked to numerous mountains of the Alps – the Saint Bernard, the Glacier (Théodule Pass), the Furka pass where the Rhône begins, the Saint Gotthard where the Ticino, the Ursa, and the Rhine flow, the Lukmanier with the middle Rhine, the Urschler (also called Splügen), the Septimer and others besides – and also went through the foothills on both sides several times.[19]

Where reliable coordinates were unavailable (the legend complained that the lack of longitudinal data for so many places had compromised the accuracy of the map), Mercator had fixed places using nautical charts, and the reports of travellers and navigators. In a process which Mercator noted as being *'laboriosissimum'*, questionable places were shifted around until they accorded with known locations, distances, directions and latitudes.

An epic of spatial harmonization, Mercator's new Europe corrected errors which had been repeated over fifteen hundred years. Continuing a corrective process that he had begun on his globe of 1541, when he had tentatively reduced Ptolemy's width of the Mediterranean by 4 degrees, to 58 degrees of longitude, Mercator now sliced off another 5 degrees to bring Europe's sea down to 53 degrees.

The new map of Europe was published in Duisburg in October 1554.

Perronet's support was recognized within the grand dedicatory cartouche off Finisterre, and the cardinal – whose name now adorned the largest, most accurate map of Europe ever engraved – paid Mercator a sum of money appropriate to his 'magnanimity and exceptional generosity'.[20]

'This work', reported Ghim of Mercator's Europe, 'attracted more praise from scholars everywhere than any similar geographical work which has ever been brought out.'[21]

Mercator had finally produced a map that had significant commercial potential, and he knew it. For the first time, he obtained two privileges. As well as protecting himself for ten years within the Empire, he applied for a privilege of the same duration from the Senate of Venice. By the end of the year, copies of the map had reached Italy, where they appeared for sale bearing the copyright of 'Gerardo Rupelimontano'.

With the completion of the map of Europe, Mercator was able to consider the next in his incredibly slow-moving sequence of cartographic productions. Sixteen years had passed since he had announced on his world map of 1538 that his readers could expect a series of regional maps of the world, of which Europe would be the first. But now his mind had changed and Europe's successor would not be America, Africa or Asia, but another map of the world, a map which would erase the memory of the flawed *'Orbis imago'* of 1538.

Like Europe, it would be a long-term project, a definitive work which harmonized ancient and modern sources. His intention was announced as the parting shot in the main legend of his map of Europe. Just as Münster had appealed to his map-users to contribute geographical information, Mercator concluded his address to his 'kind-hearted reader' with an appeal for cartographic sketches, astronomical coordinates and distances, to further his aims of publishing a new map of the world. To guide his readers in their voluntary research, he promised to publish instructions concerning cartographic transcription.[22]

18

Frankfurt Fair

It was a reinvigorated Mercator who took the boat trip upstream to Frankfurt in 1554.[1]

Each spring and autumn, this town on the Main became the bibliographic capital of the world as printers, book dealers and publishers poured through the city gates to buy and sell the latest printed works.[2] All flocked to the part of the town between the river and Leonardskirche, and in particular to Buchgasse. Beneath signs advertising each dealer, barrels of arriving books were unpacked, inspected for damage and then displayed in booths and shops. For the past few years, catalogues promoting new publications had begun to appear, and these were perused for potential bestsellers.

The thrill of Frankfurt infected anyone with ink in their blood: the year before Mercator moved to Duisburg, a young bibliophile called Josias Maler had stopped off at the fair en route from England to Zürich. To Maler – a step-brother of the Froschauer who had printed the first Bible to contain a map – Frankfurt was 'the world famous and in all countries famous city'.[3] Maler arrived on 8 September 1551, having travelled up the Main:

> There we found the honourable gentleman Christoph Froschauer, the senior, citizen and printer of Zürich, who lodged us ten days in his own quarters, and because I was not useless in his bookstore, as I was as much brought up from childhood in a bookstore and could very readily answer strange peoples in Latin and French, and could give information of many sorts, he would not let me leave till the end of the fair. I had an evil time in carrying books back and forth, and could not escape any time to see the city, as usually there are many things to be seen at the annual fairs. Intense thirst at last drove me to the great stone well; I also saw the faubourg Saxenhausen and the enormous crowd of drivers, wagons and carts. On board a boat in the Main I

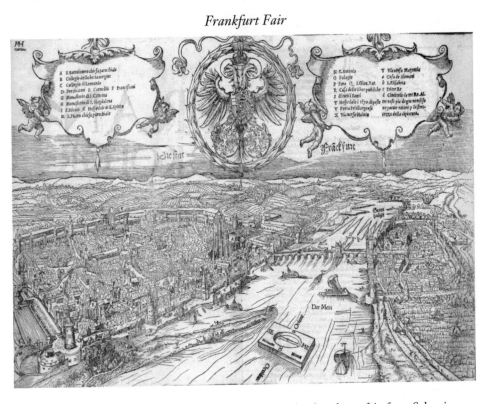

Fig. 17. Frankfurt, the home of the then twice yearly book and map fair, from Sebastian Münster's *Cosmographia* (1544). (Private collection)

finally got a good glass of beer, and having refreshed myself, I went again to the bookstore. Herr Froschauer highly appreciated my services, and on the fourth Friday after Ember Week, when we had breakfasted, he let us journey on with recommendations.[4]

While Maler was chained to the bookstore, escaping only for a sip of well-water or a snatched beer, the publishers, printers and dealers lubricated their minds in taverns where negotiations for new works, translations and revised editions took place alongside conspiratorial deals to smuggle bibles to England and to break authorial privileges by issuing unattributed or pseudonymous versions. Should they require a rest from their tireless transacting, publishers could enjoy the diversion of the rope dancer who pushed a boy in a wheelbarrow along a hemp hawser stretched from the Nikolai Thurm. An elephant, an ostrich and a pelican had been displayed at earlier Frankfurt fairs, and a couple of years after Mercator came to Frankfurt, a woman with no hands astonished the publishing world with her performing feet. Less conspicuous with their

163

consumption were authors on perennial quests for publishers, printers, patrons and dedicatees; Calvin's dedication of *harmonia Evangelistorum* to Frankfurt town council earned him forty gulden.

Mercator's name was known among the booths of Buchgasse long before he strolled the street in person. An instrument-maker, an engraver of globes and maps, a victim of the Inquisition, he was also the man who had brought italics to the north; whose 'few brief rules' in *Literarum latinarum* had helped to change the printed sheet. Since the 1540s, Low Country artists such as Cornelis Floris, Hans Collaert, Hieronymus Cock, Hendrik Goltzius and Jan Sadeler[5] had all turned to italic when requiring text. In Liège, a poet-architect-engraver-printer called Lambert Suavius had been demonstrating his new italic skills within three years of Mercator's manual appearing. Shortly afterwards, Pieter Coecke van Aelst – the father-in-law of Breugel the elder – published a book in Antwerp in which he substituted the earlier typeset captions with engraved italics. In instrument-making and geographical works, the influence of *Literarum latinarum* had been especially profound. Gemma's later editions of Apian's *Cosmographia* had for years been a showcase for italic annotation, but by the 1550s it was becoming common practice to use italics on instruments too. The standard of annotation on maps and globes produced by northern European cartographers had been set by Mercator, whose hand – or influence – had most recently been employed on Anton Wied's new map of Russia.[6] Mercator's influence had also been spread by those who had known him in Louvain: Gemma's pupil Juan de Rojas Sarmiento had published in Paris in 1550 a book on the astrolabe illustrated with sixty-three woodcuts which had been labelled in italic. Since leaving Louvain, Thomas Gemini had established himself in England as an ethically compromised editor and publisher: his *Compendiosa totius anatomiae delineatio* of 1545 had borrowed heavily from Vesalius' *De humani corporis fabrica* of two years earlier; Gemini's forty anatomical engravings were a dramatic vehicle for italic lettering which the plagiarist had learned in Louvain. Now in its third edition, *Literarum latinarum* had crept across the continent to the very borders of Italy, where, in 1549, a Zürich schoolteacher called Urban Wyss had printed a writing manual which plagiarized (without acknowledgment) the pages of *Literarum latinarum*.

One of those who wanted to meet Mercator was a young coin collector and map colourist called Abraham Ortelius. Tall and slender, with grey eyes, hair 'of a yellow colour'[7] and a broad, bony forehead, Ortelius was a striking, congenial presence. Ortelius and Mercator had dissimilar backgrounds, but related histories. Followers of the Reformed Church, the Ortels' family house in Antwerp had been ransacked in 1535 for banned books after Abraham's uncle Jacob had fled to England to evade the Inquisition. Four years later, aged thirty-nine, Abraham's father had died. Then twelve, the boy had been brought up by his uncle Jacob, now back in Antwerp. Educated in Latin, Greek and mathematics, Abraham and his two sisters had taken work colouring maps. The boy was skilled, rather than scholarly, well connected and ambitious. Admitted to the Antwerp painter's guild – the Guild of St Luke – in 1547, Ortelius had begun dealing in coins and antiques, books and maps. At the time that they met in Frankfurt, Ortelius was probably a collector as well. And in Frankfurt there were particularly rich seams of geographical works to be discovered. Fifteen years younger than Mercator, and aware of his elder's back-catalogue, Ortelius had much to gain by cultivating a friendship. Mercator saw in Ortelius an unorthodox young map dealer and colourist who was ambitious to travel. A traveller with a topographical eye was also a cartographical source. The friendship could only bear fruit.

As if Mercator needed any reminding that he belonged to a generation apart from eager young Ortelius, his old friend and master Gemma finally succumbed to 'stones' within a few months of that visit to Frankfurt.[8] He was forty-seven.

Mercator owed more to Gemma than to any other individual. Gemma's work on triangulation had made possible accurate surveying and thus cartography. Gemma's globes and his development of instruments had given Mercator a template for his profession. Had the lives of these two paupers' sons not intersected at Louvain in the early 1530s, Mercator would not have been handed the mathematical key to cartography.

Just as Gemma had initiated Mercator's plans, the new generation began to nurture them. Ortelius would play a leading role in Mercator's future cartography. So would Ortelius' young friend and neighbour in Antwerp, Christophe Plantin. Born near Tours, Plantin had worked in the print shops of Rouen and Paris before realizing that the future lay on the Schelde: 'From my point of view', he wrote later to Pope Gregory XIII, 'I could have easily assured myself of the greater advantages offered me by other countries and cities, but I preferred Belgium and, above all other towns, Antwerp. What chiefly dictated my choice was that in my judgement no other city in the world could offer more facilities for practising the trade than this one.'[9]

Plantin had arrived in the city in 1548. It was a measure of Plantin's ambition – and Antwerp's opportunities – that the eager bookbinder was registered as a burgher by 1550 and as a member of St Luke's Guild by the following year. He received his printing permit in 1555, and published his first book the same year. In 1557, he placed his first order with Mercator, buying four copies of the wall-map of Europe.[10]

Faces were changing, and places too: the year that Mercator received his first order from the new printing house in Antwerp, Hieronymus Cock printed a new view of the city. As long as Mercator could remember, Antwerp had been pictured from the water, as a bustling, dishevelled riverport. Cock's viewpoint was high above the opposite side of the city. Antwerp had become a disciplined urban network of streets and civic symbols surrounded by massive geometric defences. Waggons entered landward gates, bound for fleets of patient ships on a distant, placid Schelde. Cock's Antwerp was a celebration of mercantile might, and a working diagram of a modern city. Mercator's longevity would enable him to participate in the new functional geography; the celebration of God's creation was beginning to extend beyond mere description. Geographers were beginning to concern themselves with the mechanics of landscape.

19

Spies and Cardinals

On 13 February 1558, Mercator and his family moved to a large house on Duisburg's best street, Oberstrasse. The house came with a collection of outbuildings and a slice of the Duisburg forest.

In a letter which Mercator wrote later in his life, he may have been thinking of the Oberstrasse house when he alluded to 'plans' to create a 'comfortable and grand dwelling' suitable for 'an honest citizen of modest fortune'.[1] The order of work demanded that the essentials 'for maintaining [the] household' be provided first, 'such as a kitchen, a store for food, bedrooms, cisterns, or perennial wells. The rest, which concern entertaining, the amenities of life, ornament, and magnificence, such as porticos, halls, courts, dining rooms, a third floor, pleasure gardens, and orchards,' could be added 'as time passes and opportunity and convenience arise, according to the plan of the whole work as it was drawn up at first'.[2] More than a house, this building would become a space for independent scholarship; a place where humanist minds could meet.

Such were Mercator's domestic dreams (the same letter suggested that his edifice on Oberstrasse was never completed). But, for the first time in his life, he could assemble without fear the 'most amply furnished library',[3] a body of books, treatises, manuscripts, letters and, of course, maps, which would form the foundation of his future labours.

'I saw a lot of him,' wrote Walter Ghim, 'because we were friends and neighbours, but I never found him idle or unoccupied; he was always busily engaged in reading one of the historians or other serious authors, of whom he had a fine stock in his library, or in writing or engraving, or was absorbed in profound meditation.'[4]

To one side of Mercator's new home was the house of the burgomaster, Otto Vogel, and to the other, that of a well-known Duisburg trader Diedrich Berck. Further neighbours were respected Duisburg burghers such as the families of Redingchoven and Tybis'; one of the latter (the

Fig. 18. Sixteenth-century Duisburg, from *Civitates Orbis Terrarum*, 1575. Mercator's centrally located house lay near the foot of Oberstrasse, which ran from the main square towards its gateway in the town walls. (Private collection)

merchant Deryck) had been painted by Hans Holbein in 1533.

Ghim remembered his friend as a convivial and considerate neighbour who never forgot his humble origins:

> Although he ate and drank very little, he kept an excellent table, well furnished with the necessaries of civilized living ... He always did his best to help those who were poor and less fortunate than he and, throughout his life, he cultivated and cherished hospitality. Whenever he was invited by the magistrates to a banquet or by friends to a dinner, or if he himself invited friends, he was invariably cheerful and witty, and adapted himself to the company of others in so far as his bodily constitution and his respect for decent living permitted.[5]

The neighbourhood was familiar with incoming humanists in search of a congenial refuge. The previous year Johannes Oeste -or Otho – had arrived from Ghent, where he had run a private school until the Inquisition had raided the place and seized his books. Duisburg had welcomed the arrival of an educationalist, and authorized Otho to open a new private school for twenty-five pupils. Among those to enroll in Otho's classes were the younger Mercator children.

The town had also been chosen by Mercator's old friend from his Castle days, Georg Cassander and his companion Cornelius Wouters.

Having settled in Cologne, they also bought a Duisburg house known as Oversnest, four doors down from Mercator on the corner where Jörisstrasse met the Knuppelmarkt. Cassander was beginning to make his name as a controversial advocate of Church reform, publishing books that would make their way onto the prohibited index, and in 1561, an anonymous work intended to promote peace between Catholics and evangelicals.

Ghim provided an insight into the table talk at Oberstrasse:

Conversing in the company of friends, he was easy and good-humoured; and, when he was among scholars, nothing gave him more pleasure than a friendly discussion about general philosophical, physical, and mathematical questions; about the preservation of mental and physical good health; about the settlement of religious controversy; the achievements of famous men; geographical and astronomical problems; and the customs, laws, and statutes of foreign nations.[6]

At the time, Mercator was encouraging another Louvain refugee to settle in Duisburg. Twenty years younger than Mercator, and from Flanders, the 'scholar and noted poet'[7] Johannes Molanus had been teaching history at Louvain before his views on religious reform had led him to flee to Bremen. There, he had found a post in a school for orphans. Although Molanus was barely twenty when Mercator left Louvain, the two had been closely acquainted. Mercator heard from Molanus shortly after the move to Oberstrasse: 'A few days ago, my beloved brother Gerard,' wrote Molanus, 'my dear, dear wife, my only solace in this unpleasant period abroad, [has] died, worn out by the considerable and long-lasting troubles of her husband as much as those of her own.'[8]

The 'long-lasting troubles' Molanus referred to were understood by Mercator to mean the continuing price Molanus was paying for his religious beliefs. For some time, Mercator had been persuading him to move to Duisburg, and the death of his wife had finally won Molanus to 'the attraction of the location' and to 'spending time with you, my old friends'.[9]

Mercator had his own reasons for urging Molanus to move to Duisburg: in February 1559, a group of Duisburg's citizens had sent a letter to the municipal council proposing the establishment of an *Akademisches Gymnasium*. Offering a course which would correspond with that of a Faculty of Arts, the new school would prepare students for the proposed university. The plan was enthusiastically endorsed by the town's burghers and Mercator volunteered his services as a teacher of mathematics. By

July, posters were raised announcing the opening of the school and premises were found in the old market hall. But the appointment of a rector – often an issue of political and religious divisiveness – proved controversial. Mercator encouraged Molanus to take up the post, but his old friend, bereft and still responsible for his twenty orphans, wrote that he didn't have 'sufficient knowledge, either of mind or of prudence' and that he was 'worn out by [my] difficulties and by the long misfortune of my wife' and 'scarcely up to dealing with small private matters, let alone a public function'.[10] The post went to Heinrich Castritius – or von Geldorp – another refugee from the Low Countries.

Molanus meanwhile reassured Mercator that he was 'of a mind' to move to Duisburg and 'rent a place by you next autumn'.[11] If, speculated the widower, 'my life lasts that long, then finally I can join you whole-heartedly for whatever I am worth, to offer myself in assisting the studies of those in early youth as far as my meagre learning allows, a service not of the sort you might hope for, but of the sort that the simple reality will offer'.[12] Only twenty-six, Molanus was clearly suffering from a chronic bout of melancholia.

Courses at the new school began on 28 October 1559. Without the restrictions forced upon the professors of Louvain, Mercator's curriculum included the cosmography of Sacrobosco and the geography of Pomponius Mela. The texts of Johannes Vögelin and Gemma Frisius were used for geometry and arithmetic. (A runaway hit with European schoolteachers, Gemma's *Arithmeticae practicae* of 1540 had already been issued in twenty-five editions from Paris to Wittenberg to Leipzig – all of them in Latin.) Mercator also used Gemma's books (and those of Oronce Fine too) for comprehensive 'practicals', often outdoors. These included the practice of surveying and advanced astronomy, in which the location, movements and relationships of celestial bodies were studied.

Mercator's course appears to have been a success (the council gave him three fattened pigs in lieu of payment), but the school's administration suffered a serious hiccup in its second year, when the rector Geldorp resigned following a dispute. His replacement was none other than Johannes Molanus, who had finally succumbed to Mercator's persuasions and accepted that there was no better place for a reformer to live than Duisburg.

The arrival in Duisburg of the widowed poet-scholar Molanus revealed that Mercator may have had an ulterior motive for bringing his friend up the Rhine: 'He was so fond of Johannes Molanus,' noted Mercator's neighbour Walter Ghim, 'that he gave him his eldest daughter in marriage.'[13]

While Mercator's cares regarding Emerentia were to see her married to a suitable husband, his concern with his eldest son Arnold was to groom the young man for cosmography. 'As soon as he had acquired some knowledge of the liberal arts,' observed Ghim, 'Mercator taught him and kept him at work studying mathematics.'[14] Already showing promise as an instrument-maker, Arnold had married the daughter of the public school rector in Düsseldorf. Arnold's first exercise in cartography was undertaken around the time that the family moved to Oberstrasse. Dated 1558 and bearing his name, Arnold engraved a neat little map of Iceland. Evenly lettered and decorated with ships, sea-beasts and a winged putto wielding splayed dividers, the map was delightful in every respect except one: having copied the outline of the island from his father's recent map of Europe, Arnold had failed to engrave it as a mirror image; printed, Arnold's Iceland was a reversed form of his source.[15] With his instrument-making and engraving, Arnold had recently established himself as a surveyor. Following survey work in the west Eifel region, he produced in 1560 a manuscript map for the Archbishop and Elector of Trier.

Of Mercator's other two sons, Bartholomeus was showing promise as a scholar, while the youngest lad, Rumold – just into his twenties – was away in Dordrecht being taught philosophy by Humanus Caesarus.

The price Mercator paid for freedom in a German duchy was more than a part-time teaching obligation. As Duke Wilhelm's most notable mathematician, Mercator could only wait for the inevitable call. And come it must: ever since his ill-judged quarrel twenty years earlier with Emperor Charles v over Gelderland, the duke's preoccupations had been strictly internal, reforming education, upholding his father's 'Erasmian' church ordinances, consolidating his authority. During Mercator's brief teaching career in Duisburg – in the summers of 1560 and 1561 – Duke Wilhelm commissioned him to survey the disputed boundary between the county of Mark and the duchy of Westphalia.

Alongside the teaching and surveying, Mercator was also trying to fulfil orders for his globes, but the strain was beginning to tell, and a slip-up in 1561 led to an irate letter from Rumold's teacher: 'I am rather upset', wrote Humanus Caesarus, 'that you did not read through my letter carefully enough', before repeating testily that he had ordered a celestial globe, only to have delivered a terrestrial globe. In an attempt to rectify the mistake himself, Caesarus intercepted a globe ordered by the mint-worker Peter Zanders, hoping that the two had been mis-directed, but Zanders' globe was terrestrial too.

In the meantime, the plans for Duisburg's university remained on the drawing board. In 1561, papal bulls authorizing its foundation were issued, only to be rescinded, and in 1562, Mercator resigned his teaching post at the *Gymnasium*, handing the responsibility to his second son, Bartholomeus. As the dream of a university receded, the *Gymnasium* reverted – in 1563 – to a *scola grammatica*.

Duisburg would never be the same again. The town's profile had been elevated by the influx of influential humanists, and by the continuing efforts to establish a university. Mercator's contribution to Duisburg's status was far from over. At around this time, a young man from Breda – a town in Brabant not far from 's-Hertogenbosch – came to live in Mercator's house on Oberstrasse. From Mercator, Johannes Corputius learned mathematics, instrument-making and engraving, and by 1563, he was working on a town plan of Duisburg. Inspired by Mercator, the plan would advertise the town's security, tranquillity and excellent location; its suitability as a centre of learning. At around this time, in 1564, a papal bull authorized the creation of the university, but the lack of an Imperial licence, and funding, continued to delay its foundation.

The failure of the university to materialize removed Mercator's last chance to occupy his remaining years in peaceful academic dignity. He would never belong to a conventional body of learning, any more than he had ever been able to claim nationality or citizenship. In Louvain, this independence had become a prison; in Duisburg it promised freedom. Unconstrained by Imperial pressures and local theologians, Mercator's streams of thought were free to take lateral excursions, changing course, bursting the banks to flood the fields of conventional wisdom. In his renewed exclusion from established hierarchies, Mercator preserved his greatest asset: the ability to reconcile apparently disparate concepts, and to articulate the result with unprecedented graphic eloquence. Some would describe his gift as one of originality; he viewed it as an exercise in harmonization.

Buffeted again by contrary winds, Mercator returned to his workshop, his books and his correspondence.

For long a source of succour, these letters were both a window on the world and a means of discreet discourse, the worlds they hid weightier by far than the paper and sealing wax which wrapped them.

Occasionally, a letter arrived from England, for Mercator was still in touch with the eccentric young beetle-builder who had come out to Louvain to seek mathematicians. In the years since John Dee and Mercator had enjoyed their 'sweetly protracted cooperation in phil-osophizing',[16] Dee's fortunes had replicated those of Mercator's, a spectacular rise of fortune preceding a terrifying and ignominious fall. Returning from Paris to Cambridge with claims that his mathematical lectures in France had aroused an even 'greater wonder'[17] than the levitating scarab of Trinity Hall, Dee had presented the boy king Edward with the two astronomical works which he had written in Louvain. Patronage followed, and Dee had been set for Mercatorial acclaim in Protestant England when Edward had died and Catholicism had been restored under Mary. As heretics began to burn, Dee – like Mercator – had found his name on a list. Charged with witchcraft and accused of endeavouring 'by enchantments' to 'destroy'[18] the queen, his house had been sealed and books seized. Suspected of being part of an underground Protestant group which had gathered about Princess Elizabeth, Dee had remained under arrest for three months. On release in the summer of 1555, he had lost the rector's post that had provided a living and had been due for further interrogation on the grounds of heresy. Dee had avoided the stake by magically transforming himself into a Catholic chaplain to the Bishop of London. Released, he had applied himself with obsessive zeal to collecting books, manuscripts and curiosities for a 'Library Royal'. He had also begun compiling a great life work, *Propaedeumata aphoristica*, which would explain through a series of maxims how terrestrial phenomena were influenced by the heavenly bodies. Mercator had been enthusiastic, urging Dee in one of the many letters[19] to pass between the two, to publish 'most urgently' his 'great demonstrative work'.[20] With a view to his own reading, Mercator had also asked Dee for a catalogue of the Englishman's writings.

When Mercator eventually received a response to his request, he learned that the English mathematician was dying. Dee had been struck down by an epidemic which had swept through England in 1557 and again the following year.[21] Addressing Mercator as the 'renowned philosopher and mathematician',[22] Dee informed his old master that his health had been 'dangerously shaken for a whole year now' by an 'extremely dangerous illness'.[23] To make matters worse, the doctor had 'also borne many other inconveniences (from those who, etc.) which have very much hindered my studies',[24] a reference to the spell in prison he'd served, for the 'lewd and vain practices of calculating and conjuring'.[25]

As Mercator had hoped, the letter did contain a list of Dee's works. Among them was the astronomical pair Dee had written in Louvain, works on perspective and 'two books' which Dee described as 'Concerning a New System of Navigation'.[26]

Preparing for his own death, Dee had delegated the completion of his *Propaedeumata aphoristica* to 'that most learned and grave man who is the sole relic and ornament and prop of the mathematical arts',[27] Pedro Nuñez, the Portuguese mathematician and geographer who had brought to Mercator's attention the nature of rhumb lines.

Dee didn't die, but Mary did. Back in favour as Queen Elizabeth's 'intelligencer', Dee crossed to the continent in 1562 and embarked on a quest for rare texts, an extended journey which took him from Antwerp to Zürich and over the Alps to Italy, then east to the Danube, to Pressburg and back to Antwerp.

During his great tour, Dee passed through – or close to – Duisburg. The two men had much to share.[28] Dee was still working on *Propaedeumata aphoristica*, and Mercator – having resigned from the *Gymnasium* – was becoming embroiled in a pair of troublesome maps. One of these maps was of particular interest to Dee; the other would come close to killing Mercator.

The maps, of the British Isles, and of the duchy of Lorraine, were both tainted by the house of Guise.

Related to the ducal family of Lorraine, and descended from Charlemagne, the Guises were one of the most powerful aristocratic houses in France. Catholic to the core, their interests extended from Italy to Scotland, the latter an alliance that had been forged through the marriage (in 1538) of Mary of Guise (the widow of Louis of Orléans, Duke of Longueville) to James v. At the head of the family was the Duke of Guise, Francis 'The Scarred', and his sibling Charles Guise, Cardinal of Lorraine. Unlike his mutilated brother, the cardinal had the irresistible beauty (and questionable genes) of Adonis: he was tall (like all Guises), with piercing blue eyes, neat teeth and a personality which could be described generously as enigmatic, but more accurately as cruel, cowardly and insincere.

Unsurprisingly for a family motivated by power and territorial gain, the Guises keenly appreciated the utility – and symbolism – of maps. When the Duke of Guise took Calais and Guines in 1558, he had removed an English bridgehead on the continent which had existed for 220 years. Within weeks, the French geographer and spy Nicolas de Nicolay had produced a copper-engraved map of the region and presented a print to the cardinal.[29] *Nouvelle description du pais de Boulonnois, comte de Guines, terre d'Oye et ville de Calais* celebrated the recovery by France of a long-lost stronghold. The map also publicized the humbling of England. Nicolay had followed this by copying a pilot book of Scotland,[30] which he had dedicated to the cardinal, an act which had been interpreted by Dr Nicholas Wotton (the former English ambassador in

Paris) as being tantamount to supplying the French king with an invasion map. Since French forces at that time were being sent to Scotland to support the cardinal's sister, Mary of Guise, Wotton's suspicions had been justified.

Along with Nicolay, the Guises had cultivated another mapmaker-spy. Born at the furthest reach of northern Scotland, John Elder had trained as a priest and visited Rome. Affecting a strongly anti-papal attitude, he had been paid an annuity by Henry VIII, to whom he had presented a map of Scotland showing 'every port, ryver, loigh, creke, and haven',[31] together with a letter outlining 'a proposal for uniting Scotland with England'.[32] But Elder came to change his mind about the Pope, and while the papal legate Cardinal Reginald Pole had sought to reconcile England with Rome through official channels, Elder had connived towards the same end by manipulating print. When the new (half-Spanish) Queen of England had married the King of Spain, Philip II, Elder had released his effusive *Letter* celebrating the 'calling home' of England to Catholicism; England had rejoined 'the unity of Christ's religion'.[33] A couple of years later, the Scotsman had taken a boat for France. There, he had fallen in with Cardinal of Lorraine, who offered him a pension. The spectre of a Scots cartographer cosying up to the rabid cardinal was too much for the English ambassador in Paris, who wrote in alarm to Sir William Cecil, warning that Elder was 'as dangerous for the matters of England as any he knew'.[34] Elder, added the ambassador, should be watched.

With Guise behind him, Elder returned to England around the end of 1561, having persuaded Cecil that he could provide the English with 'some good service'.[35] Once in Scotland, Elder turned blackmailer and informed the English that they could choose between repaying his stripped pension, or standing by as Elder 'be welcome to the Scottish Queen'.[36] In England, Cecil was advised that the Scotsman had 'the wit to play the spy where he list'.[37]

The English had reason for concern. In his continuing mission to facilitate a united Catholic island, Elder had produced a provocative new map of the British Isles.[38] It was this map which found its way to Duisburg.

The map had arrived at Oberstrasse a year or so before Mercator had resigned from his teaching post at the *Gymnasium*.[39] It had been accompanied by a request that Mercator should engrave the work onto copper.

Mercator, who had by now engraved the outline of the British Isles at various scales onto maps, globes and astrolabe plates, realized that he was looking at a map which depicted the northern islands in unprecedented detail.

Scrawled across the map in secretary hand were the names of over two thousand places. There were important changes to outlines too. The square protrusion of Wales had changed into a pair of pincers separated by a gulf. Scotland was shown in freshly wrinkled form, with a newly coherent island chain marked 'Hebrides'. The territories of earls and chieftains were named, along with a couple of mythical figures: on the northern shore of Loch 'Nessa' ran the lines 'Here dwelleth Ihan the Grand'. Close by, another legend claimed that 'In Petty leyth the bons of lyttell Jhon, he hat ben xiij [14] fote en heyght'. There were other signs too that the map's creator not only favoured Scotland but knew its features first hand: off the islands of Orcades a warning read, 'Here be daungerous rockes called Petlant Skyrres'. Off Elder's birthplace, Caithness, individual tide races were labelled 'The Swell', 'The hopper' and 'The boyer'.

Across the entire island group, principal rivers appeared with a precision Mercator cannot have seen before. And the map was packed with chorographic features too small to appear on Mercator's map of Europe: the southern promontory of 'Purbek' and its off-shore island 'Corf castell'; 'Snowdon hylle' in Wales; the western rocks marked 'The bishop and hys clerks'.

One of the few regressive elements on the map was the outline of Ireland, which appeared to be less accurate than Mercator's own 'Hybernia' of 1554.[40] This was most pronounced in the north, where the towns, rivers and headlands of Mercator's Europe map had been replaced by a smooth coast with a blank hinterland. Ireland appeared as an off-shore colony, the cartographic effort focused around Dublin and the Pale.

Had the geography of the map been surveyed by triangulation, Mercator would have been looking at a truly remarkable document, for the Low Countries – thanks to Van Deventer – were still the only extensive area in the world to have been mapped mathematically. But the British map had been created by more conventional means. Local surveys, regional maps, itineraries and observations had been amalgamated and then plotted onto a framework of places derived from Ptolemaic coordinates.[41] As an editorial exercise it was impressive; as a work of applied mathematics it was horrible. Elder had not even supplied the map with a grid of latitude and longitude.

A map of little mathematical integrity, which described the British Isles –

albeit subtly – from a Scots, Catholic, anti-Elizabethan perspective, was hardly the kind of commission Mercator could accept without anxiety. Among others who might find such an image of England awkward was John Dee. But Mercator was in no position to refuse the commission: the request to engrave it had come from a friend, 'a certain singular friend'[42] whose name Mercator would be unable to reveal.

A possible – indeed likely – candidate was Mercator's old student friend from Louvain, Antoine (now Cardinal) Perronet de Granvelle, to whom Mercator had dedicated his previous map.[43] Now that it was clear that Spain – and France – had more to lose through insolvency and heretical insurrection than they could gain by continuing their territorial wars, Granvelle had been entrusted with the task of exploring the possibility of peace. In May 1558, he had travelled to Péronne to meet with the French monarch's representative, one Charles, Cardinal of Lorraine.

The negotiations of 1559 had taken place at a derelict château in the neutral territory of Cateau-Cambrésis. With broken windows papered over, the delegations from Spain, France and England had occupied respective corners of a single freezing room, periodically dispatching spokesmen to argue points in their common language, Latin. Cateau-Cambrésis could almost have been a colloquium of map collectors: facing Guise, the French patron of the mapmaker-spies Nicolay and Elder, had been Mercator's patron Granvelle, abetted by the bigoted Viglius of Aytta (appearing in his capacity as a member of the State Council). By now, Viglius owned perhaps the largest private collection of maps in the Low Countries. As the most able Latinists in the room, Granvelle and Guise were the principal conduits during the negotiations. The outcome was that the English lost Calais, their last toehold on the continent, while the French kept Metz, Toul and Verdun, in the duchy of Lorraine.

Guise and Granvelle were about to become two of the most powerful men in northern Europe. In the celebratory tournament which followed Cateau-Cambrésis, the French king, Henry II, was pierced above the eye by the shattered stump of his opponent's lance. 'Vesalius, the great surgeon,'[44] was summoned from the Low Countries, and Mercator's university acquaintance attended to the vomiting, sweating, convulsing king for the final seven days of his life. Civil war erupted, Catholics aligned behind the Guises, and Huguenots behind France's other great aristocratic house, the Bourbons.

Granvelle's rise was less bloody but equally fated: following a secret agreement made in May 1559 between the Pope and King Philip II, fourteen new bishoprics had been created in the Low Countries, with an archbishop to control them. Antoine Perronet de Granvelle had become Archbishop of Mechelen, primate of the Low Countries and a cardinal too. Granvelle's role in Cateau-Cambrésis, and his primary role in the new bishoprics of the Low Countries, had given him particular reason to press Mercator to create a glorious, copper-printed version of a map which depicted a 'Catholic' British Isles, an aim shared by Philip II and by the Cardinal of Lorraine, whose acolyte Elder could provide such a map. If it *was* Granvelle who had asked Mercator to engrave the map, Mercator would have found it hard to refuse.

Mercator may also have had more prosaic reasons for accepting the commission: during 1562, the promising orders of maps and globes from Plantin's new business in Antwerp had come to an abrupt halt after the Frenchman had been accused of printing heretical works and forced to flee the city for Paris.[45]

So Mercator engraved the British Isles. But by the time he had finished, circumstances had changed for the friend who may have sent Elder's original. Having risen meteorically to be archbishop, primate, cardinal and the most influential member of Margaret's Council of State, Granvelle was outmanoeuvred by nobility intent on seizing power. Mocked in song as a papist parrot, and in dress as a 'red devil',[46] the red-capped cardinal slipped out of Brussels on 13 March, 1564, bound for Burgundy and exile. To the door of his old home somebody pasted a sheet of paper which read 'I am for sale.'[47]

Angliae & Scotiae & Hibernie nova descriptio was published in April 1564 – a few days after Granvelle's humiliating departure from the Low Countries. Assembled from eight sheets, this was the largest, most detailed map of the British Isles that had ever been printed. For continental readers used to cramped, crooked Ptolemaic derivations, this was an astonishing transformation. The pair of uninviting islands moored off northern France had changed into a sprawling archipelago surrounded by seaports. Multiple river systems hinted at extraordinary fertility; scores of settlements, church towers and castles told of lands which were prosperous, pious and well protected.

Mercator had been unsparing in his artistic execution. The quality of the engraving was better even than the 1554 map of Europe, the letter forms more fluid and the place-names more comfortably spaced. Where mountains and hill ranges had appeared on the wall-map of 1554 as improbable ropes of even-sized bumps, they now gathered in realistic clusters and were scaled from diminutive 'mole-hills' such as those beside the rustic southern town of 'Croydon', to gigantic northern excrescences, the largest of all being a towering pile overhanging 'Cokermouth' in the county of 'Comberland'. While none of Scotland's mountains could compete in sheer stature with Comberland's solitary pile, the ranges north of Argyle were engraved in a more restrained version of the riven denticular style which Mercator had used to dramatic effect nearly thirty years earlier in the deserts of Arabia. Compositionally, the map was pleasing too, text panels being placed on blank areas of ocean to create a sense of symmetry.

Angliae & Scotiae & Hibernie nova descriptio was, of course, more than a geographical description. While the engraving was identifiably Mercator's, aspects of the map's design and detail were intended to promote the Catholic aspirations of Guise and Granvelle. To those who knew Mercator, it was perfectly clear from the face of the map that its engraver did not regard this as an exercise in mathematical cartography. Just as he had done with his one other 'political map' – that of Flanders in 1540 – Mercator had omitted the grid of latitude and longitude. The British Isles did not even have a compass rose. The implicit message to map-users was that *Angliae & Scotiae & Hibernie nova descriptio* should be regarded as a disoriented truth.

Mercator had also tipped the British Isles onto their side, with west at the top. In revolving the islands through a quarter-circle, he was following George Lily's map of 1546, and Lily's various medieval predecessors, whose images of the world were oriented eastward. But in doing so, he was also creating a Scotland which was England's spatial equivalent: instead of receding off the top of a tall, thin map, Scotland occupied the right-hand half of a wide map. Lavishly castled, *this* Scotland had a healthily populated east coast and an interior decorated with lakes, forests and hills. The impression, of a secure, prosperous paradise, was not unintentional.

The map's main text panel had also been engraved at the expense of the English. While the origins of the Scots and Irish – and the wonder of their lands – were described in lengthy, glowing terms, England was subtly assaulted with a reference to the Italian historian Polydore Vergil, generally despised (notably by the English antiquary John Leland) for the doubts he had cast on the mythical ancestor of the Britons, Brutus. The tale of Brutus – as the Italian exile who'd sailed with his Trojans 'beyond the setting sun' to Totnes in Devon then marched to the Thames to found Troia Nova, Trinovantum, London – had been relayed by the twelfth-century Welsh chronicler Geoffrey of Monmouth in his *Historia Britonum*. Geoffrey's fantasy also contained the adventures of Brutus' follower Corineus, the Cornish giant Gogmagog, and King Arthur. For Leland and his fellow Elizabethans, Brutus and Arthur were the blood link between modern England and classical civilization. Geoffrey of Monmouth had facilitated a royal descent for Elizabeth, who had passed beneath statues of Gogmagog and Corineus on the eve of her coronation. In repeating the aspersions of Polydore Vergil (whose English detractors accused him of burning cartloads of manuscripts to cover his tracks, and of stealing boatloads of books from English libraries) on his map of the British Isles, Mercator knew that he would be offending the English, especially his friend John Dee – to whom Brutus and Arthur (and Prince Madoc and King Edgar) were pillars of British history.

There were other coded taunts. While a panel at the top of the map listed the bishoprics and archbishoprics of the British Isles, and designated each with its own symbol, care had been taken to omit the six bishoprics created by Henry VIII following his break with Rome and his proclamation that he was Head of the Church in England. With the monasteries dissolved, Henry had thought it 'most expedient and necessary'[48] to empower his new, secular Church with the erection of sees at Westminster, Oxford, Chester, Gloucester, Bristol and Peterborough. The omission of Henry's home-made sees from the new papal map of the British Isles bore the fingerprints of the Catholic primate of the Low Countries, Mercator's friend Granvelle.

The seas bobbed with discreet messages too. Midway across 'Oceanus Britannicus', a lone ship towed an empty vessel on a symbolically point-less mission between France's Calvinists and England's Protestants. That the ship had its rigging set square to the prevailing westerlies may have been a reference to the old saying that it was 'easy to sail with a following wind'.[49] The pair of ships off the Hebrides reinforced the message: strengthened by number, these were the vessels of the future, their sails trimmed to catch the side-wind that would drive them north to the passage to Cathay. The only two fish on the map reiterated the point: off the Orkneys, a muscular beast surfed through the map's methodical puncta, the black disc of its iris focused on an unseen north-east passage; southbound for Dieppe at the far side of the map, a dim cod drowned. Intended primarily for a French readership,[50] the map rejected the cause of European schismatics, favouring instead a positive, universal future.

Aware that the printing of such a magnificent map, with its Catholic, Hispano–Franco–Scots viewpoint, could cause him acute embarrassment (and complicate relations with the likes of John Dee), Mercator used the opening lines of his principal legend to deny responsibility for the entire contents of *Angliae & Scotiae & Hibernie nova descriptio*:

A certain singular friend offered to me this representation of the British

Isles, accumulated with truly a good deal of conscientiousness and with the highest accuracy, asking that I expand it, made proportional according to my scale of measurement, into many copies, which, since I was not willing to refuse flatly this friend and I judged it of no advantage to draw away my hand from an overview of worthy workmanship so perfect and of such learned men, I am showing it to you just as I received it, illuminated however with expositions of those things which, to a fellow geographer are most greatly relevant for detailed knowledge of areas...[51]

On a map riddled with riddles, it was the omissions which were most instructive. The missing bishoprics, the deleted heroes of England, the blanks in Ireland, the lack of grid lines and compass were all functions of the map's non-mathematical purpose. But one omission glared brighter than them all: for the first time in Mercator's life as a mapmaker, he had felt unable to name a dedicatee.

Could it have been that the suddenly vulnerable Granvelle, exiled in Burgundy and waiting anxiously for Philip ii to reinstate Habsburgian order to the Low Countries, could not afford to be associated with such a geo-religiously loaded map? For his part, Mercator may have decided that a dedication to a disgraced minister would be likely to render the map unsaleable.

In the time since Elder had devised his propagandist description of the British Isles, Granvelle's world had been turned topsy-turvy. Only five years before he'd been driven from Brussels, one of the cardinal's favourite artists, Pieter Bruegel, had caricatured the new inverted world of scorn and chaos by painting a fool defecating on an upside-down globe. Granvelle understood that a terrible retribution must soon befall the Low Countries. In the mean time, his interests were not best served by appearing in name on a map that had been produced on behalf of those forces of retribution – the armies of Spain which must surely punish the land of his adopted homeland.

20

René's Domain

The British Isles were barely off the copper before Mercator was precipitated into the map which pushed him to the brink. Again, there was a Guise connection. And again, Mercator's motives remained unstated.

This was a place which had been plotted on Mercator's mental map of the universe since his early days in Louvain. Lothairingia, Lothierrègne, Lorraine, the remnant of the kingdom which had been inherited by Charlemagne's grandson, Lothaire, was also the cradle of modern cartography. For it was in this duchy, an ambiguous buffer between the warring giants of France and the Empire, that Duke René II of Lorraine, the titular King of Jerusalem, had fostered his school of cosmographers in the century's opening decade.

The cosmographers had come to the town of St Dié, set in a cleft of the Vosges. Hidden by forest, and fenced each winter by snowfall, St Dié had grown from the monastic seed planted by St Deodatus eight hundred years earlier. Despite its mountain site, the town was well connected and served as a way-station on the road from the upper Rhine to Nancy and Paris. A two-day ride away lay Basle and Strasbourg, where merchandise and ideas from Italy and Germany and France met at one of Europe's most invigorating crossroads.

In this secluded yet accessible sanctuary, a group of scholars had gathered around one of René's secretaries, a canon called Walter Lud. Mathias Ringmann, a Heidelberg contemporary of Gregor Reisch was here, and Martin Waldseemüller. Encouraged (and financed) in their cosmography by Duke René, whose extensive library had recently received copies of the Vespucci letters, various nautical charts and a manuscript planisphere, the St Dié 'school' had set up a press to print scholarly books. The first had been the *Cosmographiae introductio* of 1507, the book which named America. This had been followed by the globe and the map of the world, and by Waldseemüller's treatise on surveying and perspective, his booklet on globes and his map of Europe and eventually, the great work which would ensure his cartographic immortality – his new edition of Ptolemy.

Waldseemüller had been working on his Ptolemy for many years – indeed it was meant to have been the first item off the St Dié press. In words which anticipated Mercator's later difficulties, Waldseemüller had written in April 1507 to the Basel printer Johann Amerbach describing the problems he faced with manuscripts which did not agree. There was, explained Waldseemüller, 'a Greek manuscript of Ptolemy' in the library of the Dominicans in Basel, 'which I deem to be as correct as the original. I beg you', implored the mapmaker, 'to procure [it] by any means possible, either in your name or mine, that I may have this book for the space of a month.'[1]

Delayed by the death of René, the Ptolemy had appeared in 1513. The twenty-seven woodcut maps, tables of coordinates, index of over 7,000 place-names were a marvel of scholarship, and with Jacopo d'Angelo's Latin translation of Ptolemy's text, the St Dié geographers had produced a definitive edition. This was the edition which broke new ground by including for the first time a supplement containing a systematic set of modern maps.

The last of the *tabulae novae* in the supplement honoured Duke René with 'a map of his domains most carefully printed'.[2] It was Waldseemüller's Lorraine on a single page. With south at the top, Waldseemüller had depicted a land whose very form seemed to be derived from biblical imagery, the trunk and dendritic branches of the Moselle reaching upward to a frothing crown of mountainous foliage where the circular symbol of 'S. deodatus'[3] hung like a ripe apple. Waldseemüller's map of Lorraine could claim another distinction too, for it was the first map to be printed in more than two colours.

To Mercator, St Dié was the cartographer's shrine; the place where Duke René's scholars, 'in the recesses of his land ... among the crags of the Vosges',[4] had consigned Ptolemy to history.

In the half-century between René's death in 1508 and Mercator's near-fatal map, Lorraine had struggled hard to maintain its precarious neutral status, both France and Spain claiming enclaves. From René's second son, Claude, had sprung the House of Guise, who had intervened so spectacularly in 1552, when they broke the siege of Metz and inflicted 30,000 casualties on Emperor Charles v.

When René's grandson Francis i had died of apoplexy in 1544 after just a year or so trying to rule the duchy, he had been succeeded by his son, Charles iii, 'the Great'. Declaring his sovereignty and neutrality, Charles had begun a programme of reforms intended to assure Lorraine's internal stability and independence. Mines and saltworks would be reorganized, armament manufacture brought under ducal control and plans were made to improve the fortifications of Nancy. But Charles lacked the most essential ducal tool of administrative reform and military defence. He had no decent map of the duchy.

Waldseemüller's *Tabula Nova* of Lorraine in the 1513 Ptolemy had located only seventy or so towns, virtually all of them on accessible rivers. Neither was there a body of regional cartography which could be assembled to form a complete picture of Lorraine. Charles Estienne's *Guide des chemins de France*[5] had covered parts of north-west Lorraine, and Sebastian Brant's somewhat older chronicle had included information on routes over the Vosges and along the Sar. But these were mere fragments of the territory which required mapping. A practical map of Lorraine would require a survey of the entire duchy, and a surveyor.

Enter Antoine Perronet de Granvelle. Between 18 and 20 March 1564 – a matter of days before Mercator's British Isles map was published – the cardinal was in Nancy visiting the duke and his mother, Christine of Denmark. Christine, who was the niece of Emperor Charles v, was in the habit of airing Lorraine's problems to the emperor's trusted minister. In his turn, Granvelle 'encouraged her the best I could, giving her advice on whatever she put before me'.[6] In all probability, it was Granvelle who convinced Christine that Lorraine required an accurate, up-to-date map, and that the man for the job was a surveyor/cartographer called Mercator.[7]

Mercator must have been informed almost immediately. But at fifty-two, he was in no state to undertake what would amount to the longest, most arduous journey of his life. The Duke of Lorraine's lands stretched all the way from Luxembourg to the wet, vertiginous forests of the Vosges, in sight of the Alps.[8] Covering an area greater than Flanders and Brabant combined, much of Lorraine was occupied by vast tracts of forest and hills, cut through with steep-sided valleys. The terrain and weather conditions were unlike anything Mercator knew in the Low Countries. Furthermore, the survey area was threatened by Lorraine's unstable borders – external and internal – and by the presence of marauding bands. Mercator was not only elderly, but his regional cartography had never depended upon his own surveys.

Leaving the safety of Duisburg for the dangers of Lorraine was unjustified on all counts bar one: a cartographic survey on behalf of René II's great-grandson would open to Mercator the duchy's fabled map-cabinets, cabinets whose untold treasures must include the copy Waldseemüller made of the Greek Ptolemy lent him by Amerbach, the Vespucci letters,

Caveri's manuscript planisphere (one of Waldseemüller's principal sources for his outlines of Africa and the continent beyond the Atlantic), various portolans and much more besides.[9] Mercator's motives for seeing Duke René's library went far beyond curiosity, for he was incubating a project which would benefit enormously from such a visit.

But the risks of such a survey were appalling. Decades spent accumulating lists of coordinates from sources which ranged from Ptolemy, Oronce Fine and Apian, had already provided Mercator with a basic framework of Lorraine's principal places. But between the major towns were innumerable villages and smaller towns, rivers, hills, forests and borders, all of which Mercator would have to plot, using triangulation. Since the wording of the duke's commission stipulated that Mercator was to 'make a map and description'[10] of the duchy, there was no alternative but to travel exhaustively. Nothing in Mercator's past suggests that he could possibly have accepted the commission with anything but the deepest trepidation – and highest expectation.

With his son Bartholomeus, Mercator left Duisburg that spring. They travelled up the Rhine to Koblenz, then turned up the narrower Moselle to Trier. From Trier, they took the high road over the hills of Hunsrück to the river Sar and so to the northern border of Lorraine. After the great forests of Warnet, the Sar would have led them to the town of Sar Alben, and it was here that they may have met the officer who had been charged by Duke Charles with caring for the surveyors' needs. On 21 May, Mercator was paid the first of two instalments of 400 francs. Bartholomeus was paid 26 francs.

'Mercator made a survey,' Ghim reported later, 'town by town and village by village, measured most accurately by the triangulation method.' A more difficult landscape to triangulate would be hard to imagine. Tree-cover and deep valleys would have impeded sightings while the roads of Lorraine were far from safe. The pair must have worried continually about the roving bands of militant Germans, bound for France to fight the Catholics of Guise. Vagabonds had become such a recurring menace in Lorraine that Charles III had been obliged to include edicts against them as part of his reforms. Some areas were worse than others. In the regions of the bishoprics of Metz, Toul and Verdun, the Mercators would have had to contend with towns and villages which had been won over to Calvinists who were unlikely to be cooperative with surveyors travelling in the name of the House of Guise. Many around Lupstein and Scherweiler still remembered the annihilation in 1525 of the Anabaptist peasant *rustauds* (boors) by the first Duke of Guise.

It was while Mercator was surveying the northern regions of Lorraine that he would have had the opportunity to visit Nancy, and René II's fabled library. Unfortunately for Mercator, the keeper of the Trésor des Chartes was a functionary called Thierry Alix from the Chamber of Accounts. A young, fervent patriot of unlimited ambition, Alix – in common with Lorraine's nobility – shared a deep distrust of outsiders. He also had cause to involve himself personally in Mercator's anticipated map. As one of the duke's most trusted administrators, Alix had already travelled the length and breadth of Lorraine gathering information on the duchy's borders. Along with a register of 2,290 place-names, he had also made a number of descriptions. On at least one occasion, he also drew a map. The man who held the key to René's cartographic treasures had his own ideas about the form of Mercator's anticipated map. While Alix could concede that the visiting surveyor was 'a most learned and sufficient geographer',[11] he also expected that Mercator's map should include a hymn in thirty-five couplets celebrating Lorraine, a description of Lorraine (in French) and Alix's list of 2,290 place names. Given his connections with Granvelle, Mercator was in no position to collude in a map which portrayed the graveyard of so many Imperial soldiers in such a celebratory, Guisian light. Later events suggested that Mercator and Alix did not become the best of friends.

Nancy lay on the line of latitude which divided Lorraine in half, and which Mercator would later use to separate the two halves of his map of Lorraine.[12] The southern half of the duchy hid extra concerns for the surveyors. In the south-east were the mountains of the Vosges. Here, the roads were much worse than in the north, and the weather was colder, wetter and more volatile. Where the road from Nancy to Basel cut through the Vosges, a chapter of warlike canonesses from the abbey at Remiremont – whose rights and privileges as a free Imperial town had been confirmed by Charles v as recently as 1554 – had been at the centre of sporadic fighting for at least a year. Yet Mercator and his son pushed their survey into the most remote recesses of Lorraine. In the mountains to the east of St Dié, they diligently recorded necklaces of tiny hamlets which were tucked away in valleys which led nowhere. The residents of Herbeaupiere, Lusse, Le Perriere, Le Merlus and Trois Maysons[13] would have been amazed to see their communities sharing a map with Nancy and Metz. 'St Piere mont',[14] the mountain above their valley, was also recorded by the visiting surveyors. Just as they had in the north, Mercator continued to survey with extraordinary diligence the duchy's rivers and tributaries.

But as they bumped along Lorraine's atrocious roads from one sighting point to the next, Mercator and his son were bound for inevitable disaster.

At some point that summer, probably in the deep south,[15] something terrible happened. Perhaps father and son were beaten or robbed by a band of thugs, or perhaps it was the plague that ravished some of Lorraine's border areas during 1564. Maybe they fell to the warring canonesses of Remiremont. Mercator, an unwilling traveller by nature, may have found the fears and discomforts of the road too much to bear. Whatever the cause, the curse of Guise finally struck, and Mercator cracked.

Hunters in the Snow

'This journey through Lorraine,' recalled Ghim, 'gravely imperilled his life and so weakened him that he came very near to a serious breakdown and mental derangement as a result of his terrifying experiences.'[1]

Mercator came home to the Rhine without Bartholomeus. It wasn't until the onset of autumn that Bartholomeus returned from Lorraine, collecting his additional 26 francs on 2 October.

From notes and measurements, Mercator made 'an exact pen-drawing' of the duchy and 'presented it to his Grace in Nancy',[2] who was reported[3] to have expressed pleasure.

The map was never printed. The temptation to produce a printed image of Lorraine as a unified, Catholic entity, territorially distinct from France and the Empire, may have been outweighed by the risk that such a detailed, accurate map was – in the words of Charles IX of France in 1564 – a 'map for war, useful to an enemy, who with a compass and quadrant could lead an army through the whole country'.[4] Squeezed between two belligerents, Lorraine could not afford to circulate a document which described an invading army's source of food and fodder, and every river too.

To meddlesome Thierry Alix, the whole episode had been deeply unsatisfactory. His hymn, his French description and his interminable register of place-names had failed to grace the legends of Mercator's Lorraine. He also criticized Mercator for failing to mark various monasteries, châteaux and villages, and for omitting the border up in the Vosges.[5]

For Mercator's part, the exercise in Lorraine had been catastrophic to his health. Unpublished, his map was secreted in the Trésor des Chartes, a relief perhaps since its classification as a discrete source saved Mercator from being associated with a work which could be read as a triumph of Guise propaganda. But Lorraine may have rewarded Mercator in a way which he felt unable to acknowledge: did he (despite Alix) get at René's library, and perhaps at the archives of St Dié too? If so, he may have been able to read, and to copy, a collection of sources which would soon lay the foundations for his life's work.

Depressed, and in what must have been a state of near invalidity after his 'serious breakdown and mental derangement',[6] Mercator entered the longest winter of his life.

Outside the windows of Oberstrasse, the world was getting colder, each successive winter bringing increasingly extreme freezes. In 1559, Breugel had been able to paint skaters outside Antwerp's gate of St George. But it was the winter which followed Mercator's escape from Lorraine which turned seasonal fear to omen. In November 1564, the temperature plummeted. By December, frosts were clamped across northern Europe. The Scheldt froze over, and Breugel painted hunters in the snow and below them a valley in black and white where nothing could grow. The Maas ports became blocked by bergs. Not until February did the ice thin to let Breugel scrape his palette with brown and black for a river mouth wrecked by broken dykes.

It had been the coldest winter that anyone could recall and when it was finally through the rain ruined the harvests of 1565. With the Baltic closed due to a war between Denmark and Sweden (which had dragged in the Hanseatic towns, and Poland and Russia) grain imports were cut and bread ran short. Between March and December 1565, the price of wheat trebled. Riots erupted at corn markets, and the following winter the prices were pushed up to famine levels. In December, a Brussels government secretary warned exiled Cardinal Granvelle that the worsening grain shortages would lead to an uprising in which 'the religious issue will become involved'.[7]

In Bremen, Mercator's daughter Emerentia had suffered a bad winter. 'There was difficulty with the corn supply,' wrote Emerentia's husband, Molanus, 'but it has been tolerable ... The Lord has thus far given us the necessities for life and I trust that he will continue to do so.'[8]

Molanus had heard that Mercator was dead. Since returning prematurely from the mapping expedition to Lorraine, Mercator had been so ill that rumours began to circulate in Antwerp that he had 'flown across to heaven'.[9] But by March, Molanus had been reassured that his father-in-law was still alive: 'I give thanks, to Our Father ... that he has preserved you for the amplification of his name.'

Mercator's wife Barbara was also showing the strain. 'For, unless I am mistaken,' Molanus warned Mercator, 'she is gradually being worn out by your troubles.'[10]

Mercator was being assailed by more than ill-health, for in the same letter, Molanus referred to his friend's 'great courage' in withstanding 'terrible attacks' from 'troublesome men'.[11] Risk of the letter's interception precluded elaboration. Who were these troublesome men? Exposed in open ground between evangelists and Catholics, Mercator could expect hostility from local Calvinists as well as distant bishops. Publicly bonded to Cardinal Granvelle through his European map dedication, he could also anticipate antagonism from the followers of the Prince of Orange. And Mercator's recent maps of the British Isles and Lorraine were especially susceptible to malicious interpretation.

Beleaguered by misfortune, Mercator's name was omitted from the elaborate plan of Duisburg which his ex-student Johannes Corputius completed for printing in 1566. Dedicated to Duke Wilhelm, the plan's main legend was embellished with the names of Georg Cassander, Cornelius Wouters and Karel Utenhove, but it failed to refer to Duisburg's own Gerard Mercator. Once again, Mercator had shrunk from an association: all three of the men on the map were prominently associated with religious reform, and Utenhove – who had settled in Friemersheim, just across the river from Duisburg – had been one of the Ghent rebels of 1539–40.[12] Mercator's house could be seen on the new town plan, but its owner had withdrawn from sight.

Mercator now took the view – like Luther and Melanchthon – that the Holy Roman Empire had become the dying beast of the fourth and last world monarchy predicted in the Book of Daniel. And the end had been brought even closer by a revised view of time itself. Instead of the old Augustinian notion of six world ages, one for each of the six days of Creation, the view among the reformers was that there were only three world ages: the 2,000 years from the moment of Creation to the giving of the law; two thousand years under the Law; and two thousand years of 'grace', from the Messiah's coming to the Last Judgement. ('Two thousand wanton. Two thousand the law. Two thousand the day of the Messiah,'[13] Melanchthon had written.) That should have left over three centuries to run from the 1560s, but a passage in St Matthew ('And except those days should be shortened, there should no flesh be saved')[14] made it clear that the final two thousand years were going to be truncated.

But by how much? Mercator (and Melanchthon, among others) believed that the time of the Last Judgement could be derived by adding the seventy-one years that the population of Judah spent in Babylonian captivity to the date that Luther and Zwingli had risen from the masses. Apocalypse could be expected in 1588.[15]

Evidence of the approaching Endtime could be seen to all points of the compass. Impregnable hereditary monarchies which had enjoyed half a century of unequivocal support from their subjects were embattled by revolution and civil war. In France, where Henry's jousting accident had already precipitated a monarchical crisis, civil war and the assassination of the Catholic leader, François, Duke of Guise, there were rumblings of worse to come as Huguenots rallied themselves for an expected strike by Catholic monarchies against the continent's reformers. Closer to Duisburg, the Low Countries' government had been gravely weakened by Granvelle's departure, and Calvinists were now talking of following the Huguenots into armed struggle.

'The town of Ieper, among others', wrote the map collector and president of the Privy Council, Viglius, 'is in turmoil on account of the daring of the populace inside and outside who go to the open-air services in their thousands, armed and defended as if they were off to perform some great exploit of war. It is to be feared,' added Viglius perceptively, 'that the first blow will fall on the monasteries and clergy and that the fire, once lit, will spread, and that, since trade is beginning to cease on account of these troubles, several working folk – constrained by hunger – will join in, waiting for the opportunity to acquire a share of the property of the rich.'[16] The monarchy was going down with the economy and the nobility.

The storm broke in August 1566, when Calvinists erupted with unbelievable violence from the mayhem caused by grain riots. The biblically minded could point to the Book of Hosea, and to the inevitable fate of people who select false monarchs: 'They have set up kings, but not by me: they have made princes, and I knew it not: of their silver and their gold have they made them idols ... For they have sown the wind, and they shall reap the whirlwind.'[17]

The whirlwind, predicted Hosea, would 'send a fire upon his cities ... break down their altars ... spoil their images'.[18] Never had the old Flemish proverb 'Eat fire and you'll shit sparks,'[19] been recalled with such ghastly presentiment.

From churches and monasteries across the Low Countries, Calvin's mobs tore images, smashed sculptures and windows, ripped paintings. Richer residents hurled decorative carvings to the street to prevent mobs scrambling over their houses. Ortelius was in Antwerp: 'Wherever these iconoclasts, armed with sticks, axes and burning torches, ran from one church to another everybody fled ... Next day all the churches looked as if the devil had been at work for some hundred years.'[20]

'S-Hertogenbosch and Ghent were next, then Breda, Mechelen, Turnhout, Lier, Bergen, Amsterdam, Haarlem ... Utrecht. Over the next days, churches and shrines in hundreds of towns and villages were stripped. Louvain was one of the few places to escape the Calvinist purge. 'All churches have been closed', reported Ortelius, 'and the Catholics forbidden to preach.'[21] In Brussels, the government appeared helpless.

For Mercator in distant Duisburg, news of the 'iconoclastic fury' could only confirm the approaching Day of Judgement.

22

A Study of the Whole Universe

In the depths of pre-apocalyptic depression, Mercator conceived the monument which he wished to bequeath to humanity.[1]

Looking back, the sum of his works had been as unstructured as his life. In the thirty or so years since he had forsaken philosophy for mathematics, he had produced a miscellaneous succession of instruments and globes. And occasionally, maps. Some were commissions, others the offspring of Christian or mathematical conviction. Some had been guided by political expediency. Mercator's plan to map the world by regions had progressed no further than his introductory – and flawed – world map of 1538, and the wall-map of Europe.

The metaphor now forming in Mercator's mind was that of a building, a monument.[2] Three decades of labour had produced no more than a partially accumulated heap of building material. He had not even plans. But now, with the Endtime imminent, Mercator stood back from the disordered accomplishments of his past and began to conceive a structure for the future. As yet they were not exactly architectural drawings, but he could, for the first time, see the form of the building whose erection would bring reason to his existence.

The plan was stupendously ambitious.

This would be a work whose scope encompassed far more than geography: 'I will prophesy', he wrote, 'that something more august will have been produced if I do not keep myself to a description of the lands of the earth alone, but I should undertake a study of the whole universal scheme uniting the heavens of the earth and of the position, motion and order of its parts (insofar as it agrees with these aims).'[3]

In his mid fifties, Mercator was proposing nothing less than the book of the universe. There was of course a circularity in his intent, for he was returning to the concept – albeit extravagantly – which had been introduced to him over thirty years earlier by Gemma Frisius: 'I believe,' he continued, 'that cosmography is of the first merit amongst all of the principles and beginnings of natural philosophy.'[4]

Mercator's melancholic condition produced a clarity of vision. His seemingly unwieldy field of interest could be reduced to a series of manageable parts: 'I had decided at the beginning', he recalled, 'to investigate thoroughly the two parts of the universe, the celestial, of course, and the terrestrial.'[5] The revolution of heavenly bodies would be examined, and their distances, and sizes and interrelationships. In the second part, he would examine the size of the earth, its weight, 'its disposition towards sea and heavens and more things of this type'.[6]

But as Mercator contemplated the elements of his planned cosmography, he came to see that the celestial and terrestrial were united by, and preceded by, history: 'I realized that such a work was nothing other than the history of the first and greatest parts of the universe, moreover that history keeps hold of the premier place and rank in every application of philosophy.'[7] If, he considered, he were to 'trace out thoroughly, by diligent contemplation, the first origin of this mechanism and the genesis of particular parts of it',[8] such a historical approach would ease his 'judgement about the nature and power of individual things'.[9] This in turn would open the door to 'philosophizing correctly about the universe'.[10]

With his revised, historical approach, Mercator anticipated that he would now start 'from the beginning of the universe', and thus 'probe into the origins of the nature of the universe from the first chapter of Genesis and take the whole thing, as it were, from an egg'.[11]

From egg to sphere, Mercator's cosmography would describe the history and disposition of God's creation.

The cosmography would consist of five parts. The first would describe the creation of earth, the second would describe the heavens, while the third would be 'a representation of the land and the sea'.[12]

As the geographic element of the cosmography, it was this third part which would demand Mercator's mapmaking skills. And it was the geography which spawned the cosmography's fourth part: 'I saw that geography could not be separated, indeed I did not even comprehend it correctly without the period, the order and the succession of kings who founded cities and kingdoms.'[13] The fourth part would therefore consist of a *Genealogicon*.

It was while Mercator was investigating the history of states in the early phases of the *Genealogicon* that he realized that a fifth part would be necessary 'lest any ignorance of this might deceive me in the remaining parts'.[14] That fifth part would be a 'fixed doctrine of periods',[15] a chronology of world events from Creation to the present day.

Thus did Mercator formulate his great cosmography, a work which would attempt to describe the history and disposition of the entire universe.

The unfinished churches of the Low Countries were proof enough that a massive, mathematically precise, intricate, symbolic monument required more than a brilliant architect if it was ever to rise above the plain.

Mercator had the draughtsman's gift of reducing multi-dimensional human disorder to systematic plans and geographical outlines. But that very process of reductive simplification involved a denial of chaos. Mercator's cosmography was planned as if he was not beset by the daily unpredictabilities of old age, war, plague, money and family tragedy. The cosmography's schema came from the mind of a visionary. And while nobody in their right mind would ask a visionary to build them a house, it took an unconstrained imagination to visualize a truly great work.

So the cosmography began to take shape in the variable airs that existed between the dreams of its architect and the chaotic realities of late-sixteenth-century Cleves. Of the work's five parts – Creation, heavens, earth, the genealogy and history of the states and the chronology – various elements already existed. Mercator had been thinking – and probably writing – about Creation for at least thirty years. And he had been accumulating material on the world's geography for as long. At least two major mapmaking projects were already under way: the regional maps of the world which he had announced in 1538, and the new world map promised in 1554.

With such a vast and disparate archive of thoughts to marshal, Mercator began by establishing the work's mathematical references. No questions could be answered until he had 'attained the true definition of time and place'.[16]

'Time' would demand a chronology – a table of world events which would provide a temporal framework for the entire cosmography. 'Place' would require a modern, mathematical representation of the world: a map. Or rather *the* map, for this would be the long-promised world map.

Having established his 'true definitions' of time and place, Mercator could proceed with the detailed elements of his investigation. Taking precedence here would be the description of the earth in two parts. First to appear would be Mercator's definitive edition of Ptolemy's *Geography*, partnered by a modern geography of the world, equipped with maps and descriptions. The account of Creation could wait, not least because its inevitable interpretation as a heretical work would interfere with, or even preclude, the completion of the cosmography.

In an appointment which cannot have been unrelated to Mercator's plans for the greatest cosmography of all time, he was made cosmographer to the Duke of Jülich-Cleves.

The end of 1566 brought heavy snowfalls, and as another terrible winter settled on the charred work of the iconoclasts, plague returned.

In Duisburg, Mercator's and Barbara's worries were worsened by an inexplicable failure of Molanus to write from Bremen, where Emerentia was due to give birth to their second child.

When a letter eventually arrived from the north, it began with the dread words 'pressing danger' and 'bitter calamities'.[17] It was a long letter. With chronological exactitude, Molanus described in great detail the events which – in one terrible week – had destroyed his family.

The school rector began by explaining that he had returned with his 'sweetest wife and little son'[18] to Bremen, where the Senate had at last lifted the obstacles to doctrinal freedom. If only the letter had continued in this happy vein.

The plague struck ('raging behind us and in front of us')[19] just a few days after Molanus began teaching on 18 April. Mercator's grandson had been one of the victims. For four or five days the boy suffered. Disregarding the risk of infection, Molanus and the heavily pregnant Emerentia tended their son until he died in their arms. 'Finally,' wrote Molanus to the boy's grandfather, 'when he was scarcely breathing, I tore my wife away.'[20]

Ordered by the doctor to change Emerentia's clothing and to move her to another house, Molanus located an empty dwelling and installed her with two maidservants. Then he buried his son. Having returned to the house, he was cleaning himself of contagion in the steam room when a madman waving a dagger burst in. Behind him strode a servant drawing a sword. Roaring and 'roused with frenzy and drink', the intruder threatened the 'terrified'[21] women, then demanded that they leave for fear of introducing the plague into the neighbourhood. While the maidservants 'begged for the help of the neighbours', Emerentia slipped into the steam room 'and closed the door, as her whole body shook'.[22] After the madman had been dragged away by neighbours, Molanus and Emerentia sought shelter in the house belonging to the widow of Cornelius Bacher, where, in front of the hearth, Emerentia suffered 'a cold dread'.[23] But it passed. For eight days they remained in the house while plague engulfed Bremen. On 2 May, Molanus decided to remove his schoolboys to the safety of

Emden. Emerentia and the two faithful maidservants joined the boys in the waggon, while Molanus remained in Bremen.

But the plague reached Emden too. On 8 May, it claimed a friend of Emerentia's. On 10 May, Mercator's eldest daughter gave birth to 'a thin boy who lived in the light for a mere few moments'.[24] On the evening of the eleventh, 'with her mind still intact, she asked that a prayer be made; at that same time lifting her face, when suddenly she lent onto the couch, struggling briefly with the pain of death, she fell asleep in the Lord'.[25] One of the maidservants died the same day, and the other – a girl called Greta whom Mercator had known well – was claimed by the plague five days later.

Too distressed to write further, Molanus closed 'in great grief' by willing the Day of Judgement: 'This sinful nature evidently must be destroyed, so that the spirit may be safe for the arrival of Jesus Christ.'[26] Emerentia had lived for nearly thirty years.

The news of Emerentia's premature death was followed a year later – in 1568 – by the death of Mercator's second son, Bartholomeus.

The young man who had rescued the survey of Lorraine had always held – as Ghim later wrote – 'great promise'.[27] Shortly before that fateful surveying expedition, he had published *De Sphera*, a collection of his father's lectures, and in May 1567 – as Emerentia lay dying – Bartholomeus had matriculated at the university of Heidelberg, where he intended to study ancient languages, philosophy and theology. The capital of Palatinate, Heidelberg had become a centre of evangelical learning, and Bartholomeus could have looked forward to a far more liberal curriculum than his father had endured at Louvain. And as the son of the recently appointed cosmographer to the Duke of Cleves, the costs of his university education were being borne by the ruler of Palatinate, the Calvinist-inclined elector, Frederick III. But before he could finish his second year of studies, Bartholomeus was also claimed by illness. He was only twenty-eight. All surveying duties now passed to Mercator's eldest son Arnold.

News Mercator received at this time was mostly bad. Beyond Duisburg, the monarchies and the Catholic Church were fighting for survival. France was now engaged in its third civil war in six years, with the Huguenots in the ascendant. And in the Low Countries the iconoclastic furies of 1566 had been followed by a massive, terrifying response from King Philip II, who had unleashed Granvelle's old friend and court ally, the Duke of Alva. At the head of an army of Spanish and Italian troops, the elderly, gouty, captain general descended on Calvin's deltaic breeding grounds with disproportionate fury: the anti-Granvelle league were rounded up, 12,000 were tried and over 1,000 executed.

It could only be a matter of time before Alva and his Catholic crusaders swept across the rivers to rub out the heretics of the Rhinelands – and Jülich-Cleves. In Cleves, Duke Wilhelm v's heart faltered and paralysis crippled the left side of his body.

23

Time . . .

The great cosmography got off to a dramatic start. In the space of a year, Mercator completed his history of the universe, and his new world map, a map which would redefine the planet for all time.

In tackling chronology first, Mercator returned to a subject which had vexed him since the 1530s. Then, his compulsion to investigate Creation had led him up a philosophical cul-de-sac. Now, he needed to enumerate Creation before embarking upon the main part of the cosmography.

Mercator's universal history posed its author with challenges which he did not have to confront as a geographer. While 'place' existed in the observable present, elapsed time was a function of record: geographers plotted their coordinates from contemporary data; chronologers divined their dates from inconsistent histories. Compiling a date-list that spanned time's entirety presented difficulties of verification and continuity. And although a life spent reconciling sources as disparate as Ptolemy, Marco Polo and the Cabots had turned Mercator into a master of the amalgam, even he would struggle to resolve the gaps and contradictions inherent in time. It was a task which involved consulting the works – many of them multi-volume – of 123 authors.[1]

Not only did Mercator have to create a single continuous chronology from a number of partial histories – without contradicting the Scriptures – but also he had to devise a method for reconciling different methods of time-keeping: Babylonian years were counted from the region of Nabonassar; Roman years from the foundation of Rome; Greek years were measured in Olympiads, from the year in which Coroebus won the Olympic games. Christian years began with the birth of Jesus. A supplementary complication was introduced by the differing criteria for year-changes: Greek years changed at the summer solstice, Hebrew years at the spring equinox, Christian years at Christmas (although the Low Countries still followed the Julian year-change at Easter).

Mercator's solution was ingenious, and without precedent. He synchronized the major calendar systems against a single chronology which began with the Creation of the world in Year Zero. He likened this method to a trunk and its branches. The early pages of the work were simple enough, with a single column to carry the chronology Mercator had derived from the Hebrew Bible. But the columns soon multiplied. By the crucifixion of Christ, six parallel calendar systems allowed the reader to date each historical entry in Babylonian, Roman, Hebrew, Greek and Christian years, and to read off the elapsed years since the moment of Creation.

But the chronology was not as seamless as he would have liked. The Hebrew Bible provided dates for events up until the reign of the Babylonian King Nebuchadnezzar, when Mercator switched to Ptolemy's *Almagest*. (In doing so, he was following the astronomer Copernicus, who had believed that Ptolemy's Babylonian King 'Nabonasser' was the biblical King 'Salmanasser'.) The *Almagest* carried Mercator forward in time to the Roman Emperor Antonius Pius, at which moment he switched to dates gleaned from Onofrio Panvini, an Italian monk and antiquary with whom Mercator had been corresponding. One of several continuity glitches was caused by a lack of information on the history of the Trojans. Somewhat rashly, Mercator turned to a source he acknowledged as 'fable'[2] – the notoriously flaky Johannes Annius of Viterbo, whose seventeen-volume, regularly reprinted *Antiquitates* of 1498 included various forged texts and assorted scraps from pre-Christian writers. (Mercator was not alone in failing to resist Annius: Erasmus had used the *Antiquitates*, despite admitting that he was 'not really satisfied', and despite referring to the Italian as 'rash and pompous, and in any case a Dominican'.)[3]

Mercator's sources ranged across the full spectrum of historical fidelity and theology. Contrasting with Annius, Johann Sleidan was a modern, reliable, revisionist historian. A contemporary of Mercator's, he had been born in the hills of Eifel to the south of Gangelt. Educated at Liège, Cologne and Louvain, Sleidan had served Francis I in France, but was dismissed and became the official historian of the Schmalkaldic League, dying young (felled by a wave of pestilence in the autumn of 1556) to leave a monumental history of religion, the 940-page *Commentariorum de statu religionis et reipublicae, Carolo V. Caesare*. One of the leading adherents of the four-monarchies view of history, Sleidan had also written a book on 'the four chief empires'. The Inquisition had raised Sleidan to 'most banned' status, recording him (in their Rome index of 1559) in their expanding list of *Auctores quorum libri & scripta omnia prohibentur*. Attacked by Lutherans for his moderation and by Catholics for his partisanship, he was a historian who deplored value judgements and who argued that every event should be recorded 'as it actually happened'.[4] His *Commentariorum*, impartial and packed with quotes and state abstracts, was as austere and devoid of authorial intrusion as Mercator's recent maps. They shared the same ideals, of *candor*, of *veritas*.

Sleidan's history of the religious cataclysm provided Mercator with the *Chronologia*'s climax – the sequence of apocalyptic events which would soon culminate in universal upheaval. After 334 pages of kings, popes and eclipses, Mercator reached the year 1517, and the entry ('*Martinus Lutherus contra Indulgentiarum* . . .') which had triggered the revolution.

Over the final pages of the *Chronologia*, Mercator chronicled the build-up to apocalypse: the *colloquium religione* of Berne in 1528; the submission of the *Protestatio* in 1529 by the Empire's evangelical states, princes and Imperial cities, and the subsequent emergence of the 'Protestant'[5] Schmalkaldic League; the deaths of Zwingli and Luther; the publication in 1548 of Charles v's *Interim* law for Protestants. Mercator's historical record of man on earth culminated in August 1566 with the pre-Apocalyptic desecration of the Church in 'Belgium templis' and a warning of the great rebellion to come.

Apart from a couple of eclipses which Mercator had observed in Duisburg, there were no further entries. So it only remained for him to conclude with one predictive date, 1576, the '*Initium cycli decemnovalis*' – the beginning of the ten-year cycle of fallow.

In the '*decemnovalis*', Mercator alluded to his own vision of the apocalypse. There would be no Antichrist, deluge or all-consuming fireball, but instead, Hosea's rain of righteousness. For a decade, exhausted earth would lie unseeded. The call to renewal could be read in Hosea's instruction to 'break up your fallow ground: for it is time to seek the Lord'.[6] In this sense, the end point of the *Chronologia* was a return to Mercator's first map, of the Holy Land, on which he had engraved Micah's lament for backsliders. Hosea completed the circle: 'I will heal their backsliding, I will love them freely . . . I will be as the dew unto Israel.'[7] So the Lord would bring peace and prosperity back to burned 'Belgium': 'They that dwell under his shadow shall return; they shall revive as the corn, and grow as the vine: the scent thereof shall be as the wine of Lebanon.'[8]

Included in the *Chronologia* were three supplementary chapters. In one of these, Mercator examined the controversial issue of Jesus' ministry. To demonstrate that Jesus had preached for four years, rather than the commonly accepted three, Mercator devised another table in which he compared the four Gospels. He also attempted to define the period of Christ's Passion by analysing solar eclipses.

Mercator knew that he had written a work which would be categorized as heretical. Drawing on ancient works and banned books, the *Chronologia* addressed humanists who sought a path between extremes. The absence of any entries relating to voyages of discovery, or to mathematical breakthroughs, might have seemed strange to those who knew Mercator's work. But Mercator had not intended to mirror Sleidan's religious history with a chronology of human achievement. The *Chronologia's* principal characters were those entrusted by God to govern the world: kings and popes. (The only contemporary humanists from the world of mathematics to merit inclusion were Copernicus and Gemma Frisius, both credited with providing the dates of astronomical events.) Mercator's closing inclusion of the low-born protagonists of the religious schism was a means of anticipating the apocalypse. The *Chronologia* was an urgent warning.

For Mercator, it was also a deeply personal reflection on his own lost ideals. The monarchs and ministers who had been his earthly leaders were gone: Charles v and his sister Maria of Hungary had been dead since 1558. Mercator's first patron and Erasmian ally on the Mechelen Grand Council, Frans van Cranevelt, had died in 1564. Granvelle had been sent by King Philip ii to Rome. The paralysis that had attacked Mercator's own Duke Wilhelm of Cleves in 1566 had progressed to dementia.

The *Chronologia* was dedicated to 'the most celebrated gentleman, Henricus Oliverius,[9] the honourable Chancellor of Cleve'.[10] Educated at Cologne, Orléans and Bologna, the elderly law professor was admired among Mercator's circle for his religious moderation.

A buzz of anticipation preceded the book's appearance, Ghim recalling that the *Chronologia's* author had to be 'pressed by his printer and his friends'[11] into getting it printed.

The sight of Martin Luther, Zwingli and various other heretics taking their due place in a printed world chronology can only have excited the blood of those who had been confined to secret histories. More than that, the *Chronologia* was an innovative attempt to apply the laws of mathematics to theology. The effect of Mercator's rigorous temporizing was to verify the Scriptures as historical fact.

Among those who welcomed Mercator's *Chronologia* was an old friend from Louvain, the French antiquarian Johannes Metellus.[12] (A close friend of Ortelius' also, Metellus would move to the Rhine, make maps and write admiringly of Mercator setting 'in order the times of the whole world in annals, so that the history of the immense orb might be known'.)[13] Shortly before the *Chronologia* was completed, Onofrio Panvini had written to Metellus of his shared interest in Mercator's latest work: 'But concerning Mercator,' enthused the Italian monk,

> I freely admit that, of all those whom I have come across in the department of learning – and I have read them all – there is no one by a long chalk whom I have put before your Mercator, whether you look to his matter and arrangement or examine his judgement and industry or consider his observations of the heavenly bodies, in so far as they are relevant. So I beg you by the bonds of our friendship that, when you meet him or write to him, you will introduce me to him as a friend of reason of our common interest in chronology.[14]

Christophe Plantin was also impatient to see the book. Somehow, the Antwerp printer managed to get copies by October 1568, several months ahead of the 1569 publication date which the book's Cologne printer had typeset onto the title page. Within days of receiving them, Plantin had sold his twelve copies to a buyer in Paris. Since the *Chronologia* was bound to be added to the index of prohibited books, there were merits in buying it in before it could be examined by the Inquisition.[15] By the end of 1569, Plantin had sold another twenty-four copies.

24

... and Place

After the 'true definition' of time came the 'true definition' of place.

In resolving 'place', Mercator considered the greatest cartographic riddle of the day: how could the course of a ship following a constant compass bearing be represented as a straight line on a map which had been constructed on a grid of latitude and longitude?

On the ocean, such a course took the form of a continual curve. The compass always sought to maintain a constant angle from the imaginary meridians that ran north to the magnetic pole. And since those meridians were not parallel, but converging, a constant compass bearing would – if maintained for long enough – describe a gentle spiral over the surface of the earth. In 1541, Mercator had drawn attention to the properties of these curving 'rhumb' lines by becoming the first to mark them on a globe.

If these rhumb lines could be straightened and marked onto a map with a latitude/longitude grid, the navigational techniques of mariners would be revolutionized: a mariner wishing to sail between two known points on, say, opposing coasts of the same ocean, could lay a ruler between those two points and measure the angle between the ruler's edge and any intersecting meridian. That angle would be the constant compass bearing he needed to follow in order to sail between the two points. (In fact, such a course would not be the shortest route to follow, since in reality it curved, but nobody had yet devised a method of sailing along a 'great circle' course, with the continual changes of direction it would demand.)

Mercator had never been to sea (and never would). His desire to devise a map projection which straightened rhumbs had arisen from decades of studying conflicting maps. Because mariners were unable to mark their courses of constant compass bearing as straight lines on charts which were ruled with latitude/longitude grids, the records of their voyages could not be integrated accurately with the projections used by cartographers. A projection permitting rectilinear rhumbs would – for the first time – allow mariners and cartographers to work from the same map. Thus reconciled, sailor and humanist would be able to unify their investigations. Uniquely, Mercator's new projection would harmonize the geography of globes and maps; the spherical with the planar; the three-dimensional with the two-dimensional.

Mercator had been struggling with this issue of cartographic incompatibility since his earliest days as a cartographer: 'Whenever I examined nautical charts,' he had once written to his friend Antoine Perronet,

> I had to wonder, how it could be that ship-courses, when the distances of the places were exactly measured, at times show their difference of latitude greater than it really is, and at other times on the contrary, smaller, and again frequently upon a correct difference of latitude for the places in question. Since this matter caused me anxiety for a long time, because I saw that all nautical charts, by which I was hoping especially to correct geographical errors, would not serve their purpose, I began to investigate carefully the cause of their errors, and found them chiefly to rest on an ignorance of the nature of the magnet.[1]

That was 1546. His new projection had been at least twenty years in the making.

Perhaps the moment of discovery occurred one day as he prepared another set of gores for his perennially popular terrestrial globe.[2] With the geography of the world laid flat on his work-surface, prior to being glued onto the sphere, the meridians on the gores would have appeared as straight, equidistant lines. Broken by the separations between each gore, the parallels and rhumbs would have appeared as interrupted curves.

And that may have been the moment, the once-in-a-lifetime moment of recognition, when two familiar, indissoluble patterns suddenly dissolved into a singular, universal truth: if the parallels – the lines of latitude – were straightened too, then moved further apart, the rhumbs must also straighten. The trick would be to judge just how far apart to pull the latitudinal parallels. And in this the gores on the work-surface already told him the answer. Since each gore tapered progressively towards each pole, it followed that the latitudinal parallels would have to be spaced progressively too. A few minutes with scissors would have produced the empirical evidence.[3] The result, exclaimed Mercator, corresponded 'to the squaring of the circle so perfectly that nothing ... seemed to be lacking except formal proof'.[4]

With the parallels correctly spaced, Mercator was looking at the projection that had eluded so many for so long. Parallels, meridians and rhumbs were straightened into a rectilinear trinity. Mercator had ruled an immutable framework for global mapping, a planar grid which would prove as timeless as the planetary theory of Copernicus. In seeking the essence of spatial truth, he had become the father of modern mapmaking.

Mercator's projection appeared in the form of a wall-map of the world which he titled: 'New and more complete representation of the terrestrial globe properly adapted for use in navigation'.[5]

Writing twenty-six years later, Mercator's slightly perplexed neighbour Walter Ghim merely recorded that the cartographer had 'set out, for scholars, travellers, and seafarers to see with their own eyes, a most accurate description of the world in large format, projecting the globe on to a flat surface by a new and convenient device'.[6]

Composed of eighteen separate sheets, this was the largest map Mercator had ever produced.[7] The density of detail, the precise lettering, the numerous legends and the compositional symmetry of the huge map created a cartographic spectacle. But the geography was bizarre.

Instead of the relatively skeletal continent which Ortelius had depicted only five years earlier, North America was now a bloated giant occupying nearly half of the northern hemisphere. And Mercator had done something peculiar to the poles. In what appeared to be a Ptolemaic regression, he had created polar landmasses which ran the full width of the map; landmasses which dwarfed even the latitudinal girth of North America. That this was a work which required some explanation was suggested by no less than fifteen framed legends.

The panel which explained the map's purpose occupied the vast unexplored interior of North America. In it, Mercator outlined his intention 'to spread on a plane the surface of the sphere in such a way that the positions of places shall correspond on all sides with each other both in so far as true direction and distance are concerned and as concerns correct longitudes and latitudes'.[8] Anticipating incredulity, he reassured his readers 'that the forms of the parts' had been 'retained, so far as is possible, such as they appear on the sphere'.[9]

Readers (and of course they had to be readers of Latin) who found themselves no wiser after persevering to the end of the entire 750-word panel could only turn back to the map, where examination of the graticule demonstrated graphically how and why the shape of the world had changed for ever. Through some extraordinary trick of mathematics, neither the parallels nor meridians curved. The entire planar face of the planet was divided into neat, symmetrical rectangles. And somehow, Mercator had taken the spiralled lines of constant bearing from his globe, and contrived to straighten them too.

Mercator's only previous world map had been shaped like mirrored hearts; a projection which had interrupted (and inverted) land and ocean so severely that it was near impossible to interpret the geography. On the new projection, there were no geographical interruptions, but the progressive distortion of space towards the poles required the reader to allow for an elastic scale.

A map of this size and mathematical ingenuity could only reflect the most accurate description of the world.

In a further effort to refine his mathematical model, Mercator had abandoned the old Ptolemaic prime meridian through the Canary Islands in favour of a line further west, which he believed to coincide with zero magnetic deviation. The decision to fix his prime meridian on the Cape Verde Islands had arisen from information gleaned from a 'skilled pilot' of Dieppe, who had reported that his compass had pointed to 'true' north while in the islands. Conflicting with 'Franciscus' of Dieppe was evidence from other navigators who had reported zero deviation at Corvo, in the Azores. Unable to suppress potentially useful information, even when it came at the expense of clarity, Mercator explained in one of the map's legends that there were in fact *two* magnetic poles, one on a rock in the polar ocean north of the Strait of Anian (the terminal point of the Cape Verde meridian) and another magnetic pole on an island to the north-west (at the terminal point of the Corvo meridian). Mercator believed that the true magnetic pole must lie between the two islands.

Revisiting the geography of the entire globe for the first time in three decades, Mercator found himself trying – just as he had with his *Chronologia* – to reconcile conflicting sources.

Mercator's cartography was dictated by his conviction that truth could be sought through universal harmony; that by harmonizing his sources he would discover the true geographic form of the world. But such an approach was demanding increasingly prodigious research. As opportunistic editors bundled together new and rediscovered travel narratives and maps, the business of harmonizing them into a single world picture was becoming more complex. And Mercator's reverence for ancient authorities was increasing the tensions between sources.

While the outlines of Europe and Africa could be engraved with some level of confidence, the western shores of the Americas and the eastern shore of Asia were open to interpretation. And the Antarctic landmass had yet to be visited in modern times. But 'Magellanica' could wait; more urgent by far was a definitive description of the Arctic coasts which might open a short-cut to the Indies.

Mercator had never been able to make up his mind about the Arctic. On his first world map of 1538, he had followed Gemma in depicting a northern extension of Asia covering the north pole. Three years later, on his terrestrial globe of 1541, the Arctic landmass still existed, but now it was joined to North America rather than Asia. The implications for navigators in search of a sea route to the Indies were considerable. On Mercator's first map, there had been a north-west passage to the Indies, but no north-east passage, then on his globe there was a north-east passage but no north-west passage. On his new world map, there was a north-east passage *and* a north-west passage. This final position was a conviction that Mercator had been suppressing for some time, protecting perhaps any residual Imperial dream to open up a shorter sea route to Cathay and India. (In a few years, Ortelius would claim to an Englishman that 'if the wars of Flanders had not been, they of the Low Countries had meant to have discovered those parts of America, and the north-west Strait'.)[10]

Mercator's doubts concerning the true nature of the Arctic lands were clear enough from his cartographic vacillations. He also knew that his concept of the Arctic pole as an extension of Asia or America was not universally accepted. As long ago as 1492, Behaim's globe had depicted an aquatic north pole surrounded by islands. Fifteen years later, the Rome Ptolemy had appeared with Ruysch's revolutionary map showing four islands orbiting an aquatic north pole, a depiction that had been developed by the Frenchman Oronce Fine.

Time had run out. Mercator's great cosmography required a certain Arctic. Many awaited his decision, especially the English, and Ortelius, who for some time had been urging seafarers to explore northwards. The failure of the English to push north and to return with new coordinates threw Mercator at the mercy of sources that he must have suspected were dubious. Two of these sources had appeared since the issue of the 1541 globe; both had been given the appearance of freshness through resuscitation.

Comprising three volumes of travel narratives and maps, *Delle Navigationi et Viaggi* had been published between 1550 and 1559 by the Venetian editor Govanni Battista Ramusio. Mercator owned the set.[11] Well connected through his role as secretary to the Venetian Senate,

Ramusio's collected geographical gems included material from a new book written by his fellow Venetian, Nicolò Zeno, a well-born geographer and mathematician who had served on the Venetian delegation to Charles v in 1543.

Zeno's *Commentarii* described how two of his ancestors, brothers who had been Venetian ship-owners, had sailed the shores of Iceland and Greenland in the late 1300s, writing home of their trials and observations. Zeno included in his book 'a navigating chart which I once found that I possessed among the ancient things in our house, although all rotten and many years old'.[12] The map revealed astonishing new truths about the Arctic: appended to the north of Europe was the peninsula of Greenland, marked with the monastery described in detail by the Zeno brothers, while two sections of coastline labelled 'Estotiland' and 'Drogeo' suggested a previously unknown landmass to the west of three, new, northern islands – 'Frisland', 'Estland' and 'Icaria' – anchored between Iceland and Scotland.

Nicolò Zeno became a cartographic celebrity and his Arctic revelations were adopted in Venice editions of Ptolemy's *Geography*. For Mercator, this was confirmation enough. He had seen the recent, 1562, Moletius edition, and its statement that Zeno was an authority in history and geography who was 'universally held to have, at this day, few equals in the whole of Europe'[13] would have been enough to convince him that the strange tale of the rotting map was true.[14]

The Venetian's new Greenland was complemented by another source which Mercator had come across since publishing his globe of 1541.

For some time,[15] Mercator had been aware of a travel account which cast extraordinary new light upon the dusky lands of the Arctic. The account had never been printed, and seldom read. Indeed, it was so rare that Mercator did not own it himself, but had borrowed a copy from 'a friend ... at Antwerpe'.[16]

The author of this revelatory account came originally from Mercator's school-town of 's-Hertogenbosch, and appeared to have travelled during the latter part of the fourteenth century 'throughout all Asia, Africa, and the North'.[17] According to Mercator, Jacob Cnoyen's account included 'certain historical facts of Arthur the Briton', but that 'the most and the best information' had come 'from a priest who served the King of Norway in the year of Grace 1364'.[18]

From this priest, Cnoyen had learned of 'an English minor friar of Oxford, who was a mathematician', who in 1360 had 'reached these isles' having 'described all and measured the whole by means of an astrolabe'.[19] The geography that the minor friar had passed on to the priest, and so to Cnoyen, bore a remarkable resemblance to the Arctic depicted by Ruysch in the 1507 Ptolemy.

Even Mercator, so used to divining truth from ancient sources, must have wondered at the man from the Maas. In adopting Cnoyen's geography, Mercator would be trusting the virtually unknown account of a mysterious traveller whose observations were based on hearsay from an unnamed priest, who had got *his* geography from an unnamed minor friar, two centuries earlier.

If Mercator was discomforted by such a tenuous chain of truths, the temptation was extreme. Cnoyen provided Mercator with documentary justification for completing his Arctic. And Cnoyen had the news which navigators – particularly English navigators – most wanted to hear. Both the north-east and the north-west passages existed.

In scrapping his earlier polar landmass in favour of Cnoyen's description, Mercator found himself creating a far more detailed version of the islanded Arctic than had been seen on Behaim's globe, or achieved on the maps of Ruysch and Fine.

So Mercator's new map showed four islands in a ring around the pole. Around the outer, southern limit of each island, he had depicted a rim of mountains. Each of the four islands had been separated from its neighbour by a narrow torrent of north-flowing seawater, braided at its southern mouth. According to Cnoyen (whose account agreed with Ruysch in this respect), two of the islands were inhabited. On the island closest to Norway, Mercator had added a legend which read: 'Here live pygmies whose length in all is 4 feet, as are also those who are called Screlingers in Greenland.'[20] The neighbouring island to the west was 'the best and most salubrious of the whole of Septentrion'.[21]

Inside the ring of four islands was the roughly circular polar sea, where waters from the inrushing torrents congregated before being 'absorbed into the bowels of the Earth'.[22] The site of this cosmic plughole was marked at the pole by a tall rock 33 leagues in circumference.

Because the new projection caused extreme distortion in the far latitudes, Mercator had depicted the Arctic in a dedicated panel. It was the only cartographic inset on the map and its prominence gave added weight to this final, definitive description of the northern sea-routes.

Anybody admiring the map could not help but notice that the inset also performed another function. While the world map upon which it was superimposed was a geographical plot which appeared to be entirely without pattern, the polar inset had a beautiful, unearthly symmetry. The four islands were of similar size and equally distanced from the central, polar rock. The peripheral mountain chains followed the southern edge of each island and thus formed a ring around the globe. The widest of the four inward-flowing channels was almost aligned with the prime meridian.

With its quartet of rivers and perimeter wall, Mercator's polar inset bore a curious resemblance to the future 'Eden', the paradise promised to the righteous after death; the 'third heaven'[23] of St Paul. Just as the mind confirmed that Mercator's world map was a work of mathematical wonder, the heart urged the inset to be the place which knew no earthly troubles.

25

'liketh, loveth, getteth and useth'

If fate (plague, war, famine, fire ... the Inquisitor's knot) had intervened and caused Mercator's projection to be his dying work, he would not have been denied his place in the cartographer's Pantheon. He had expressed his Big Idea; had his Copernican moment; made his Vesalian mark. Nothing more could be expected of a sixteenth-century cobbler's son. But Mercator's projection was intangible. Few of his contemporaries would understand its purpose, and its vast, wall-map format and copious Latin legends restricted the map's circulation to a small circle of wealthy humanists with very large walls. In a community which valued design, craftsmanship, erudition and ownership, the projection circulated as a nebulous concept. Mercator's tangible contribution to his oeuvre, his one coherent body of work, would be the sum of the parts of his great cosmography. The *Chronologia* and the projection were the prologue; the historical and spatial scaffolding.

But Mercator was fifty-seven. And as he sat in the light of his Duisburg window, he knew that his last adversary would be time itself.

The next phase was the cartographic component of the cosmography: the 'representation of the land and the sea'.[1]

There would be two parts. The first would describe the world as it was known to the ancients. This would be Mercator's definitive edition of Ptolemy's *Geography*. For the second part, he would publish a systematic set of modern maps describing the entire planet.[2]

Although maps were nothing new to Mercator, he would be exploring fresh territory. In the past, his mapmaking had been sporadic and unsystematic: ranging from small regional surveys to gigantic wall maps of the world, they appeared almost random in their subject and format. Only the Holy Land map and the wall map of Europe conformed to any kind of pattern. In thirty or so years, Mercator had produced less than ten maps; now, as he approached his sixties, he intended to produce well over one hundred maps, in two systematic sets, one Ptolemaic, one modern.

Ptolemy had to come first.

Ptolemy's lunar observations had shared the same page of Mercator's *Chronologia* as the crucifixion of Christ;[3] Christ and Ptolemy came from the same century; the geography of the Christian era began with Ptolemy.

Ptolemy had created the template for modern geographers, and the Wittenberg professor of mathematics and astronomy Georg Rhaeticus had spoken for a generation when he had written in his *Chorographia* of 1541 that only 'the mathematician following in the footsteps of Ptolemy can reform geography'.[4] Mercator's entire adult life had been spent in the Alexandrian's shadow.

Mercator knew better than most that in passing through the centuries, Ptolemy's *Geography* had accumulated corruptions such that 'in the entire work there is not one part which is not riddled with errors'.[5] Ptolemy's interminable lists of coordinate numbers had always invited transcription slip-ups, and the strange-looking names of places and peoples had also led to misunderstandings. Further mistakes had been introduced when Ptolemy's text had been used to create new editions of the maps, and when the maps had been used to 'correct' existing texts. The discrepancies, felt Mercator, had discredited Ptolemy: 'Greatly concerned by this abuse of this famous antiquity', he considered 'it necessary to restore and reconstitute the most ancient Geography of Ptolemy, passed down since the beginnings of this art, and reproduce it as closely as possible, following the intentions of the author'.[6] Mercator intended his *Geography* to be an accurate historical resource; a compendium of geographical knowledge as it was known to the earliest Christians.

Mercator's Ptolemaic mission was one of homage, of editorial scholarship, of re-creation. It was also a coming of age: only by creating a just epitaph for the master could Mercator move on; move out of Ptolemy's shadow.

Initially, he would concentrate on re-creating authentic versions of Ptolemy's twenty-seven maps: the single 'Universalis tabula' describing the known Ptolemaic world, the ten maps of Europe, the four maps of Africa and the twelve maps of Asia.

The challenge was daunting and familiar. Once again, Mercator would be engaged in a reconciliation process between truth and fabrication: 'The most difficult thing has been distinguishing the false from the original,'[7] he would write. 'You will not find two editions of this famous work which are identical in every respect.'[8]

The solution, as always, was to scrape away the accumulated historical sediment: 'To fulfil this task one has recognised the necessity of comparing the different editions of the work of Ptolemy and the original pieces of different manuscripts which are available.'[9] By comparing 'the many different derivations of one source one can arrive at a more certain, more studied and more true definition of that primary source'.[10]

Mercator had fewer versions of the *Geography* to work from than he might have wished. He had access to three printed editions: Rome (1490), Lyon (1535) and Venice (1562), the last having Pirckheimer's translated text. Backing up the three editions which contained the maps of Ptolemy, Mercator had access to two versions with the text alone: a printed Latin edition which had appeared in Cologne in 1540, and a manuscript which had been commissioned over a hundred years earlier by Nicolaus Cusanus. This had been lent to Mercator by the new Prince Elector and Archbishop of Trier, Jacob von Eltz, for whom Arnold Mercator had recently undertaken a triangulated survey of the upper Moselle.

Each of his five principal sources had been chosen carefully. Since Cusanus had been close to the Pope, Mercator could assume that his manuscript had been copied from a *Geography* in the Vatican library, while the printed Rome edition – which had been lent or given to him by a friend in Cologne[11] – was understood to have been corrected from a Greek manuscript.

The Lyon edition of 1535 also came with an encouraging provenance, having been carefully corrected from Greek manuscripts by Michael Servetus.[12] Mercator turned to Servetus in the knowledge that the radical Spanish humanist had been burned at the stake as a heretic twenty years earlier; controversially, Servetus had held the view that man could rise to God through love rather than belief, a view which required the dismantlement of the Trinity. One of the pieces of evidence thrown at him at his trial was his edition of Ptolemy, which had included a map of the Holy Land bearing on its verso a statement that it was largely infertile.[13] The flames which had consumed Servetus had been fuelled by copies of his own, heretical Ptolemy.

The third and most recent printed Ptolemy which Mercator used had been revised, edited and annotated by Josephus Moletius in Venice. With sixty-four copper-plate maps and Pirckheimer's translated text (again corrected from Greek manuscripts) this was a particularly handsome edition.

Less reliable was Johannes Noviomagus' Cologne edition of 1540, which contained no maps, while its lists of coordinates seemed at greatest variance with the Ptolemaic norm.

*   *   *

Fig. 19. The Antwerp printer, Christophe Plantin (1514–89). (Science Photo Library, London)

The process of reconciling the Alexandrian's coordinates and compiling the twenty-seven Ptolemaic maps was to take Mercator far longer than he had anticipated. For nearly ten years, he would pore over his ancient sources, and during that decade, the market for modern maps took off without him.

Even as he embarked upon his huge mapmaking programme, sales of his earlier maps were falling. Plantin had returned to business in 1563, and his ledgers recorded that orders for Mercator's works resumed briskly at the end of 1564, when fourteen copies of the Europe map and six of the new British Isles map were delivered to Antwerp. Just over a year later, on 19 January 1566, the ledger recorded a single order for 208 copies of Europe. And the figures kept climbing.

A comparison of the first five years of Mercator's trading relationship with Plantin, from 1558 to 1562, with the five years after Plantin resumed operations (September 1564 to September 1569) revealed that orders for Mercator's maps had increased by over 2,500 per cent, from 25 maps to 644. Globe orders over the same two periods had gone up from 3 single globes to 22 pairs.[14] But sales of a map tended to peak soon after publication, then plummet. The European wall-map had been hugely successful, but by 1568, Plantin was reducing his order. Sales of the British Isles map were falling too. The heart-shaped map of the world which Mercator had engraved in 1538 was not in Plantin's ledgers – perhaps because Mercator disliked it too much. The Flanders map had long ceased to earn money. In 1568, Plantin made an isolated purchase of 107 uncoloured copies of the Holy Land map.

The figures suggested a future fall in earnings. Mercator's combined map sales had declined irreversibly, and the new map projection was not going to produce more than a three- or four-year blip in earnings: Plantin had sold 41 copies by December 1569 (more than half of them to Paris), but in the whole of 1570, he sold only 39 copies; down to 30 in 1571, then 10 in 1572...

Solitarily, Mercator took his rocky path towards cosmographic knowledge, while opportunistic wholesalers, dealers, publishers – even cartographers – trooped the highway to financial reward. Leading the field was Abraham Ortelius. With a web of correspondents from Lisbon to London and Bologna, Ortelius (unlike Mercator) was also a traveller, regularly visiting Frankfurt, and taking trips to Paris and Italy. When the precocious bibliophile and graduate of Wittenberg Johann Sambucus (whose own circle included Peter Apian and Nicholas Olah) wrote to Ortelius from Ghent in 1563, the letter was addressed to 'Cosmographo Antuerpiensi amico suo',[15] the reference to Ortelius as a cosmographer anticipating by a year the antiquarian's metamorphosis from coin collector to cartographer. Ortelius had announced himself to his fellow mapmakers with a vast wall-map of the world. Engraved on eight sheets of copper, the map was cut with great skill and – following the example of Mercator's map of Europe – Ortelius had added the names of winds in Latin, Dutch and Italian. The map included a panel listing the sources of European imports such as gold and silver, precious stones and spices. Copied from Giovanni Battista Ramusio, two other panels depicted views of Cuzco and Mexico City. Mapped on a heart-shaped projection, the map bore a close resemblance to the geography of Mercator's globe, although Ortelius significantly removed Mercator's Arctic landmass and left open both the north-east and the north-west passage. Ortelius' world was a reflection of his ambitions and his connections.

Ortelius and Plantin quickly became a hub of cartographic production and sales. Plantin was exporting maps and globes from Antwerp to booksellers in Spain, Italy and England, Germany and France. At the Frankfurt fairs, buyers could expect to find the latest maps from Rome, Venice and Bologna, maps of Africa from Forlani, Lily's Britain or Van Deventer's Brabant. There was even a fair catalogue listing new publications. Started by the Augsburg book dealer Georg Willer in 1564 as his own stock-list, it had become the official *Messkataloge*, complete with a section for maps. By the 1570s, Willer was listing so many maps that he devised separate sections for wall-maps, for maps from Venetian publishers, and for the cornucopia of historical, astronomical and military maps.

Willer's lists promoted a more formalized market for maps. Alphabetically ordered by region with a note of author, engraver or publisher, the output and interests of the cartographic community could now be reviewed on a regular basis. Mapmakers and publishers could keep abreast of cartographical coverage.

The map boom also bred a new species of systematic collector, though few who began in the 1570s could hope to match the extraordinary collection now owned by the 'passionately obsessed' Viglius van Aytta,[16] whose two hundred or so maps included diverse works by Italy's leading cartographer, Giacomo Gastaldi, the English traveller Anthony Jenkinson, Mercator's near-neighbour in Kalkar Christian Sgrooten, Apian and now Ortelius ... The three maps by Mercator which Viglius owned – the wall-map of Europe, the world map of 1569 and the map of Flanders – would have been among the more impressive in his collection.

Three decades after Viglius and Cranevelt became hooked on maps, John Dee described the rising fervour among English collectors who sought geographical descriptions:

> Some to beautify their Halls, Parlers, Chambers, Galeries, Studies, or Libraries with; other some, for things past, as battles fought, earth-quakes, heavenly firings, and such occurrences, in histories mentioned: thereby lively as it were to view the place, the region adjoining, the distance from us, and such other circumstances: some other, presently to view the large dominion of the Turk: the wide Empire of the Muscovite: and the little morcel of ground where Christiandom (by profession) is certainly known, little I say in respect of the rest, etc.: some other for their own journeys directing into far lands, or to understand other men's travels ... liketh, loveth, getteth and useth, Maps, Charts, and Geographical Globes.[17]

The sheer scale of Mercator's new enterprise, and his limited resources, precluded him from sharing the spoils of the booming map market.

While he accumulated material for his maps, ancient and modern, his friend Ortelius was poised to launch the next generation of spatial illustration; a new method of modelling the planet which would offer an alternative – perhaps even a replacement – for wall-maps and globes.

For Ortelius, the issue was one of utility. Existing maps were unman-ageable:

> For there are many that are much delighted with *Geography* or *Cho-rography*, and especially with Maps or Tables containing the plots and

descriptions of Countries [... who ...] would very willingly lay out the money [for them], were it not that by reason of the narrowness of the rooms and places, broad and large Maps cannot so be opened or spread, that every thing in them may easily and well be seen and discern'd. For ... those great and large *Geographical* Maps or Charts, which are folded or rolled up, are not so commodious: nor, when any thing is peradventure read in them, so easy to be look'd upon. And he that will in order hang them all along upon a wall had need have not only a very large & wide house, but even a Princes gallery or spacious Theatre.[18]

From his commercial throne in the heart of northern Europe's most powerful economic metropolis, well-travelled Ortelius knew that he was on the brink of a breakthrough. In faraway Duisburg, Mercator was working towards the same goal, for different reasons. His was not an immediate need to replace the flapping map but a longer-term dream to produce a method of representing three-dimensional space in two dimensions, whilst retaining the 'reality' of a globe and the detail of a map.[19]

Both men had come up with broadly the same solution: a uniform collection of modern maps. Bound volumes of maps had, of course, been around since Ptolemy's *Geography*, portolan charts and a host of other thematically assembled maps had been clamped between book-boards. Waldseemüller had introduced a systematic organization of map leaves by separating his Ptolemaic and modern maps in his edition of 1513. But nobody had taken the trouble to engrave to a uniform pattern a methodically selected spread of modern maps, and to market them with minimal text as a generically novel product.

Through compression and selection of geographical data, the world's regions would be presented in a format which would allow near-instant access to any part of the world in a format which would fit on a lap. The essence of the project was miniaturization. Ortelius wanted to market a predictable utility, an indispensable map-book 'that every student would afford a place in his Library, amongst the rest of his books'.[20]

The two cartographers had devised very different approaches. Ortelius had seen that he could create such a product by employing an engraver to copy at reduced scale a choice selection of other cartographers' maps, which he would then bind in a logical order as a book. Mercator's purist approach demanded that he alone should research the cartography for his maps, and that he (or his sons) should engrave them. By 1570, Ortelius was over a decade ahead of Mercator.

Fig. 20. Abraham Ortelius (1527–98), Mercator's friend – and competitor (Science Photo Library, London)

Since his friend Mercator was working to an unrelated timetable, Ortelius might have expected to publish without competition. But an opportunistic Antwerp printer, Gerard de Jode, was also working on a bound volume of modern maps. A well-known presence at Frankfurt – where he had a stand managed by the Nuremberg merchant Cornelius Caymocx – Gerard de Jode was one of those who had ridden the wave of Antwerp's print boom. Born three years before Mercator, de Jode had been admitted to the Guild of St Luke in 1547 and by the late 1550s was dealing regularly with Plantin and running a team of engravers that included the three Wierix brothers and the two van Deutecums, Jan and Lucas. It was de Jode who had published Ortelius' highly successful world map in 1564, but the two men had fallen out. Younger, better connected and dedicated to maps in a way that de Jode was not, Ortelius used his influence to make sure that de Jode's printing licence would be delayed for ten years.

Ortelius' *Theatrum Orbis Terrarum* (Theatre of the World), was published in Antwerp on 20 May 1570, just nine months after Mercator's new projection was lifted off his press in Duisburg. Buyers familiar with title pages that represented the three continents with three human figures were surprised to find two additions: a naked savage representing America, and a woman's bust representing 'Magellanica', or Terra Australis, her hands and feet truncated 'because she is scarcely known'.[21]

With a Ptolomaic sense of order, Ortelius opened his collection with a map of the world, followed by maps of the 'four quarters or principal parts ... to wit, Europe, Asia, Africa and America'.[22] Following these overviews were maps of smaller geographic and political regions, from Britain, Spain, Portugal and Gaul through to Russia, Persia, Palestine and Barbary. The engraving – much of which had been undertaken by Frans Hogenberg – was attractive and the folio size immediately struck a fine balance between convenience and clarity. As if there was not already enough to admire, Ortelius had included in the volume a *catalogus auctorum tabularum* – a list of the eighty-seven cartographers whom he had used as sources. Together with his Rhineland neighbour Sgrooten, Mercator had the longest entry, as well as receiving an honorary plaudit in the introduction to the world map. To Ortelius, Gerardus Mercator was 'the Prince of modern Geographers'.[23]

The *Theatrum* was rewarded with salvoes of epistolary applause. Petrus Bizarus responded with a laudatory poem in which he hailed Ortelius as 'the everlasting ornament' of his country, his race 'and the universe'.[24] From Speyer, Emperor Maximilian's leading physician, Johannes Crato von Krafftheim wrote to Ortelius of the 'relief' that the *Theatrum* had provided from his 'oppressive cares and labours at Court',[25] an understandable reaction given that those cares included the Imperial haemorrhoids. Such were the palliative benefits of the maps that Crato handed the *Theatrum* 'to his Imperial Majesty ... as soon as I received it'.[26]

Among the names of cartographers which the stressed physician noticed in Ortelius' *catalogus auctorum tabularum* was that of Gerard Mercator, soon to be another recipient of Crato's kind words.

The extraordinary praise which was lavished on the *Theatrum* was an unnecessary reminder to Mercator that his habitual perfectionism worked against his commercial interests. At the end of November 1570, Mercator also wrote to congratulate Ortelius:

> I have examined your Theatrum and compliment you on the care and elegance with which you have embellished the labours of the authors, and the faithfulness with which you have preserved the production of each individual, which is essential in order to bring out the geographical truth, which is so corrupted by mapmakers. The maps published in Italy are especially bad. Hence you deserve great praise for having selected the best descriptions of each region and collected them into one manual, which can be bought at a small cost, kept in a small space, and even carried about wherever we please.[27]

In praising Ortelius for selecting the best sources and successfully compressing them into the portable *Theatrum*, Mercator discreetly defined his friend as an editor rather than as a cartographer.

Mercator made no mention in his letter of his friend's claim to have been the originator of the map-book. In the *Theatrum*'s address to the reader, Ortelius wrote that it was his cogitations upon the disutility of existing maps that had led 'at length' to the discovery that 'it might be done by that means which we have observed and set down in this our book'.[28]

Mercator's distance from commercial map publishing was illustrated by a commission which he undertook shortly after the publication of Ortelius' *Theatrum*.

While Ortelius basked in the success of his editorial compilation, Mercator painstakingly assembled a one-off book for a client who wanted a portable collection of European maps.[29]

With some ingenuity, Mercator created fifteen maps by cutting up four copies of his European wall-map and three copies of his British Isles map. Also included were two of the sheets from his recent world map. Since the client had a particular interest in northern Italy, Mercator added to the collection a pair of manuscript maps of Tirol and Lombardy, at a larger scale. Both were drafts from the growing series of maps which he was preparing for the cosmography's 'representation of land and sea'. The two sheets were exactly the same size, and fitted together with a slim, accurate overlap. Following his normal practice, Mercator must have supplied these maps on the basis that the new owner return from Italy having revised their geography.[30]

Work on this one-off collection of European maps coincided with another job which took Mercator away from his great cosmography.

Responding to demand, he was in the final stages of completing a new edition of the wall-map of Europe that he had published back in 1554. Of the fifteen copper plates which made up the map, six had been modified, a process which required each plate to be hammered out, repolished and then re-engraved with the revised geographical information.

A new legend on the map explained that it was the 'most famous navigation of the Englishmen by the Northeast sea' that had given Mercator the opportunity and the 'certain direction'[31] to reform his map. As well as new discoveries in the northern parts of 'Finmarke, Lapland and Moscovie, laid out according to the just elevation and the quarters of the world', the English had returned with the 'true observation of the latitude of the city of Moscow', information which had yielded Mercator 'an infallible rule, for the correcting of the situation of the inland countries'.[32] It was, wrote Mercator, his 'duty to exhibit to the world' a 'more exact and perfect' map 'than hitherto had been published'.[33]

After so many years of waiting, the English – through the activities of the Muscovy Company – were at last beginning to provide coordinates for the northern lands. Mercator was able to reshape Russia, modifying the northern coast and moving Moscow south onto the adjacent copper plate. Scandinavia was updated, and to the north of the Black Sea, mountains and river systems were rearranged.

But this was not merely an exercise in revision. Unnecessarily from a geographical point of view, Mercator revisited the map's appearance. Where the sea had been blank on the original map, he now separated it from land by stippling the aquatic parts of the plates with thousands of burin pricks. New sea creatures frolicked, spouted and gasped in the Mediterranean, and Mercator added a maritime vignette: off the Mauritanian coast, three ships foundered in turbulent seas.

And in a coded call to English seamen, Mercator removed the solitary ship which had been sailing on the 1554 map for the north-east passage, and instead engraved a fleet of three bound for the north-west passage. Having brought home the shape of Russia, the English – implied Mercator – should turn their attentions to the coordinates for Cathay.

Mercator knew that the English were listening: within weeks of the publication of his world map of 1569, John Dee had penned a treatise (published the following year) in which he urged 'our English pilots' to take advantage of the island's 'situation most commodious for navigation to places most famous and rich'.[34] Dee recalled that the 'courageous Captain' Sir Humphrey Gilbert had been poised to take such advantage by sailing 'either westerly to Cape Paramantia, or easterly above Nova Zemla', but that he had been called from this momentous quest by 'the Irish rebels'.[35] Nevertheless, continued Dee, 'some one or other should listen to the matter' and 'little and little win to the knowledge of that trade and voyage . . . which I should be sorry . . . should remain unknown and unheard of'.[36] Without mentioning Ortelius and Mercator – although these were the 'learned' men that he had in mind – Dee concluded that the English were 'half challenged by the learned';[37] they were duty bound to continue exploring the northern lands.

Dated March 1572, Mercator's modernized map of Europe was ready in time for the spring fair at Frankfurt. It enjoyed a promising launch. Plantin bought fifty-three uncoloured copies at the end of 1572 and from London, Nicholas Reynolds wrote to Ortelius asking for two copies of the new map. Willer listed it in his autumn catalogue the following year, but the only other item of Mercator's – in a list dominated by Italian maps – was his 1569 world map. No less than three versions of Ortelius'

Theatrum were on Willer's list: the original Latin edition, a *second* German edition *and* a special supplement of eighteen maps. Not only that, but the *Theatrum* had been placed at the head of the map section.

Such were the benefits of employing others to draw one's maps.

Mercator must have wondered how long he would have to wait for the coordinates of the north-west passage to filter from London through to Duisburg. A description of the true disposition of the northern lands was critical to the cosmography's modern maps. Indeed, Mercator had originally intended the northern lands to occupy the first part of the work.

While the northern lands held some of the greatest problems, work on Gaul and Germany was relatively advanced. Age and his aversion to travel increased the difficulty in obtaining the maps required as sources. The peripatetic Ortelius was one of several friends who helped the increasingly isolated Duisburg cartographer: 'Many thanks for sending me the map of Bavaria,' wrote Mercator in spring 1572,

> which I was not able to buy at Frankfort. I have likewise failed to obtain the description of Bavaria by the same author. As regards the description of Moravia, you are the first to inform me of its being published, as we have here no merchants who procure such wares from Frankfort, and the booksellers of Cologne, occupied with their books, neglect geographical maps. Send me, therefore, the map of Moravia and the description of Hainaut, with the prices.[38]

The letter was an example of Mercator's epistolary symmetry, the passage relating to the assistance offered by (and requested of) Ortelius, being balanced by reciprocatory advice from Mercator. The barter was explicit: Ortelius' maps for Mercator's scholarship. In the letter he wrote that May, Mercator repaid his debt by advising Ortelius how to create a new map of America by combining the best elements of his own world map with a recent map of 'Nova India':

> Arnold Mylius was with me a few days ago ... from whom I learnt that there are in existence certain detailed descriptions of Nova India, which perhaps you will obtain and publish. I recommend however that in publishing them you observe the longitudes and latitudes of certain more important places as they are on my Universal Map, unless by chance any better can be produced from faithful observations. It will be sufficient to do this for St Domingo Hispaniola, for Cape Razo,

Canada and Hochelago in Nova Francia, for Mexico, for Cape St Augustine in Brazil, and likewise for some promontories and towns in the Western and Peruvian Sea, and the rest may be inserted according to the proportions of the new descriptions.[39]

Anticipating perhaps that Ortelius may have wondered why he was not making use of this information himself, Mercator added: 'I ask this because, having been occupied for some years in correcting Ptolemy and the recent maps, I should wish in the first place you to profit by a description of Nova India, and further myself to be freed from this labour, as I have more than enough to do in other ways.'[40]

Ortelius had been busy too: in the couple of years since he had finished work on the *Theatrum*, he had published a wall-map of Spain, a map of the Roman Empire, a new edition of the *Theatrum* containing an additional twenty-eight maps, and a book on coins illustrated by the engraver Philip Galle. Unmarried and unconstrained by domestic commitments, Ortelius was a master of opportunity. And praise for the *Theatrum* still poured in: 'You have laid me under obligations to you,' enthused Daniel Rogers of Windsor, 'by the gift of the monuments of your genius'.[41] (Rogers promised to send Ortelius 'a very accurate geographical sketch' of Ireland which, he wrote, would 'greatly embellish'[42] the *Theatrum*.) Meanwhile, a rare note of disapproval had arrived via Brussels from Spain, where Cardinal Espinosa had discovered on consulting the *Theatrum* that his own birthplace had not been marked. The cardinal sent a message to the cartographer that the map of Spain should be altered forthwith and that 'two coloured copies' of the revised *Theatrum* 'with the altered name, bound in leather and gilt' be sent with the fleet on 'the first favourable wind.'[43] Ortelius 'corrected' his map immediately and the copies were dispatched.

Others were joining the Ortelius waggon train: in Cologne, Georg Braun moved briskly to capitalize on the enthusiasm being showered upon Ortelius' *Theatrum*. At Frankfurt's autumn fair of 1572, dealers were introduced to the first edition *Civitates Orbis Terrarum*, a book of city views whose title bore more than a coincidental resemblance to the phenomenally successful *Theatrum Orbis Terrarum*.[44] Braun's collaborators had been two men of Mechelen: Frans Hogenberg and Simon van den Neuvel.[45]

While Ortelius and de Jode, Braun, Hogenberg and 'Novellanus' created their collaborative bestsellers in the mercantile honeypots of Antwerp and Cologne, Mercator laboriously continued to construct his cosmography virtually single-handed, researching his own cartography, engraving many of his own maps. Because there was no commercial printing press in Duisburg the text required for the Ptolemaic maps would have to be typeset in Cologne. Now past sixty, Mercator knew that his own chronology had little time to run. Already he had outlived many of his contemporaries. Gemma was long gone, and Vesalius had been dead nearly a decade. Mercator had lived for many more years than his parents; had lived longer than two of his children and at least one grandchild.

Mercator's humble origins, his undistinguished academic record, his imprisonment and his reluctance to travel had isolated him from various circles of humanist exchange. Steadfast in his Duisburg redoubt, he had however generated his own geographical aura. Thirty years after they had been educated together at the university of Louvain, Andreas Masius (who had graduated first among the Masters of Arts in 1533) got in touch with Mercator when he needed geographical information for his commentary on Joshua. To Masius (by now an accomplished linguist and orientalist) Mercator had become 'the learned mathematician, who is most active and experienced in the representation of the countries of the earth and who is so dear to me on many scores'.[46] To his friend Cassander, Masius expressed a wish to acquire all the maps Mercator had published.

To Ortelius, Mercator had risen to be a figure of immense stature. In later editions of the *Theatrum*, 'the Prince of modern Geographers',[47] gained an even more onerous crown when he also became 'nostri saeculi Ptolemaeus',[48] the Ptolemy of our age. In England, John Dee had dedicated his improved edition of *Propaedeumata aphoristica* to Mercator, and referred elsewhere to his old collaborator as 'the expert and grave Cosmographer'.[49]

From Antwerp, Johannes Vivianus ('a Merchant, but a great lover of learning')[50] wrote to Mercator asking his friend for a contribution to his *Album amicorum* – a book of friends which he had begun compiling in 1571. A quick platitude would have done the job, but Mercator judged it 'not at all honest to write foolish trifles to a friend',[51] and so he inscribed a drawing of the universe. Characteristically, Mercator's universe was neither Aristotelean nor Copernican, but a compromise between the two: Mercury and Venus revolved around the sun, yet the sun (and all the other planets) revolved around the earth. The various planetary orbits were described in the accompanying letter.

Excusing himself for being 'excessively occupied and consequently hurried',[52] Mercator begged his friend's forgiveness: 'If anything has not been expressed clearly enough by me, or placed in a confused order, or even omitted, pardon me, I beg you.'[53] The stress inflicted on a perfectionist unable to honour his own standards was eating at Mercator's soul: 'space', he cried, 'was not granted me for thinking'.[54]

That was the summer of 1573. A short while later, he fell to a 'serious illness',[55] perhaps a recurrence of the deep melancholia that he'd brought home from Lorraine, or the gout that would increasingly impair his mobility.

It was in the aftermath of this latest malady that a small group of Mercator's friends gathered in Antwerp to undertake a curious journey.[56]

Conceptually ambiguous, it was a pilgrimage, an archaeological investigation, a geographical adventure, a humanist quest, a homage to the master of modern mapmaking. The route, and the sequence of observations and encounters it determined, was intended to create a narrative thread. In its final, printed form, the journey would be addressed to the humanist whose theatre of life had been played out within the definition of its itinerary. It was a journey to Mercator.

The travellers were Ortelius, Vivianus, a young Fleming called Jérôme Scholiers, and the Antwerp master painter Jan van Schille. Schille had recently undertaken land surveying in Lorraine for Duke Charles III, and his knowledge of the region would aid the group's navigation.

The journey began at the bibliographical printroom of northern Europe, Antwerp, and it ended at the continent's bibliological marketplace, Frankfurt. The streams that it followed were those of Mercator's life: the Schelde, the Dyle and the Maas, the Moselle and the Rhine. The journey's turning point and its geographical climax had been contrived to take place – where else – deep inside the forested heartland of Lorraine, the duchy that had fostered Waldseemüller and the beginnings of northern cartography; the duchy that had nearly killed Mercator.

From Antwerp, the travellers passed through Mechelen, the seat of northern geography, and then Louvain, the seat of Mercator's geography. Continuing southward to Namur and the Maas, they turned downriver to Liège, where they paid a devotional visit to the tomb of Sir John Mandeville. 'Here', read his epitaph, 'lies the noble man and Lord John of Mandeville ... born in England, lecturer of medicine, humblest preacher, most generous donator of goods to the poor, who after travelling over the total world, finished the last day of his life in Liège ...'[57] Ortelius and Vivianus recorded that the tombstone bore the carved outline of an armed man with a forked beard, stepping on a lion. The fourteenth-century English knight's book of travels had influenced many of Mercator's generation; Monachus had used Mandeville's *Travels* to create his globe, and Mercator had referred to Mandeville on his world map of 1569 as 'an author who, though he relates some fables, is not to be disregarded as concerns the positions of places'.[58] Mercator's extensive library contained a manuscript copy of the *Travels* in French.[59] Ortelius also took Mandeville for a reliable geographical source, noting that the knight's 'roaming through the total world' had been 'proved by his *itinerary*, printed and available everywhere'.[60]

Having paid their respects to the world traveller who brought home tales of long-necked 'gyrfaunts'[61] and tongueless serpents who wept when they ate men, the humanists turned south again, over the hills of the Ardennes and through Luxembourg to Lorraine.

On the bank of the river Meurthe four miles[62] south-east of Nancy, the weary travellers reached the journey's turning point: St-Nicholas-de-Port. A place of pilgrimage for five hundred years, the town's great basilica had been built on the site of a small chapel that had housed the fingerbone of St Nicholas ever since it had been brought back from Bari by knights of Lorraine. Originally the bishop of Myra (a port on the Lycian coast), Nicholas had been persecuted, tortured and imprisoned for his faith during the reign of Emperor Diocletian, but had survived to see the relatively tolerant times of Constantine. His finger had been the cause of innumerable miracles, and as the protector of children, scholars, sailors and *mercatores*, persecuted St Nicholas had a *vita* which would have chimed with Mercator.

At St-Nicholas-de-Port, the travelling humanists turned and retraced their route to Nancy, and then followed the Moselle to Trier, Koblenz and so to the book and map fair at Frankfurt.

No record survives of a call Ortelius must have made in Nancy. Before the journey, he had written to Duke Charles of Lorraine's doctor in Nancy, an avid coin collector called Antoine Le Pois,[63] asking the physician whether he could supply the map which Mercator and his son Bartholomeus had made of Lorraine in 1564. As the best map of the duchy, it would have been of enormous help to the humanists on their journey. But 'the map of our native Lorraine, drawn here some years ago by Gerard Mercator', replied Le Pois, was still 'unpublished, in the hands of the Prince'.[64] The only other map of Lorraine he had been able to offer was 'a small sketch ... in the older editions of Ptolemy', which was 'not very satisfactory'.[65] When they reached Nancy, the travellers were given a guided tour of the city and its defences by one of the duke's deputies and by Joannes Scillius, who had been 'most useful on account of his distinction and diligence in matters geographical'.[66] Since Ortelius passed through Nancy twice, on the way, and on the way back from St-Nicholas-de-Port, it is inconceivable that he failed to renew his attempt to see – and possibly copy or obtain – Mercator's manuscript map of Lorraine.

When the travellers eventually returned to Antwerp,[67] Ortelius and Vivianus prepared an account, an *Itinerarium*, in which Mercator would eventually become a virtual companion.

Commensurate with his rising acclaim, Ortelius opened an *Album amicorum* – a book of friends. Mercator was one of the first to be invited to contribute a portrait and a few lines. He was cooperative but not particularly enthusiastic, replying to Ortelius that he would send a portrait with the next batch of maps 'most willingly, if it pleases you, but reluctantly as far as I am concerned, as I feel ashamed to exhibit myself, as if I were of any importance, among famous men'.[68] There was no artifice in Mercator's modesty; he genuinely believed himself unworthy of recognition.

Perhaps it was his skirmish with mortality which persuaded Mercator to pose – before it was too late – for the 'artful' Frans Hogenberg.[69] Engraved onto copper, Hogenberg's Mercator wore a biblical beard and gazed fixedly from the portrait's frame with slightly raised eyes, the subject of his gaze a point above the earth. The eyes were set above half-moons of toil; the cheeks hollowed by austerity, the mouth a silent line. Above the brow were the furrows of old anxieties. His left hand rested on his globe, turned so that his thumb and forefinger underlined the word 'AMERICA'. In his right hand he held a splayed pair of compasses, one of the tips pricking the magnetic pole – the essence of his life-long quest and the apparent source of gravity worn by his tired face. To the viewer, this was a man whose reach was both celestial and terrestrial, a man who could see the heavens and touch the globe. The point of the mathematical instrument which linked Mercator to the earth was a symbol of truth. That it rested not on the Holy Land, but on the magnetic pole and the prime meridian, described a cosmographer who had devoted his life to measuring and mapping the sublunary world of God's creation.

This was the portrait which was pasted into Ortelius' *Album amicorum*. By the time Mercator penned his brief entry, on 1 October 1575, the 'Book of Friends' already contained contributions from the Calvinist Carolus Clusius, Georges Hoefnagle, Plantin, Mercator's university contemporary Rembert Dodoens and Johannes Crato von Krafftheim, the physician who had found the *Theatrum* such a relief from his courtly cares.

Crato had been another who had troubled himself to praise Duisburg's increasingly reclusive cosmographer. The imperial physician had never met Mercator, and had a reputation for being peremptory, but following several conversations with Jean Van der Linden, the ex-burgomaster of Louvain, Crato had written 'a most loving letter'[70] to Mercator back in 1572.[71] Three years later, Crato was still waiting for a reply to his unsolicited praise. Although Mercator may have been reluctant to open a correspondence with a conspicuous supporter of Calvin's doctrines, such neglect was probably unwise. At the time, Crato was particularly close to Emperor Maximilian, whose heart palpitations, gout and chronic haemorrhoids demanded the physician's continual exertions. Maximilian had raised his principal physician to the title of *Pfalzgraf* (Count Palatine),[72] and the two men dined together most mornings at ten. The Breslau surgeon was well placed to promote Mercator's work and to ease the issuing of the publishing licences which Mercator would soon need for his Ptolemy.

26

Ptolemy Corrected

By 1575, Mercator had set aside the modern maps in order to concentrate on completing the maps for his edition of Ptolemy. The most consistent source of distraction now became his globe of 1541.

Mercator and his globe had become inseparable through Hogenberg's engraving. Yet there was more to this than pictorial conceit: the globe of 1541 had proved to be his most enduring work. The English, who had yet to produce their own printed globe, favoured Mercator's above all other imports, despite the fact that its geography had not been updated since its issue.

Rising orders for the thirty-year-old globe were as welcome as they were distracting. Orders for previously published maps were easy enough to fulfil, either by dispatching sheets held in stock, or – if the print run had been exhausted – by pressing another set of prints from the original copper plate. Globes were quite different. Although Mercator held a stock of printed gores, each order required the construction and weighting of a new sphere, meticulous cutting and pasting of the gores, varnishing and then the assembly of the 'furniture' upon which the globe rotated. Unlike maps, which could be rolled and protected in barrels for transit, or sent individually by courier, each globe had to be individually padded with straw in containers which frequently had to be specially made. Cumbersome and fragile, the packed globes then had to be dispatched by Rhine boatmen towards their eventual destination. But the globes could – in theory – earn good money. In 1575, an uncoloured copy of the Europe wall-map sold to Plantin would earn Mercator 1 florin; while a pair of globes could be sold privately for as much as 20 florins.

But Mercator's profits from his globes were being eroded by the time required to fulfil each order; it was very difficult to retain a profit from labour-intensive, low-volume products during periods of rapidly rising costs. Each globe took so long to construct that the rising costs could eradicate the profit before it could be delivered to its owner. After delivering four globes to Plantin in June 1575, Mercator supplied him with no more for another five years.

One of several customers clamouring for globes at this time was Joachim Camerarius, the son of Melanchthon's biographer. Joachim had grown up in the shadow of a father whose 150 published works included translations from Greek to Latin of Herodotus, Demosthenes, Homer and Sophocles, and whose theological views had propelled him to the forefront of the reform movement. Through a correspondence with Francis I, the elder Camerarius had explored the possibilities in 1535 of a reconciliation between Lutherans and Catholics, and had assisted Melanchthon in drawing up the Confession of Augsburg. For the same lost cause he had been summoned at the age of sixty-eight to Vienna by Maximilian II. The younger Camerarius, Joachim, had adopted a lower profile than his father, returning from a medical education in Italy to tend his botanical garden in Nuremberg. Early in 1574, he wrote to Mercator requesting a pair of the famous globes.

'Most famous fellow,' replied Mercator, 'I finally received your letter around the time of the market at Frankfort, much too late for me to be able to get ready the globes you required. And so I pray that you will be patient until the next fair. Meanwhile I will adorn and complete as exquisitely as I can the globes I am preparing for you ...'[1]

Camerarius had to wait another five months before the globes were ready. That the wait was worth it is indicated by the care that Mercator had taken in assembling them:

> I am sending ... the globes you asked me for. From many I have selected two that are equally excellent and which come as close as possible to equilibrium. There is only one which does not settle in any position, but sometimes slips back a little, though the unevenness of its weight is slight. For it is very rare to achieve equilibrium, and though we aim to obtain it with as much skill as you like, it is only by chance that we do get it exactly right.[2]

Mercator charged Camerarius 40 florins for the globes, 20 albi for the packing and 42 albi for the shipping costs to Cologne.

As 'a small gift', Mercator included with the globes his huge wall-map of Europe. Having packed each globe in straw 'in such a way that they cannot rub against each other or suffer any harm',[3] the consignment was dispatched upstream. As an apologetic afterthought, Mercator added that his letter had been 'written hurriedly amidst all my tasks'.[4]

He was a victim of his own repute, and the distractions appeared with frustrating regularity: 'I reply to you a little late,' he began to Ortelius in March 1575, 'not because your little gift[5] was unwelcome; indeed I embrace it lovingly and with equal desire to pay favour in time. Rather,' explained Mercator, 'I have waited for a ... convenient opportunity ... and am all the more confident in this as I know you to be indulgent to the occupations of friends.'[6] Mercator admitted that he was 'sidetracked by so many occupations' that he was 'really making quite slow progress in accomplishing the Ptolemy work ... I am doing what I can ...'[7]

The letter from Ortelius concealed yet another distraction, for Mercator's friend had felt obliged to draw attention to a passage in a recent work on the philosophy of history by the French jurist Jean Bodin. 'No less does Mercator slip up,' Bodin had written of the *Chronologia*, 'who thinks that the Sun was in Leo when the universe was coming into existence.'[8] With his 'foundation badly laid', Mercator's entire chronological edifice was – continued Bodin – liable to crumble: 'the other things which he set down about the movement of stars and the trustworthiness of history threaten his downfall'.[9] Bodin refrained from criticizing Mercator's other 'trivial conjectures' because they were 'too unimportant to be worth rebutting'.[10]

Stung, Mercator replied to Ortelius that he had missed the passage when he had read Bodin's book himself; 'I see a learned man and one of much reading,' wrote Mercator, 'but in this work an immature judge.'[11] Condemning Bodin's views on the origin of the world, Mercator added that he was 'amazed that a learned man makes judgements so glibly about matters either not understood or not clearly examined'.[12] Bodin, concluded Mercator, could expect a rebuttal 'when there is time'.[13]

But time Mercator did not have.

A few weeks later, Camerarius was writing to Mercator for more globes. But the letter got delayed and did not reach Mercator until 13 August. Thankful no doubt that he had an excuse not to drop his cartographical work again, Mercator informed Camerarius that his letter had arrived too late for the globes to be completed in time for the forthcoming Frankfurt fair. Blaming the courier, Mercator explained that the letter 'fell into the hands of some not sufficiently trustworthy man, who carried it round with him in his little sack for too long and only saw to its delivery when it was dirtied and worn, although it was obviously taken to Cologne in clean condition (as one would expect as it was wrapped with others)'.[14] Mercator nevertheless promised that the globes would be ready for the next fair, and that he would send them to Camerarius via the bookseller Andre Wechel.

The principal reason for Mercator's inability to fulfil his orders for globes (and to respond to Crato's adulatory letter) surfaced in his reply to Camerarius: 'I am working diligently on the old maps of Ptolemy, but working on my own I am progressing quite slowly, as other tasks, one after the other, interrupt this one ...'[15]

Typically optimistic, Mercator added that he hoped to finish correcting the maps of Ptolemy within a year. Many of the plates for the maps were being engraved for Mercator by Arnold's eldest son Johann, and his brother Gerard.

One year later, Ptolemy's maps were still being engraved, but Camerarius had got his globes:

> As promised [wrote Mercator in April 1576] I am now sending the globes you were waiting for last fair ... I have selected seven of the best bodies which I had among many and with singular care I have put them together so that I trust they will be pleasing; as you had ordered there are two each of the celestials and of the terrestrials and the price of all of these together is 40 florins and 20 sous.[16]

Taking care to account for the minutiae of the bill, Mercator explained that there would be a slight increase in the shipping cost because the Rhine was in flood, and the globes would have to be conveyed on a smaller-than-usual vessel which would have to be poled upstream to Cologne, where the globes would be trans-shipped to an even smaller vessel onward to Frankfurt.

It was a bad year for the Low Countries: the rain and floods which had caused havoc in spring, were followed in the autumn by the worst atrocity of the century. In preparation for a Spanish assault led by Philip II's brother, Don John of Austria, forces of the States-General of the Low Countries took to Antwerp's defences but were surprised by a dawn attack. Over the following days, 8,000 were killed and 1,000 of Antwerp's houses destroyed. A resident who escaped death by burying himself under the turf in his own cellar reflected that Antwerp would 'never be again as it was'.[17]

By 1577, Mercator had nearly completed the maps for his Ptolemy. But the sense of urgency which infected his correspondence turned to frustration as he applied for the vital publishing licences.

Two licences were required, one from the Imperial court in Vienna (which would protect Mercator from copyists in the Imperial cities, and in particular Frankfurt) and another from the court of Philip II of Spain at Brussels (which would protect his copyright in the Low Countries).

But Mercator had picked the century's worst moment for obtaining Imperial favours. In the Low Countries, the Spanish had lost control to Calvinist war councils, while Brussels itself was still reeling from the holocaust that had consumed Antwerp. And Vienna was distracted by the death of the emperor; having attended to Maximilian's exotic catalogue of ailments for his entire twelve-year reign, Johannes Crato von Krafftheim (whose letter to Mercator still awaited a response, *four years on*) and his fellow physicians could do nothing to save their patient. On 12 October 1576, Emperor Maximilian II had died, having refused the last sacraments. The end of an era of religious conciliation came with his unstable son Rudolf II drawing his dagger during a brawl at the funeral. A convinced Catholic with a record of erratic behaviour, Rudolph's arrival was not good news for those Germans who had grown used to Maximilian's years of tolerance.

With his direct lines of communication to the Emperor severed, Mercator had to apply for his new Imperial licence through the secretary to the Duke of Jülich. But here too there had been an unfortunate change, with a new secretary – Paul Langer – taking office in 1576.

Langer was a man whom Mercator felt obliged to approach with caution. However, through the court chamberlain, Jacob Wichius, Mercator learned that Langer was 'kindly disposed' to Duisburg's cosmographer and his 'efforts in the field of geography'[18] and in April 1577, Wichius offered to take up the matter of the licences with Langer, requesting that the secretary arrange for a letter to be written 'to His Imperial Majesty by command of the Most Illustrious Prince'.[19] The letter would be accompanied by a petition written by Langer. Once the licence had been granted by the emperor, Mercator could repeat the procedure at the court of Philip II of Spain at Brussels.

Mercator furnished Wichius with the information required for the petition, but by June, he had still not received the Imperial licence. Inquiries revealed that the petition had indeed been polished by Langer and dispatched to the new Holy Roman Emperor, Rudolf II. Mercator – eternally optimistic – still thought that he would complete the Ptolemy in time for the autumn fair at Frankfurt. Bypassing Wichius, Mercator wrote directly to Langer on 22 June. In words which can have left Langer in little doubt that the court cosmographer had much to lose by further delays, Mercator reminded the new secretary that 'the Frankfort Fair is now at hand'.[20] All depended upon Langer: 'I should gladly retail [at Frankfort] this work of Ptolemy', continued Mercator; 'I am most eagerly awaiting the granting of my petition and license; especially since, once it has been granted, I must petition the Court at Brussels in the same manner, for I am required to obtain a license from each of them before I can publish.'[21]

As he urged Langer into action, Mercator must surely have thought back to his early days in Louvain, where he had the ear of the emperor's chancellor, and the court lay a short walk away in neighbouring Mechelen. Perronet would have been able to expedite the licences in a trice. Not only did Mercator exert little influence over Langer, but the secretary appeared invisible to the Imperial court in Vienna.

'I therefore beg you,' concluded Mercator,

> in case you do not yet have word of the granting of my license, to urge upon His Imperial Majesty's Council at the earliest opportunity that it be done as soon as possible. With the hope that you may be mindful of me and my cause, I am sending you a map of Europe, mounted on cloth and coloured, for I shall be the more deserving of your kindness toward me, when I have brought this work to the light.'[22]

A more undeserving recipient of such a gift would be hard to imagine.

During the summer of 1577, work on the Ptolemaic maps picked up pace as Mercator made the most of a rare offer of assistance with his engraving. Nothing could stand in the way of this opportunity to make progress. Writing on 31 August 1577 to Theodor Zwinger in Basel, Mercator explained (in terms familiar to Camerarius) that he was too busy completing his Ptolemy and that he could spare no engravers to work on globes which Zwinger had ordered months earlier. 'For certain artisans', he wrote, 'were then offering me their services and I feared that, unless I started them quickly on my work, they would leave me and take employment with others.'[23] Apologetically, Mercator added that his time had been consumed in keeping ahead of the engravers 'during this entire year and the greater part of last year'.[24] Zwinger was informed that he would not have to wait much longer: 'When I have finished and published the Ptolemy maps (which I hope will be within a month) I shall proceed with the globes you desire and, God willing, shall complete them for the next Fair.'[25]

Having rallied his engravers to finish the Ptolemaic maps, Mercator was already thinking ahead to the next phase of his cosmography – the modern maps: 'For I have ready enough maps of the more recent geography (for the engraving of which we are now making ready) to keep the workmen occupied for some months and permit me to turn my attention to other things . . .'[26]

By September 1577, the Ptolemaic maps were nearly finished. And when Camerarius – fresh from a plant-hunting trip to Bavaria and Suabia – attempted to extract yet more globes from Mercator, he was asked to be patient. The Ptolemy, wrote Mercator, would be complete 'within a month or a little more . . . all is very nearly prepared for putting the finishing touches to the printing; and nothing is delaying me except the printing of the license which I have not yet received'.[27]

Characteristically, Mercator was underestimating the amount of time required to complete the job. Far from taking a month or so, the Ptolemy would continue to occupy him through the entire winter – a winter which must have seen Mercator waiting with increasing anxiety for the critical Imperial licence to publish.

The Ptolemy was completed before the buds had opened on Duisburg's trees.[28]

'Finally,' sighed Mercator, 'through God's kindness, I added the colophon.'[29] The dedication to the Duke of Cleves was dated February 1578.

'For three years,' wrote Mercator, 'I hoped every six months to complete that work, but more troubles and labours lay in wait for me than I had been able to foresee.'[30]

Immediately, Mercator wrote to the patient Camerarius, announcing that at last he was released from the 'vexations of Ptolemy's Geography'[31] and that he now had time to complete the latest batch of globes. These had been assembled 'more ornately than usual, with oil laid over them liberally so that they have a more splendid appearance and their colours are rendered more clear and durable'.[32] Camerarius was left in no doubt that he was a favoured customer: 'I am sending you two choice pairs,' wrote Mercator, 'but at a very slightly higher price than the ones you have had from me before.'[33] The cost of the metal fittings had risen, explained Mercator, and so had the wages of workers. 'All things are now nearly twice as expensive as they were when I fixed the price at which I have sold thus far, and for this reason I have made this meagre increase.'[34]

Although Mercator's costs had virtually doubled, he could not bring himself to add more than 20 per cent to the original price of the globes, up from 20 to 24 florins a pair. (When Plantin next ordered,[35] the price had risen to 36 florins – twice the amount he was paying a decade earlier.)

But Mercator was not entirely free from Ptolemaic vexations, for although he had at last received his licence from Philip II,[36] the all-important Imperial *privilegium* had not arrived. Having given Langer the coloured, cloth-mounted wall-map of Europe, the duke's secretary had failed to influence the Imperial court – a failure which produced a rare burst of indignation from Mercator, who blamed Langer for pressing his application 'too negligently'.[37]

Rather than miss the spring Frankfurt fair, Mercator was compelled to publish without the licence, knowing that he would have to print a new issue of the Ptolemy as soon as the Imperial *privilegium* could be added to the volume.

The delayed Imperial licence – dated 26 May 1578 – was finally issued just over one year after Mercator had made his initial approach to Wichius. Under the terms of the licence, he dispatched three copies of the Ptolemy to the Imperial Chancellery.

One of those to receive a copy of the maps was Johannes Crato von Krafftheim. Maximilian's old physician had waited six years for a reply to his kind, supportive letter to Mercator. Now he got a reply, and as a bonus, a set of the most exquisite Ptolemaic maps ever to be printed. 'It is the sixth year now,' wrote Mercator, 'from when your Most Excellent dignity thought it worth honouring my meagre soul ... with a most loving letter'.[38] He had, he apologized, been seriously ill, and 'continually overwhelmed by various and many occupations'.[39] The 'small gift' of the Ptolemaic maps was a symbol of 'gratitude' and 'friendship offered'.[40] Dilatory by force of circumstance, Mercator knew that it was never too late to send a thank you.

Mercator's Ptolemaic maps were not only the most beautiful of their kind, but the truest to Ptolemy's original intentions. In a five-page preface, Mercator described the sources he had employed, and the projections he had used. His intention, he explained, had been to arrange the graphic surface of each map in such a way that the reader could manage without much of the usual supporting text. The title page of each map included information such as the lengths in hours of the longest day for prominent places, and their time differences at midday from Alexandria. Observations for which there was insufficient space could be found on the versos of the maps.

Following Ptolemy's system, the maps were ordered so that regions appeared from west to east, starting with the lands of the north; 'Europa Tabula 1' described the British Isles. And Mercator had thrown in an extra map – an enlargement of the Nile delta – making twenty-eight in all. Unlike the densely labelled modern maps that he was also preparing, his Ptolemaic maps were elegantly spacious pictures from the past, framed by their scales of latitude and longitude, with exotic cartouches bearing fruit and peacocks ... a squatting frog. There were beaked sea creatures and ships with arabesque sterns; a winged dragon roaring in the deserts of Cyrenaicae, and a club-tailed beast having its spine gnawed by a lion; an ostrich in Aethiopiae; grazing camels; elephants; a bagpiper serenading a flock on 'Asia ii tab'. On the eighth regional map of Asia, nomads in northern Serica stood by their conical tents while a group of Anthropophagi beyond the Annibi mountains hacked at a headless human corpse beside a suspended cooking cauldron. To the west, a tribe of fleet-footed Scythians chased horses with drawn bows and spears. And Mercator had included the ten little islands of the cannibalistic Manioli, 'from which', Ptolemy had written, 'they say that boats, in which there are nails, are kept away, lest at any time the magnetic stone which is found near these islands should draw them to destruction'.[41]

Coloured, the Ptolemaic maps brought to life the world as it had been at the dawn of Christianity. The first of the maps in the volume was Ptolemy's world map: his *Universalis tabula*. With its immense Africa and absent Americas, it was a reminder of the extraordinary progress made by Mercator's generation of cartographers. The maps were virtually flawless; years had passed while Mercator had carefully plotted Ptolemaic coordinates, marking each with a precise circle, whether it be a town or the source of a river. Only one map revealed the stress Mercator had recently endured: on the coast of the Holy Land ('Asiae iiii tab') he had accidentally transposed 'Ascalon' and 'Gazeorum'.

Tabulae geographicae Cl: Ptolemei came with a lavish title page framed by the standing figures of Ptolemy and Marinus of Tyre, a terrestrial globe at their feet and a celestial globe resting on a twin-columned entablature above their heads. Mercator had finally raised his portico; now he had to erect the remainder of his monumental cosmography.

Conceived in the mid 1560s, Mercator's enormously ambitious project had been overtaken first by the *Theatrum* of Ortelius, and now – in 1578 – by another book of maps, de Jode's thwarted *Speculum Orbis Terrarum*. Although the *Speculum* did not compare favourably to the *Theatrum*, it contained no fewer than ninety maps and included reproductions of lesser-known works that had originally been printed outside the Low Countries. De Jode and Ortelius remained bitter rivals, both authors conspicuously failing to refer to the other among their listed authorities. Mercator also appeared to ignore de Jode, who had already copied – without acknowledgement – Mercator's map of Flanders for an earlier collection of maps, the erratic *Speculum Geographicum totius Germanie* of 1570.[42] Ortelius, meanwhile, had achieved near divine status: writing in 1579 from Paris, the eccentric irenicist Guillaume Postel elevated the *Theatrum* to the principal work of the world after the Holy Bible'.[43]

Mercator was probably the only one who knew that *his* modern maps would outshine even those of the *Theatrum*.

27

Adorn your Britannia!

Shortly before he finished work on the Ptolemaic maps, Mercator received a letter from his son Rumold in London.[1]

'You write great matters,' replied Mercator (without clarifying what they were), 'though very briefly of the new discovery of Frobisher.'[2]

Rumold was still working as an agent for the Cologne bookseller Birckmann, where he had made a number of influential friends. Among them was John Dee, who had been a frequent visitor to Birckmann's London shop since at least 1560.[3] Mercator's particular interest in Rumold's news – and in England – was understandable: following the established Ptolemaic sequence for a book of maps, the northern lands had to come first, and the English held the key to the cartography of those northern lands.

Expressing surprise that the English had taken so long to explore to the north-west, Mercator repeated his view to Rumold that there existed 'a straight and short way open into the West even unto Cathay'.[4] This kingdom, continued Mercator, would be found if the English took the right course. The explorers of Cathay, he reminded, 'shall gather the most noble merchandise of all the world, and shall make the name of Christ to be known unto many idolatrous and Heathen people'.[5]

The 'Frobisher' Rumold was writing about was Martin Frobisher, an English ship's captain who had spent fifteen years attempting to mount an expedition to sail the north-west passage through to Cathay and India. The Yorkshireman had finally sailed from Blackwall in 1576, with two small barks and a pinnace crewed by a total of only thirty-five men. Frobisher had taken with him the 'greate mappe universall of Mercator in prente',[6] the world map of 1569.

Frobisher's pinnace had been lost in a storm, and then one of the barks had deserted, but the survivors had sailed on to reach the coast of North America, where they had found a promising passage. Pushing on in the hope that he 'mighte carrie himself through the same into some open sea on the backe syde',[7] Frobisher had made the mistake of landing on a hostile island and losing five men to the natives. By 9 October, he was back in London with 'great rumour of the passage to Cathaye'[8] and some mysterious 'black earth'[9] which he – and his backers – had claimed was a lump of gold ore. Called to court, Frobisher had been 'greatly embraced and liked of the best'.[10] The black earth was enough to convince Queen Elizabeth that Frobisher should sail again the following year.

But what of the 'great matters' reported by Rumold? Had Rumold been referring to Frobisher, or to another momentous voyage? In England at this time there was a name which might have outshone Frobisher's in the eyes of a cosmographer's son.

Since his return from a bloody expedition to Ireland in 1575, Francis Drake had been plotting another raid on the Spanish ports of South America. His co-conspirators included Thomas Doughty and Christopher Hatton, the captain of the queen's guard. Reports also circulated that Sir Francis Walsingham was involved and that the queen herself had encouraged the idea of a raid on Spain's territories in revenge for 'divers injuries.'[11] Anyone divulging the plan to the King of Spain would, said Elizabeth, lose their head. Were these perhaps the 'great matters' which Rumold had passed to his father in Duisburg?

The Spanish were losing control of their sea lanes. Antwerp had fallen, and the Habsburgian Low Countries of Emperor Charles v had been ripped apart. Commenting on the new balance of power to Ortelius, John Dee 'grieved that [your] country is so disturbed, nay torn to pieces, that you have hardly any hope of seeing it recover its original greatness and liberty'.[12]

In truth, Dee may not have been as distraught as he implied, for he was close to concluding the 'British discovery and recovery enterprise'[13] that he had been pursuing since the 1550s. Queen Elizabeth had expelled the French from Scotland and enlarged the concept of an English England into a British Britain, and several months before Mercator had published his Ptolemy, Dee had published the first volume of *General and Rare Memorials Pertaining to the Perfect Art of Navigation*, the great work which would, he hoped, lead to the establishment of a 'BRYTISH IMPIRE'.[14] Exactly thirty years after Dee had crossed the waters to confer with the measurers of Louvain, his geographical and historical researches had reached the climactic conclusion that the future of Elizabethan England lay in her maritime resources. In Dee's view, the division of the globe under the Treaty of Tordesillas into Spanish and Portuguese hemispheres would be invalidated if Britain could claim overseas lands 'by discovery, inhabitation or conquest'.[15] Applauding the *Theatrum*, Dee probed Ortelius for the sources that he had used 'on the northern coast of the Atlantic'.[16]

Dee had played a major role in Frobisher's expedition, advising the explorer on navigation and chart-making, and he concluded his letter to Ortelius with the tantalizing information that 'our people are preparing an expedition to the northern parts of the Atlantic. Last year they only touched the Greenland Strait, but in such a way that they entertained great hopes of sailing around the entire coast and of penetrating that way as far as the Eastern Sea'.[17]

A few weeks later,[18] Ortelius showed up in London, where he met Dee and – among others – Richard Hakluyt. A recent graduate of the university of Oxford, Hakluyt had already become something of an expert on voyages and discoveries. The young English geographer was convinced that Ortelius was spying, for 'it seemed that the chief cause of his [Ortelius] coming into England was to no other end, but to pry and look into the secrets of Frobisher's voyage: for it was even then, when Frobisher was preparing for his first return into the north-west'.[19]

Ortelius told the suspicious Englishman that men of the Low Countries would have discovered the north-west passage themselves, but for the war on their own soil.

At around the time that Mercator heard from Rumold, Dee wrote to Duisburg asking for 'the very principal Authority' that Mercator had used for 'that strange plat of the Septentrionall Llands'.[20]

The 'strange plat' referred to by Dee was the inset on the 1569 world map, which showed the quartet of Arctic islands and the clear passage of water between the Atlantic and Pacific – the north-west passage. With Frobisher due to depart in May on a second voyage, Dee was extremely eager to judge for himself the veracity of Mercator's map. In asking for Mercator's 'very principal Authority', Dee was digging for Mercator's primary source; the source which would prove that Mercator's four islands were not merely extrapolated from those of Ruysch and Fine, but were based on first-hand observation.

Mercator wrote back on 20 April 1577.[21] As it happened, he did still have his primary source to hand. It was a very rare manuscript indeed.

Laboriously, Mercator transcribed the extraordinary story of northern exploration which had been related to Jacob Cnoyen of 's-Hertogenbosch.

In an exercise which Mercator cannot have known about, Dee copied Mercator's letter and its Cnoyen transcription, adding an introduction

ATLAS
SIVE
COSMOGRAPHICÆ
à MEDITATIONES
DE
FABRICA MVNDI ET
FABRICATI FIGVRA.
1651

Gerardo Mercatore Rupelmundano,
Illustrißimi Ducis Iuliæ Cliviæ & Mõ-
tis &c.ᵃ Cosmographo Autore.
Cum Privilegio.

DVISBVRGI CLIVORVM

'I have set this man Atlas,' wrote Mercator, 'so notable for his erudition, humaneness, and wisdom as the model for my imitation.' Title page, from *Atlas sive cosmographicæ meditationes de fabrica mundi et fabricati figura*, 1595 (Royal Geographical Society 264.H.4)

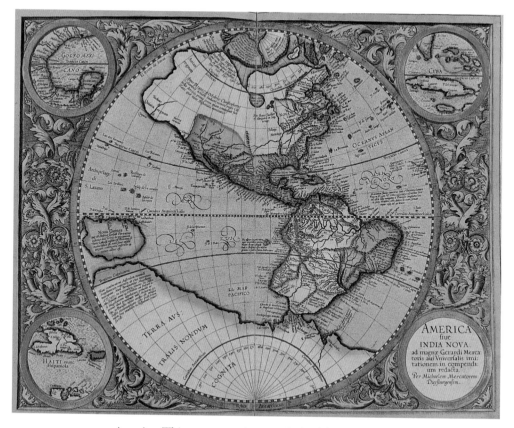

America. This representation was derived from Mercator's world map of 1569, and repeats the misinformed bulge in the western coast of South America, and the vast Antarctic continent. Such a southern landmass was believed to be essential if the earth was to maintain its equilibrium. Sixteenth century cartographers sought confirmation of its presence in the reports of Marco Polo, and from the survivors of Magellan's circumnavigation, the route of which passed through the tight straits seen here at the tip of South America. The map's engraver was Mercator's grandson, Michael. From *Atlas sive cosmographicæ meditationes de fabrica mundi et fabricati figura*, 1595 (Royal Geographical Society 264.H.4)

Asia. Mercator's globe of 1541 showed four prominent peninsulas projecting from southern Asia. By the end of his life, he had reduced the peninsulas to the two that are known today. The sailing ship draws the viewer's eye to a region of particular concern to Mercator - and to the English. Off the ship's starboard bow is the mouth of the Straits of Anian, with north-west America (now Alaska) to one side and Asia to the other. At the furthest end of the straits can be seen the solitary rock marking the magnetic pole ('Polus magnetis'), and west of the rock, the opening to the much conjectured north-east passage. Mercator's grandson Gerard engraved the map. From *Atlas sive cosmographicæ meditationes de fabrica mundi et fabricati figura*, 1595 (Royal Geographical Society 264.H.4)

The Arctic lands. This exquisite, symmetrical map was concocted from existing maps and historic travel narratives, from the voyages of English explorers and from Mercator's theories concerning the location of the magnetic pole. No other map in the *Atlas* proved to be so erroneous. The four islands around the North Pole were later found to be fictitious. The island of Frisland, seen in the upper left circular inset, also proved not to exist, having been copied by Mercator from a faked map by Nicolò Zeno. (The other two circular insets show the islands of Faeroe and Shetland.) Mercator placed a prototype of his Arctic map in a panel in the lower left corner of his world map of 1569. From *Atlas sive cosmographicæ meditationes de fabrica mundi et fabricati figura*, 1595 (Royal Geographical Society 264.H.4)

Europe, engraved by Mercator's son, Rumold. Mercator's first map of
Europe (1554) was the most advanced of its kind and 'attracted more
praise from scholars everywhere than any similar geographical work
which has ever been brought out' (Walter Ghim, 1595). Of the 107 maps
in the 1595 edition of Mercator's *Atlas*, 102 of them described regions
within Europe. Mercator's intention to extend his detailed mapping to
the other continents was thwarted by his death, aged eighty-two. From
Atlas sive cosmographicæ meditationes de fabrica mundi et fabricati figura,
1595 (Royal Geographical Society 263.G.9)

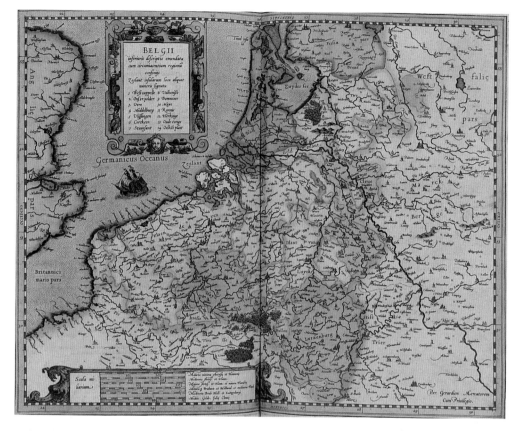

'Belgii Inferioris' (Lower Belgium), from *Atlas sive cosmographicæ meditationes de fabrica mundi et fabricati figura*, 1595. This is the map of Mercator's life. There is no hard evidence that he ever left the limits of this map, although it must be assumed that his disastrous surveying trip into Lorraine would have taken him south of Metz (at the map's foot). Mercator used this map of 'Belgii inferioris' to define a modern 'Belgium' distinct from the Roman concepts of 'Gallia' and 'Germania'. (Royal Geographical Society 264.H.4)

Opposite page 'Frisia occidentalis'. A detail from the *Atlas* map of western Friesland, illustrating Mercator's use of cartographic symbols. Note the hatched hills and pictorial depiction of water currents and salt marshes dotted with wild fowl. Throughout the *Atlas*, Mercator adopted standardised symbols for small settlements 'of no distinction' (plain circles), small towns (one tower), large towns (two towers), monasteries (a circle with a cross) and castles (a circle with a barbed line). Political boundaries appeared as pecked lines. Mercator explained in the *Atlas* that he chose symbols 'that are simple to make so that anyone can easily supply what has been omitted'. His maps, he inferred, must be regarded as a works-in-progress. The town of Dokkum, the birthplace of Mercator's tutor, Gemma Frisius, can be seen in the upper right part of this map. From *Atlas sive cosmographicæ meditationes de fabrica mundi et fabricati figura*, 1595 (Royal Geographical Society 264.H.4)

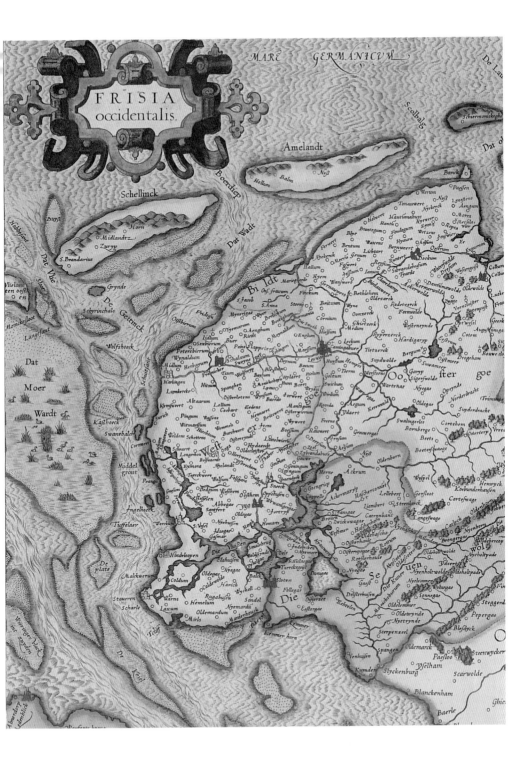

FRISIA occidentalis.

MARE GERMANICVM

De Lau...

Scolbalg

Schiermonickoghe

Dat ol...

Amelandt

Neß

Hollum   Balm

Banck

Werum   Paeßen

Boerdiep

Tonauwert   Neß   Ioestens

Hobert   Hantfmalnus   Aengum

Schellinck   Bi Idt   Hanti   Nyrerr   Morra

Branteğum   Foudegum   Hyaura   Meßlas

Oes frauwen   Blye   Wetsens   Royta wier

Horn   s.Jacob   Mariegğest   Bentum   Waxens   Bernswert   Ee

Midlandtz   Fridens   Wyer   S.Anna   Lichtart   Reythm   Subrandahufum   Dockem   Engarum...

Zwyp   Mynnyğat   Sterns   Germania   Ruñters   Iyarde   Ophterfwalde   Wertergeft   Collu...

S.Brandarius   Berlickum   Burmania   Pie wed?   Ianum   Ackerwolde   Dreßum   Collu...

Grynde   Oßterbierum   Tyemmmum   O.Aenğum   Medum   Oonzerck   Hardigaryp   Veenclafter   Buteupoft

Schyrinchals   Getrinck   Lidlum   Burr   Peins   Slappecerp   Engelum   Cornum   Ghietzerck   Feenwolde   Wyfel   Augufonfga...

Wolfhoeck   Peterabierum   Sexbierum   Mefsindum   Enğelum   Heifium   Leckum   Weftezeynde   Cotem   Su...

Wynaldum   Schalzum   Eching.   Lewaren   Tietszerck   Berğum   Drogeham   Bauwe...

Midlum   Herbayum   Deynum   Bacum   Huryum Hempr   Suydwolde   Swamęrr

Harlingen   Franicker   Txum   Wigfwert   Alfershm   Hirlum   Terms   Weftern   Garyp   Sigerfwolde   Oo   iter   goe

Lynmkercke   Hitzum   Spanghen   Romkeheru   Hune Brert   Swicham   Argum   Warteñat   Nyegae   Opeynge   Noırderdracht   Trimins...

Altaarum   Lollum   Edens   Brittwert   Oßteryeurum   Roorda   Errenwou   Ydaert   Oldegae   Swalingerlet   Suydracche

Pingum   Waßens   Hennaert   Manhipum   Frens   Groewergas   Bornberge   Betts   Oßterterp   Vree...

Kymfwerrt   Witmarßum   Wommels   Wywert   Rauwert   Srfynum   Beetesfwage

Swanbalch   Welden Schettens   Oßteryndt   Botfhum   Lutelwiern   Oldeborn   Neß   Cortehem   Oldehou...

Kaßhoeck   Schrart   Langerhou   Hydaerdt   Oldeclofterhm   Nyerbert   Zybrandahuyr   Herne   Ackrun   Wyfpel   Henryck   Koefwanderhausen

Swart   Ega   Nertwert   Wanfwer   gončhum   Goningum   Geengryp   Lellebert   Gerfloot   Cortefwage

Wardt   Ecmora   Nyelandt   ŋecrick   Groenbhr   Oßterrymgae   Hafchercomirt   Langefwage   Catick   Oldehoow

Middel   grout   Peanga   Terckwert   Wolfum   Folğaga   Sneck   Cappele   Ackmaryp   Lambert   Steenkerck   Nyenhorn   Die

Obdijum   Worckum   Weftfum   Oefthm   Oppehufen   Breuck   Zwickwagae   Catrynbant   Oßterhaule   Nyenholt   Brongregae   Oldesercop

Gaßum   Heßfen   Oldegae   Whaenfergae   Weftmer   Hafpelo   Rofferhaule   Rottum   Oßterholtwolde   wol

Tufelaer   Abbegae   Santfert   Rorreyt   Breuck   Indefuelmr   Oldezom   Oldeholt   wolde   Nyeholtpade

Engelhoeck   Workum   Nyehufum   Homarts   Vhusen   Oßterhaule   Ofterygae   Oldeholt   wolde   Ydaerin

Hindeloopen   Adegae   Hotjs   Gumle   Nyemerer   Oldelommer   Se   uen   Oldehercop

Gafmar   Smalbrug   Wolfwanga   Tjerckğaft   S.Niclasgae   Dongae   Nyegae   Die Knerr   Nyelemmer   Wohugae   Sonnegae   Pepergae

S.Vofl   Nyehufum   Woldgawick   Dele   Sloten   Ganft   Belftermafum   Oldermmer   Nyttrynde   Scerpenzeel   Bleßyck

Malckwerum   Oldegae   Harich   Balck   Follegae   Echten   Oldewrynde   Steggerda   Fin

Coldum   Oldewolde   Die   Efterzet   Nyttrynde

Staueren   Warns   Wyckell   Oefteret   Spangea   Oldemark   Steenwycker

Scharle   Rugahufen   Sindel   Lemmer   Venhufen   Paesloo   Scarwolde

Laxum   Hemelum   Nyemardii   Kuynder   Yfelham   Slyckenburg

Mirls   Oldemardum   Manderhouck   Lewmer horn   Blanckenham

Tfliff   Baerle   Ghie...

De plate

Vlielandt   een oeft   en

Dat Vlie   Sluys

Momkefloot

Langefant   De Getrinck

Dat   Moer   Wardt

Staueren   Kuil

Die   Kuil

Weranger Vlact   int zeytin

Albertsberg   domkhick   Veranges beeus

Details from two of the six regional maps of England and Wales. *Right*, the coast of eastern England from Yorkshire to Norfolk; *below*, south-west England. From *Atlas sive cosmographicæ meditationes de fabrica mundi et fabricati figura*, 1595 (Private collection)

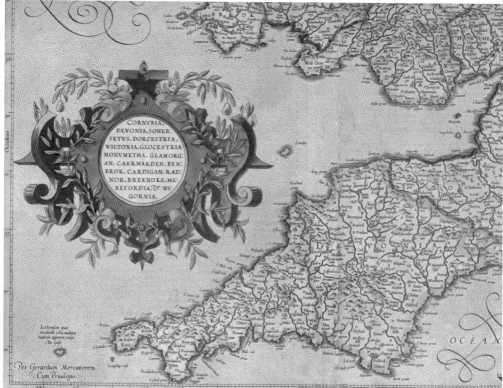

Fig. 21. John Dee (1527–1608). According to Dee, he and Mercator were at one time inseparable. Later, Mercator would become implicated in Dee's plans to establish a British Empire. (AKG, London)

in which he explained that he had, 'since the last yere', been seeking information about the 'Northerly Countries and Iles' and that one of those whom he had contacted was the 'honest Philosopher and Mathematician, Gerardus Mercator'.[22] Recalling the 'great familiarity' that he had shared with Mercator in Louvain, Dee claimed that the transcription would 'testify the honest and philosophicall Regarde that he [Mercator] had, of my earnest request to him'.[23] The regard Mercator had for Dee was, wrote the Englishman, implied by the fact that Mercator had responded so 'spedily'[24] to his request.

Dee had what he wanted: on Mercator's world map of 1569, the author had merely mentioned 'Jacobi Cnoyen Buscoducensis', together with the unnamed priest and the unnamed minor friar who had an astrolabe. Now, Dee was able to show Queen Elizabeth the words of Cnoyen himself, copied 'word for word'[25] (according to Dee's transcription) by Mercator in Duisburg.

A more portentous document could not have been written by Dee himself; here was Mercator, the most respected cartographer in northern Europe, describing in great detail how King Arthur had 'conquered the Northern Islands and made them subject to him'.[26] In a further elaboration of Mercator's remarks concerning Cnoyen and his two mysterious sources on the 1569 map, Dee's transcription identified the unnamed minor friar from Oxford as the author of *Inventio Fortunatae*, a book which 'put into writing all the wonders of those islands',[27] and which the friar later presented to the King of England, Edward III.

So Dee's transcription of the Mercator letter proved that the lands of the Arctic had been colonized by Britons one thousand years earlier, and that a reliable account of that colonization had been written by a mathematically minded Low Countries author who had 'travelled the world like Mandeville but described what he saw with better judgment',[28] an author furthermore whose work had recently been verified by Mercator himself. Bearing Mercator's name (though not of course, his handwriting or his signature) Dee's transcription was a charter to establish the British Empire.

Less than a month after Mercator had written to Dee with the Cnoyen transcription, Dee again wrote to Duisburg, asking this time for information about the narrow straits – of Anian – which separated North America from Asia.[29] Thirteen days later, on 26 May 1577, Frobisher left Blackwall with three ships and 120 men.

If Mercator thought that he was helping the British to further geographical understanding, he was mistaken. Frobisher's ships carried miners and the directive to 'defer the further discovery of the passage until another time'.[30] The British had gone north to look for Arctic gold.

On 2 November 1577, Dee was able to make the following entry in his diary: 'I declared to the Q her title to Greenland etc Estotiland, Friseland'.[31] On the back of a map of the northern lands, Dee summarized the evidence that he had provided to the queen: 'by discovery, inhabitation or conquest',[32] the British Empire extended from Terra Florida, over the north pole to Nova Zemlya, off the northern coast of Asia.

Dee was not the only English scholar looking to the continent for source material which might help to establish Britain as a noble entity.

In the first days of 1579, Mercator was removed again from his work on the cosmography by the arrival of a letter from an English schoolmaster whom Ortelius had met during his visit to London in 1577.[33]

An elegant Latinist whose fellowship of All Souls College in Oxford had been blocked by Catholic theologians, William Camden was working on an ambitious historical survey of Britain between duties as second master at Westminster school. Following their encounter in London, Camden had written to the Low Countries asking whether Ortelius could help him by copying 'the routes of Britain'[34] should he happen to have a manuscript copy of Antonine's Itinerary, the invaluable register of distances and stations for the roads of the Roman Empire.[35] Ortelius evidently could not help, and when Camden wrote again to Ortelius the following year, he confessed that his historical survey was 'under difficulties' and that he wished 'that some light might arise over our ancient Britain which is so enveloped in darkness'.[36]

Still searching for a copy of Antonine's Itinerary, Camden then wrote to Mercator, presumably on the recommendation of Ortelius.

Mercator's epistolary response to the 'most learned Camden'[37] reflected the pressures in Duisburg, and the respect that the old cartographer felt for a young, alienated humanist less than half his age.

'I hope you will forgive my delay,' began Mercator, 'and put down my belated response to your request to the fact that I have had so many things to take care of.'[38] Mercator explained that he had been further delayed by the lack of anyone in Duisburg to whom he could entrust the task of transcribing the material that Camden had requested. So Mercator had done the job himself, carefully harmonizing the miscopied Antonine manuscript of Cardinal Cusa's original with the version printed by Henri Estienne. As an exercise in reconciling conflicting sources, it was a task which Mercator was peculiarly suited for, but one which was hardly justified in view of his own, pressing commitments. But Mercator saw in Camden a fellow antiquarian and humanist who was engaged in a 'most remarkable work'.[39] With the Antonine notes, Mercator also sent Camden a copy of his new edition of Ptolemy's maps, that it 'might contribute in some way to your research, particularly in the matter of the location of places; for I have corrected this the best I could by comparing it with other works and I have provided it with degrees of longitude and corrected latitude'.[40] Mercator was unsparing in his support: 'If I can be of service to you in any other respect of your prestigious work,' he wrote, 'I am completely at your disposal'.[41] In a final entreaty, Mercator urged Camden to 'adorn your Britannia with all the beauty of antiquity'. Through a beautifully adorned Britain, continued Mercator, the English antiquary would 'procure even more eternal renown' for himself, and 'even more honour for the work which we devote to the geography of the ancients'.[42]

Camden's *Britannia* informed continental humanists that the British were about to regain the continental toehold that they had so humiliatingly lost twenty years earlier when 'le grand Guise' had ejected the Calais garrison from the continent.

By 1580, Frobisher's inconclusive quests were finished and the English turned once again to the north-east passage, the route to Cathay favoured by John Dee. Once again, Mercator became the recipient of urgent geographical inquiries.

In the spring of 1580, Mercator 'was informed from England' that merchants trading with the Russia Company were about to 'send out secretly a certain very experienced mariner' with orders to survey the coast of northern Asia 'even beyond the promontory of Tabin'.[43] Mercator's informant revealed that the mariner's name was 'Arthur Pitt'[44] and that he was equipped with a fast ship and provisions for two years.

Pet's proposed expedition to the north-east passage offered Mercator an extraordinary opportunity to plot one of the world's least-known shorelines, and it promised to prove conclusively whether the north cape of Norway marked an opening to a direct route to Cathay.

Mercator's informant was probably the young geographer whom Ortelius had met during his trip to London in 1577.[45] Shortly after Ortelius left, Richard Hakluyt had begun delivering lectures that 'shewed both the old imperfectly composed and the new lately reformed maps, globes, spheres, and other instruments of this art'.[46] Like Camden, he harboured ambitions to print, and by 1580 he was working on a book describing the voyages of discovery to America. Furthermore, Hakluyt's influential elder cousin – another Richard – had been responsible for drawing up notes concerning the trade that would arise from Pet's discoveries.

Like Dee and Camden, the two Hakluyts saw in Mercator a geographer who held the key to certain, critical British interests. Urgently, the younger Hakluyt sought advice from Mercator on the proposed route. Unfortunately, Hakluyt's letter did not reach Mercator until 19 June; Pet had sailed from Harwich three weeks earlier, on 30 May. Another five weeks passed before Mercator found time to reply to Hakluyt, on 28 July.

'It grieved me much',[47] wrote Mercator of the late arrival of Hakluyt's letter. 'I wish Arthur Pet had been informed before his departure of some special points. The voyage to Cathay by the East, is doubtless very easy and short ...'[48] Not only that, continued Mercator, but the north-east passage would open the way to the interior of Asia, by way of the 'great rivers' which flowed into the 'great Bay' which lay beyond the 'Island of Vaigats and Nova Zembla'.[49] Mercator recommended that the English merchants choose a port in one of the rivers of the great bay 'from whence afterward and with more opportunity and less peril, the promontory Tabin and all the coast of Cathay may be discovered'.[50]

With obvious concern for Pet's fate, Mercator described the 'very many rocks, and very hard and dangerous sailing' which lay beyond

Tabin, and the perils of compass variation as a ship neared the Loadstone: 'If master Arthur be not well provided in this behalf, or of such dexterity, that perceiving the error he be not able to correct the same, I fear lest in wandering up and down he lose his time, and be overtaken with the ice in the midst of the enterprise.'[51]

The outcome of Pet's explorations were of intense interest to Mercator, who had yet to commit to copper his new maps of the northern latitudes. Having provided Hakluyt with an appreciation of the north-east passage, Mercator left the Englishman with some questions which he hoped would be answered on Pet's return: how high, wondered Mercator, were the tides along the Asian coast of the north-east passage? Did the sea in the passage flow to the east or to the west, or did it ebb and flow like a tide? And, asked Mercator, could Frobisher provide the same information for the north-west passage, too?

Mercator also addressed the reason for Hakluyt's letter, for the English geographer had followed Dee in inquiring after the mysterious source which had caused Mercator to reconfigure the Arctic and to create a north-west passage.

But Mercator had disappointing news for the Englishman: the account of Jacob Cnoyen's northern voyage had gone missing. After reading it, Mercator had returned it to his Antwerp friend. 'After many years I required it again of my friend,' Mercator confirmed to Hakluyt, 'but he had forgotten of whom he had borrowed it.'[52]

Having emerged from the past into the hands of Mercator's unnamed Antwerp friend, Cnoyen had delivered his Arctic truth then slipped away into the bibliographic night. Only Mercator appeared to have seen him.[53]

Diligently answering Hakluyt's questions, Mercator wrote that he would 'most willingly communicate' further information, and urged the Englishman to send to Duisburg 'whatsoever observations'[54] arose from the voyages of Pet and Frobisher. In an acknowledgement of the value of such observations, Mercator added that the information 'shall remain with me according to your discretion and pleasure'.[55] Having suspected Ortelius of spying, Hakluyt was sharing confidential information with Ortelius' friend, Mercator.

The mistrust was mutual. By now, even Mercator suspected that the English had become agents of disinformation. Through Rumold, Ortelius had fed Mercator 'a report' about Drake's 'new English voyage', and then 'a dispatch'[56] once Drake had returned – having circumnavigated the world – in September 1580. To his evident frustration, Mercator had been unable to learn anything conclusive about the route Drake had taken: 'I am persuaded', wrote Mercator to Ortelius in December, 'that there can be no reason for so carefully concealing the course followed during this voyage, nor for putting out differing accounts of the route taken and the areas visited, other than that they must have found very wealthy regions never yet discovered by Europeans.'[57] The evidence, continued Mercator, was the 'huge treasure in silver and precious stones which they pretend they secured through plunder'.[58] Mercator also believed that Pet's secret voyage to the north-east passage had in fact been launched as a search-and-escort role to help Drake home with his riches. Convinced by the implications of his own world map (on which he had conveniently obscured the critical neck of the north-west passage with a cartouche) Mercator argued to Ortelius that Drake must have returned from the east because – as Frobisher had proved – the north-west passage 'is obstructed by many rocks'.[59]

A couple of months later, Mercator did receive some information on the northern coast of Asia, in the form of a letter from John Balak.[60] The letter had been sent from 'Arusburg upon the river of Osella'.[61] Recalling the 'exceeding delight' that Mercator had taken in their 'being together ... reading the Geographical writings of Homer, Strabo, Aristotle, Pliny, Dion[ysius], and the rest', Balak reported to his 'most dear Friend' that he had recently met a man called 'Alferius' – 'by birth a Netherlander' – who had been held captive 'for certain years ... in the dominions of Russia'.[62] Alferius had been obliged by his captors ('two famous men Yacovius and Unekius') to travel to Antwerp where he was to 'procure skilfull Pilots and Mariners ... by propounding liberal rewards'.[63] Back in Russia, the seafarers would board two ships built by 'a Sweden ship-wright' and proceed to explore 'the river of Dwina'.[64]

But this was a sideshow. The letter addressed to Mercator went on to reveal that Alferius had told Balak of the route to Cathay: 'This man's experience', continued Balak,

> will greatly avail you to the knowledge of a certain matter which hath been by you so vehemently desired, and so curiously laboured for, and concerning the which the late Cosmographers do hold such variety of opinions: namely, of the discovery of the huge promontory of Tabin, and of the famous and rich countries subject unto the Emperor of Cathay, and that by the Northeast Ocean sea.[65]

In a curious repetition of a phrase used by Mercator himself the previous year, Alferius stated that the route to Cathay by way of the north-east passage was 'without doubt very short and easy'.[66] Apparently Alferius had travelled all the way to the river Ob, both by land and by sea. The rest of the letter was taken up with the incredible plan Alferius had devised, to sail east to the Ob at the end of May with 'a Bark laden with merchandise'.[67] All being well, the Low Countries adventurer would explore the mouth of the Ob 'so that he may learn where the river is best navigable'[68] and then push inland 'against the stream' through 'the country of Siberia' to 'the lake of Kittay … whereupon bordereth that mighty and large nation which they call Carrah Colmak, which is none other than the nation of Cathay'.[69] In a sentence which might have been written entirely for English eyes, Alferius added that the people along the route had 'seen great vessels laden with rich and precious merchandise brought down that great river by black or swart people'.[70] At the mention of Cathay, these people 'fetch deep sighs, and holding up their hands, they look up to heaven, signifying as it were, and declaring the notable glory and magnificence of that nation'.[71]

Balak concluded the letter by writing that Alferius had promised to visit Mercator at Duisburg 'for he desireth to confer with you, and doubtless you shall very much further the man'.[72]

If Alferius did arrive at Duisburg that spring, he left no trace of his visit. And neither did Balak appear to write again. Both Tabin and the Ob did, however, feature prominently on the maps which Mercator was preparing for his 'new geography'.

The flurry of correspondence between the English and Mercator had run its course by 1581. It had been a fairly unsatisfactory exchange on both sides. The English had failed to receive absolute confirmation of Mercator's sources for his Arctic geography, while Mercator believed that the English were withholding information about their discoveries. But, for a few years, the vicarious voyager had come closer to the cutting edge of exploration than at any time in his life. As his 'new geography' and his life approached their respective end-times, Mercator and the English had appeared on the brink of settling the greatest exploratory and cartographic riddle confronting Europeans. In the absence of a fleet from the maritime Belgii, the English had tried and tried again to reach Cathay. Their efforts would be remembered by Mercator when the great cosmography was finally completed.

28

The New Geography

By the time Mercator had released himself from the 'vexations' of correcting Ptolemy (and from the English), work on the cosmography's modern maps had been in intermittent progress for over ten years.

To complete the cartographic content of the cosmography, Mercator had to map the entire world according to contemporary geographical knowledge. These were the maps that Mercator now referred to as his 'new geography'.[1]

In the three decades that had passed since he had worked on the geography of Europe for his wall-map, Mercator's concept of 'geography', and of the purpose of a map, had modified.

Geography was no longer simply the plotting of rivers and coasts, mountains and towns; geography had a new, post-Ptolemaic application:

> The use and utility of geography in reading and remembering histories are too apt and well known for them to need any proof or recommendation from me. But geography has another and much more eminent dignity (if it is rightly directed where it is valid), namely, in that it will contribute greatly to the knowledge of political regimes, providing that it describe not only the position of various places, but also their nature or legitimate condition, which the duty of the geographer always demands. For just as a painter who limns a man according to the proportions of his members, but, regardless of colours and physiognomical signs, does not investigate his nature and emotions, fails to satisfy the requirements of his profession, so too a geographer will fashion, so to speak, a dead geographic corpse, by simply placing locations according to their distances, without indicating their mutual political relationships.[2]

This was Mercator's geographical manifesto. It would be geography's 'eminent dignity' to contribute to 'the knowledge of political regimes'.

Geography retained its role as the arbiter of location, but now it also had to describe the 'nature', the 'legitimate condition', the 'emotions' of place. Not to do so would be to fall before – as Mercator put it – a geographic stiff.

These political relationships were far more than another tier of cartographic data; they redefined the purpose – and established the power – of maps. Mercator intended to extend the role of the map as a sublunary tool for describing the miracle of creation. Maps could also be used for plotting the weighting of human forces; maps could represent political (and religious) regions as effectively as they could display the mountains and rivers which separated contiguous landscapes. Later, Mercator would insist that it had not been his 'plan to pursue political rather than geographical studies. Rather,' he wrote, he sought only to 'disclose how the two branches of study, those of geography and of political administrations, can illuminate each other'.[3] The aim, as ever, was harmony.

But Mercator cannot have been immune to the implications of his 'new geography', for there was a fundamental and far-reaching difference between a topographical border and a political border: where mountains and rivers were cartographically unequivocal, the boundaries between human kinds were open to innumerable permutations – and thus to cartographic manipulation.

Only a man who had grown up in borderlands could have reached this moment of affirmation. Gangelt, Rupelmonde, 's-Hertogenbosch, Duisburg were each a stone's throw from another political entity. (And all, of course, also occupied sites on old-fashioned topographical borders: rivers.) Mercator, furthermore, had grown up in a part of the world which had as many political borders as it had rivers. Depending upon which border he observed, he could be 'German', 'Gallic' or 'Belgic'; he could be a subject of the Holy Roman Empire, of the Habsburgs, of Spain; of his duke. As the geographer Sebastian Münster had observed in his *Cosmography* of 1552, regional boundaries were no longer defined by rock and water, but by 'languages and lordships'.[4]

Geography's new purpose was immensely empowering. To Mercator's century of humanists, ink was definitive. Political distinctions could be created with maps; with Mercator's own burin. 'I have considered it just', he wrote, 'to make this geographical work ... as useful as possible to the commonwealth.'[5]

In practice, the depiction of political geography was an issue of selectivity: the cartographer chose which category of political border to mark (usually with pecks of the burin point); the colourist chose the complexion.

Fig. 22. The map-making process: surveyors at work in the countryside (above) and a cartographer at his table (below). From Paul Pfinzing's *Methodus Geometrica*, 1596. (Fotomas Index)

But the new geography also required complementary text which would explain the various criteria used to define each area: 'For this reason', explained Mercator, 'I have thought it of the highest importance to preface my maps in each region with a discussion, suitably distributed, of the nature and order of the forms of governance of their dependent locations, so that our work might contribute something to students of political division and the forms of commonwealths.'[6]

Mercator's 'new geography' went beyond being another book of maps. Laudable his predecessors may have been, but Ortelius' and de Jode's books consisted of maps copied from the work of various other cartographers. Standards varied and individual maps could not generally be matched up to their neighbour.

Mercator's new geography replaced piecemeal cartography with a reformed, universal standard. Researched and designed by one man, Mercator's maps had editorial consistency. They conformed to a uniform style intended to optimize accuracy, versatility and spatial clarity. (Superfluous legends and gratuitous monsters were not, for example, representative of the new geography.) The new geography's single greatest technical innovation was the creation of cartographic overlaps between sheets: 'Thus all places that lie closest to the margins of each map are repeated in the neighbouring maps, and in every case they have the same longitude, latitude and distances, so that the passages and journeys from one map to the next are equally recognizable and perceptible as if there were one continuous map containing both descriptions.'[7]

The one difference, Mercator pointed out, was 'that often the maps are not compared to the same magnitude of celestial degree, so that the space of one mile ... is larger in one than the other'.[8] In other words, the overlaps on adjacent maps were to be identical in all respects except scale.

With the continuous map, politician and scholar could roam the globe from home, crossing from sheet to sheet with constant reference to latitude and longitude – and distance.

The scale of each map would reflect the available geographical knowledge about each continent. So Africa, Asia, America and the mysterious southern continent of 'Australis' (or 'Magellanica')[9] would require fewer regional maps than Europe.

As if that were not ambitious enough, Mercator intended that certain maps should be designed so that they could be assembled into regional wall-maps. This complicated the overall planning considerably, because these 'dual-function'[10] maps had to have identical scales, and matching frames. The dual function maps also required that the non-geographical information such as the distance scale, privilege, and title cartouche were so placed that they were neither repeated nor obscured when overlapped and pasted to their neighbour. As a result, some of these dual-function maps carried no title cartouche.

In introducing the notion that adjacent cartographic regions could, and should, harmonize, Mercator would illustrate that the entire planet could be mapped seamlessly at Ptolemy's chorographic/regional scale; Mercator's modern maps would wrap the three-dimensional spherical globe in two dimensional, planar maps. In doing so, he would give every citizen access to a mathematical representation of their own panel of God's Creation.

Such a complex, ambitious cartographic enterprise had been an enormous undertaking, and one that had benefitted from a gradual evolution over twenty or so years. In 1578, when Mercator had sent a copy of his Ptolemaic maps to the marshal of Jülich, Werner von Gymnich, he had estimated that the '*nieuwe geographie*'[11] would demand at least one hundred maps (ten more than de Jode had issued with his recent *Speculum*).

The letter to Gymnich had also carried a veiled warning that the court cosmographer's plans were proving a little *too* ambitious. (Mercator admitted that he was having difficulty researching the maps, and asked Gymnich to let him know if his impending trip to Italy produced any fruitful information.) Having devised a vast, rigorous and imaginative cartographic sub-section for the intended cosmography, Mercator was finding that the task was somewhat harder than anticipated.

Gymnich learned that the new geography would not be issued as a single vast compendium, but episodically. On his low margins of profit, Mercator had little option but to break the new geography into stand-alone commercial fascicles. The twenty-eight Ptolemaic maps had consumed around a decade of his time. The hundred or so maps for the 'new geography' would be drawn from sources that were even more dispersed and contradictory than those of the Ptolemaic canon. To postpone publication until all 100 maps were complete was financially – and perhaps humanly – impossible.

The new geography would also be released in a sequence other than that originally intended. Instead of the northern regions appearing first, Mercator intended to break with Ptolemaic tradition and begin by publishing '*die nederlanden mit Vranckrijck unde duytslant*'.[12] The reason, he had told Gymnich, was that he had better descriptions of the Low Countries, France and Germania.

*   *   *

After decades of close, focused work in variable light, Mercator was finding that he was 'unable to distinguish letters in broad daylight'.[13] His own diagnosis of the condition was less that it was due to the normal declining of sharpness in vision which comes with age, but that 'thick and viscous humours' had 'enveloped the optic nerve'.[14]

The duke's personal physician, Reiner Solenander, prescribed the most reliable remedy known: the hill flower euphrasy, or 'eyebright'. Mercator was instructed to dry it and then to drink it as a solution in 'good Rhineland wine', the doses to be taken 'in the morning, before lunch, at midday, before and during dinner'.[15] The treatment, said Solenander, should be followed 'for a number of months, until there is some sign of an improvement'.[16]

After several months, Mercator did indeed notice an improvement: 'Thanks to this herb, the humours have been wonderfully dissolved, diluted and washed out of my head. However, as it is the sharpness of vision which is weakening, I think it more useful to chew fennel seeds, without neglecting to keep up a moderate dose of euphrasy'.[17]

With his weakening eyes, Mercator was also short – as usual – of skilled engravers. One of the best he knew was his eldest son, who had engraved some of the Ptolemy maps. But Arnold, rued Mercator, was 'too taken up by the affairs of our illustrious Prince'[18] to whom 'he was bound by stipend'.[19] Rumold was still with the Birckmanns. Frans Hogenberg had been able to work with Mercator on the new engravings for a while, but had produced only 'a handful of maps'[20] before being called back to his own work. Mercator's only regular assistant was Arnold's eldest son, Johann, who was just into his twenties.

By the beginning of 1583, Mercator was able to inform Ludgerus Heresbachius – a humanist friend he used to visit in Cologne – that he had 'almost finished' six of the maps of Germania and that he hoped to have finished the maps of Gaul 'by the end of the year'.[21] Heresbachius had been unable to hide from Mercator his disappointment that the maps of Germania would not be the first to appear. In phrases which suggested that Mercator felt unable to reveal to Heresbachius his reasons for opening the new work with Gaul, he wrote that he too had 'often desired' Germania to take the first place, and acknowledged the 'immense pleasure' such a decision would have provided 'a man of culture and a good friend'.[22] Mercator explained that he had been obliged to concentrate on the Gallic maps 'because of more pressing reasons'[23] – reasons which he felt unable to illuminate. The Germania maps, reassured Mercator, would eventually number 'about twenty in total, many of them previously unpublished, which I know will give you great pleasure'.[24]

Only when these were finished, would Mercator turn to 'Italy, Spain, England and other countries'.[25]

But progress on the cosmography was still being hampered by Mercator's commitments to make globes. Before the spring fair of 1583, he reported that he had sent all the globes he possessed to Frankfurt, and that he had undertaken to construct more, reckoning that it would be three months before they would be ready. He planned to sell them for 10 thalers each.[26]

When Mercator's son Arnold went with an old friend of Ortelius' – Arnold Mylius – to the Frankfurt fair that spring, one of the items on their agenda must have been the new edition of Mercator's Ptolemy. Five years after publication of the Ptolemaic maps, Mylius had revised Pirckheimer's Latin text of the *Geography*, and Mercator's edition now appeared in its complete form, with a new title page and title: 'The eight books of the Geography of Cl. Ptolemy, carefully corrected and newly published, together with maps produced and corrected according to the author's intentions by Gerardum Mercatorem'. Mylius' dedication to Ortelius was dated 1 July 1583, but the volume was not ready for publication until the following year.

Back in Duisburg, Mercator's travels were limited to the roving of failing eyes across his many maps: 'I often think of Cologne,' he wrote to Heresbachius, 'where I had the great pleasure of meeting the humanist that you are, as well as other friends, and of exploring the libraries.'[27] But Cologne was beyond Mercator's range: 'Alas, my work keeps me away, as well as the infirmity of my years and the dangers of war.'[28]

War, once again, had come to Mercator's door. A few months before Mercator wrote to Heresbachius, Cologne had converted to Calvinism. It had happened after Calvinists convinced the wayward Archbishop of Cologne, Gebhard Truchsess von Waldburg (an old adversary of Heresbachius), that he could take a wife *and* retain his see if he relinquished Catholicism. With a Calvinist archbishop controlling Cologne, the Catholic status of the entire lower Rhine was suddenly threatened. Although the Pope had deposed Truchsess by 1583, the ex-archbishop took up arms and civil war erupted. Duisburg found itself in a war-zone rampaged by Bavarian and Spanish troops sent in to rout the recalcitrant archbishop.

But Cologne was a sideshow; the main event was being played out in the Low Countries, where the Spanish had embarked upon a ruthlessly efficient reconquest under the astute military tactician Alexander Farnese, the Prince of Parma. Having gained a foothold in Hainaut, Parma had been methodically working his way north: in the summer of

1583, Dunkirk was retaken, and Diksmuide and Bergues. In October, he turned to the Schelde towns and recaptured Eeklo, Hulst and Axel. Then he moved inland, to Rupelmonde. The riverport of Mercator's boyhood – and imprisonment – was devastated.

The Spanish reconquest drastically changed the political landscape of the Low Countries. Aalst fell, and Ieper surrendered, and Bruges, then Dendermonde and Ghent. As Parma pushed into Brabant, there seemed little that would stop the apparently invincible prince from reclaiming the Low Countries for his Spanish, Catholic king. In the geographical unity of his strategy, Parma was creating a coherent political space on the map; the Flemish humanist and classical scholar Justus Lipsius would soon refer to Parma as '*conditor Belgii*', the founder of Belgium.[29]

In September 1584, Vilvoorde was captured and Parma turned at last to the greatest prize of all: Antwerp. As the Spanish military machine began to strangle the city, the Antwerp printer Christophe Plantin rushed out a book extolling the virtues of 'Belgian Gaul'. The book was dedicated to Gerard Mercator.

Itinerarium per nonnullas Galliæ Belgicæ partes (Itinerary through some parts of Belgian Gaul) bore the date October, 1584 and it described the journey which Ortelius, Vivianus, Scholiers and Schille had taken nine years earlier, whilst Mercator had been struggling to complete his Ptolemaic maps. This was the second occasion that Mercator had become the dedicatee of a printed work.

A slim, octavo volume of seventy-nine pages, the work had been written by Ortelius and Vivianus, and it took the form of a letter addressed to Mercator, the traveller who had been with them in spirit. In a preamble to their description of the journey, the authors explained in self-deprecatory tones that they had set off with no expectation of recording anything about 'our Gaul' that was not already 'made known everywhere in so many histories from all ages'.[30] More, 'it seemed better to make a note of some sort or other than to be idle on the whole journey'.[31] Referring obliquely to Mercator's forthcoming maps of Gaul, the authors talked down their 'little commentary' as being 'useful for ourselves' and 'not without pleasure for you too, who, when you scrutinize those regions carefully at some point, will be aided by the opportunity to reread them as often as you like at your own home and without the bother of a journey'.[32] Ortelius and Vivianus also took care to note the ambiguous disposition of the region they had travelled through, referring to the inhabitants as being 'peoples who, although they were of German origin, had seats however in Gaul and were counted among the Belgae'.[33]

Pared of the travellers' escapades popularized by the letters of Erasmus, the *Itinerarium* revealed nothing of the group's modes of transport, accommodation or trials. Humanists in towns along the route were met and introduced to the narrative, archaeological sites described, inscriptions transcribed. At various points, the text was supplemented by illustrations: a bas-relief at Sarpainge; aqueduct arches at Jouy; the ruins of the Baths of St Barbara at Trier. Uncluttered by anecdote, the journey had been reduced – like Mercator's modern maps – to proportional essentials.

The journey ended at Frankfurt's book fair with a concluding reminder that the region of Galliæ Belgicæ 'has always been, as it is, the most celebrated of all'.[34] Although the words were unwritten, the request was implicit: a map of Galliæ Belgicæ in Mercator's forthcoming cosmography would do no harm to the cause. In the last sentence of the *Itinerarium*, the authors delicately placed their suggestion before the cartographer: 'If in some way this will be approved by you, Mercator, it will be of great reward for us to have achieved what we wanted.'[35]

Dedicated to the greatest cartographer of Belgian Gaul, *Itinerarium per nonnullas Galliæ Belgicæ partes* appeared in print as the lands of the Belgii were being tumultuously redefined, and as Mercator prepared the press for his 'new geography' . . .

Mercator was finally ready by the summer of 1585; the first instalment of the new geography would be presented to the book trade at the autumn fair in Frankfurt.

With various supporting pages of text (a two-page address to the 'Studious and Benevolent Reader', four pages 'On the Political Status of the Kingdom of France', one page titled 'Advice on Using the Maps', an 'Index of the Maps of Gaul', the dedicatory letter to Duke Johann Wilhelm and an index to the names appearing on the maps), there would be fifty-one maps. For Mercator this was a staggering achievement; those fifty-one maps more than doubled his life's entire cartographic output.

In the dedicatory letter to the duke, Mercator explained that the great, long-awaited cosmography could not be released in its entirety, yet: 'As I began a great and laborious work', wrote Mercator, 'what usually happens to an honest citizen of modest fortune when he plans to build a comfortable and grand dwelling for himself happened to me. Namely, when rising costs or the difficulty in building the edifice as planned prolong the work over several years, he takes care first to build what is immediately needed for maintaining his household'.[36] The kitchen and food store, continued Mercator, had to come first; the pleasure gardens and orchards later, as 'opportunity and convenience arise'.[37]

For the benefit of his duke, Mercator outlined the grand plan for the cosmography, a plan which had undergone modification since it had appeared in the preface to the *Chronologia* seventeen years earlier. There were now six, rather than five parts: the first part would now 'treat of the fabric of the world and the general disposition of its parts'.[38] The second would treat 'the order and motion of the celestial bodies', and the third, 'their nature, radiation and confluence in their workings'.[39] Fourth would come 'the elements'; fifth, the 'kingdoms and a description of the whole earth'; sixth, 'the genealogies of princes from the creation of the world, in order to investigate the migrations of peoples, the first inhabitants of the earth, and the times of discoveries and antiquities'.[40] The new section on the behaviour of celestial bodies had been included 'in order to inquire more truly into astrology',[41] but otherwise the content was similar, though slightly rearranged.

The six-part sequence was, continued Mercator, 'the natural order of things, which easily demonstrates their causes and origins and is the best guide to true knowledge and wisdom'.[42]

As deftly as he could, Mercator described the difficulties which were delaying his 'best guide'. The work, wrote Mercator,

weighs on my shoulders alone, and I am able to take advantage of the assistance of no others (except those who engrave the plates) in completing it (as a person building does, who entrusts everything to the hands of masons, carpenters and others), and, moreover, since there were not enough engravers to finish the work as planned in a few years, I was reduced to the necessity of beginning from the middle of the work to be built. Nor would it have been just to deprive students of the use of that part that I have now finished, even if I could have suppressed everything until the completion of the whole.[43]

The anguish could be read between the lines. This was not the best way to build an edifice, but, under the circumstances, it was all he could do. The priority was to release 'those maps of the *New Geography*' that were already engraved and which 'might best serve the commonwealth at this time, namely, those of Gaul and Germania'.[44]

But he had at least made a start. Some forty-seven years had passed since Mercator had first announced in print that he intended to map the world by regions. Map-buyers had waited a lifetime for this moment.

* * *

The maps looked dull.

How the heads of the Frankfurt dealers must have shaken. Where were the crests and legends and compasses ... the beetling crags and cataracts? Where were the fleets of caravels? (True, there were a few ships, but they'd been dropped in as solitary tokens rather than as emblems of delight.) Where, in the Lord's name, were the *fish*? And as for the frames ... they were utterly unadorned; merely ruled and repetitively marked off in degrees of latitude and longitude. The map titles were minimally decorated, and some of the maps had no titles at all. None of the maps began to look like a *picture* ... Against the world-wide, exotically decorated, fashionably packaged *Theatrum* and *Speculum*, Mercator's unimaginatively titled *Geographic Maps of Gaul* looked like a publishing turnip.

Few of the dealers on Buchgasse would have known what Mercator was doing. Austere and incomplete they may have looked but these fifty-one maps were not another set of regional 'pictures'; they were the first instalment of a multi-function, universal scheme of cartography. Mercator's maps overlapped to create unbroken coverage; Mercator's maps came in a variety of scales which allowed overviews of entire monarchies, and close-up examinations of selected regions; Mercator's maps could be sold individually, or they could be bound as thematic sets into map-books. (Separate title pages for Gaul and Germania had been printed, in German, English and French.) Many of Mercator's maps (those were the ones without titles) could be trimmed and pasted to a neighbour, to create a wall-map. Among the fifty-one maps were no fewer than seven potential wall-maps: of Lorraine, Burgundy (a present for Perronet perhaps), Alsace, Lower Saxony and Brunswick, Hesse and Thuringia, Westphalia (a third sheet – of Berghe, Marck and Cologne – could be pasted to the bottom of Westphalia). The largest potential wall-map was a magnificent four-sheet spread of Switzerland. Except for one or two minor irregularities, the match between wall-map sheets was perfect; pasted together and coloured, the joins would be undetectable.

Mercator's maps were also accurate. Where the location of a place was called for in the text, he not only provided the degrees, but the nearest minute of latitude and longitude. Places listed in the index also carried their location to the nearest minute.

Mercator's maps were practical and accessible, their utilitarian appearance an aid to enlightenment and education – which was also the guiding principle behind the supporting text. In the section titled 'Advice on Using the Maps', for example, students and non-geographers were given a brisk tutorial on map grids: 'You will find the degrees of latitude and longitude designated at the sides of the maps; most often the degrees of latitude are at each side, and those of longitude at the top and bottom, whenever, that is, it is possible to place north at the top of the map.'[45] The exceptions were those occasional maps where 'the region to be described extends farther from north to south than east to west'.[46] In such cases, west would be at the top. And there was a mathematical reminder for absolute beginners: 'Each degree is divided into sixty parts, called minutes.'[47]

That Mercator intended to wrap the entire globe in these precise, overlapping maps seemed even more incredible now that the first examples had been revealed in print. Completing Europe would be an enormous undertaking. And after that, he could look forward to 'Africa, Asia, America, and, if it is discovered, which is our hope . . . the Magellanic or southern (*Australis*) land'.[48]

By 'beginning from the middle' of Europe, Mercator had mapped for his readers a vast tract from the Atlantic coast of France eastwards to 'Polonia' and 'Hungaria'. The entire span of this tract was described on two overview maps, of 'Gallia' and of 'Germania', each with its own title page. Mercator's definition of 'Gallia' accorded roughly with Roman Gaul, a four-cornered block of territory bounded by the Atlantic, the Pyrenees, the Mediterranean, the Alps and the Rhine; his 'Germania' was an agglomeration which extended beyond the borders of the Holy Roman Empire to include parts of Roman Raetia, Noricum and Pannonia.

Of Germania and Gaul, it was the latter – as the symbol of a functioning monarchy – which occupied the prestigious opening pages of the new geography. In 'seeking to establish the best political status' for themselves, the French ('the wisest and most warlike people') had recognized 'the desirability of turning by necessity to one person in all affairs as though to a head and origin'.[49] (Mercator did not mention the German Empire by name, but it was clear which political structure he meant when he added that 'the commonwealth' would suffer 'inconvenience and danger when several people simultaneously have command'.)[50]

France had never been systematically mapped at the regional level. In three overlapping sheets, Mercator covered the whole monarchy, with a further four larger-scale sheets of specific localities such as Bologne and Guines, and the duchy of Berry on the upper Loire. At a time when France was being torn apart by religious conflict, Mercator presented the country as a unified, coherent whole, comprised of regions with distinct geographical identities.

But Gaul and Germania were not alone. Mercator had prised from Gaul and the Holy Roman Empire a third entity, a region whose political presence had been subtly endorsed by the tip of Mercator's burin.[51] Its territory appeared on the western edge of the 'Germania' map, and on the eastern edge of the 'Gallia' map, but was identified on neither. It was however identified with its own title page, and with its own overview map.

In his description of the region's political geography, Mercator recalled that Julius Caesar had divided Gaul into Celtica, Aquitania and Belgica. 'As it happens', he continued, 'the king of Spain has possessed one half of Belgica for some centuries now', while the other half, 'namely, Picardy, Champagne, Normandy (although these last two are not entirely included in Belgium), and the remaining part of Belgium belong to the dukes of Lorraine and of Jülich and Cleve, the archbishops of Trier, Mainz, and Cologne, the bishop of Lyon, and others'.[52]

From 'that part of Gaul that the king of Spain possesses',[53] Mercator had created an entity called 'Belgii inferioris', Lower Belgium, a region which included Flanders, Brabant, Holland, Zeeland, Gelderland, Artois, Hainaut, Trier and Luxembourg. No less than nine of the fifty-one new maps described Belgii inferioris.

In the accompanying text, Belgica emerged as the most noble part of Gaul, a territory which 'always produced more notable and braver soldiers than the rest of Gaul'.[54] And although 'all of Belgica [was] wonderfully famous ... that part that has always obeyed the Catholic king (i.e., of Spain) is by far the most noble'.[55] Lower Belgium's fame came from it being 'the origin and native land of so many monarchs, kings, princes, and dukes; and then because of the populous and exceedingly wealthy cities in it, its infinite towns, and innumerable territories, as well as the marvellous multitude of its inhabitants, their riches, civility, and strength of spirit'.[56]

Neither Gaul nor Germania was described in such terms.

The implication was clear: Lower Belgium was not a political appendage to Gaul and Germania. 'They say', added Mercator (coming as close as he could to inciting cartographic secession, 'even the emperor Charles v was so moved by these [qualities] that he often deliberated whether to raise these provinces into a kingdom'.[57] Only the 'diversity of privileges, manners and laws' and the 'difficulties of the continuous wars'[58] had diverted the emperor from his plan.

It was no coincidence that the book which Ortelius and Vivianus had dedicated to Mercator the previous year had described a journey through 'Belgian Gaul'. The Roman antiquities, Mandeville's tomb, the shrine of

St Nicholas, the encounters with learned humanists were ordered parts of an itinerary whose narrative structure had been designed to illustrate the historicity and cultural coherence of Belgian Gaul.

The hand of a political cartographer could also be seen at work at a more local scale.

Among the fifty-one maps there was just one which had no title, and yet was not part of a 'dual-function' wall-map either. An anomaly within an otherwise systematic schema, it had been mapped at a different scale to its neighbours and carried no 'editorial' information beyond the normal '*Per Gerardum Mercatorem Cum Privilegio*'. The area it described was neither a single political entity nor a geographical region (although it could be said to display some level of riverine symmetry). No other map was so completely covered with place-names. There was no title because there was no 'dead space' in which to place one. Even the distance scale had been reduced to a tiny four-mile bar.

On its reverse side, the map was identified as 'Brabantia, Gulick et Cleve'. Having presented Lower Belgium as the noblest, most mercurial region of northern Europe, Mercator had cut a rectangle containing Brabant and Duke Wilhelm's duchies from the heart of the region and proffered it as the essence of Lower Belgium. Within this rectangle were the region's geographical organs and arteries: the rivers Schelde, Maas and Rhine; and Antwerp, Mechelen, Brussels, Louvain, Cologne, Liège, Aachen – the centres of commerce, of printing, of cartography, of learning, of the Church. Charles v had been crowned on this map, and Sir John Mandeville entombed.

'Brabantia, Gulick et Cleve' was also the map of Mercator's life, an autobiographical basket of the places he had known as home. On the right of the map was Gangelt, the town of his conception and boyhood, and on the far left was a sliver of Flanders, with Rupelmonde. Up at the top was the town of his schooldays, 's-Hertogenbosch, and at the foot, his university town, Louvain. Over on the far right was Duisburg. And in the centre of the map was the heart-shaped, unpopulated void of the Kempen, surrounded by the veins and capillaries of the Belgic waterways.

Mercator had spent virtually his entire life on this one map, its geography was steeped with meanings that were uninterpretable to anyone but himself. True, the map was too crowded to take a title, but did Mercator perhaps experience a moment of retributive pleasure when he denominalized the duchy which had persecuted him and so many of his friends? And how just to partner misguided Brabant with the duchy which had received her refugees in a spirit of Erasmian moderation.

And what was passing through Mercator's mind when he snapped Cologne's church tower and left it hanging over the city like the sword of Damocles? A minute twist of the burin, it was a detail so small that it would be missed by anyone without a magnifying glass or a specific interest in Cologne. One of Cologne's churches had indeed got a crooked tower, a tower which had appeared in print forty years earlier on Münster's view of the city in his *Cosmographia*. But, given the absence of representational symbolism in the new geography, what had made Mercator break his own rule? Cologne appeared on three of these new maps, and in each case, he had snapped the tower. Could thoughts of retribution have been on his mind again, for it was Cologne which had recently gone to the Calvinists with such devastating consequences for the lower Rhine. Apt perhaps that the city which had broken with the Church should be represented by a broken Christian symbol.

Curiously, it was this map, anomalous 'Brabantia, Gulick et Cleve', which was the only one of the fifty-one maps to show evidence of having been subject to a major correction, and it was very likely to have been one of the 'two folios' which Mercator had 'ordered' to be 're-engraved correctly' after 'some errors were found . . . which it seemed could hardly be borne'.[59] It was 'on the advice of several' that Mercator found it necessary to correct the plates and reprint the two maps before he could offer them 'as ready gifts'.[60] Since the corrections required knocking out the affected parts of the copper plates from behind, repolishing them and then re-engraving them with updated geography, the effort required was considerable. In spite of the pressure to complete his cosmography, Mercator was unprepared to reduce his standards. Dispatch of the gifts was delayed for several months, in one case until December. The correction to 'Brabantia, Gulick et Cleve' involved an extensive area at the foot of the plate. All the way from Namur to Liège the course of the Maas had been modified, and the location of around forty towns and villages had been altered. Since the same area occurred on two other regional maps, both of which reflected the updated information, it can be assumed that 'Brabantia, Gulick et Cleve' was one of the earlier maps Mercator completed; perhaps even a prototype for the entire 'new geography'. The corrected area was on the route taken by Ortelius and Vivianus in 1575 during their tour of Belgian Gaul, and it is not beyond the bounds of possibility that it was Ortelius – or Jan Van Schille, the surveyor in the group – who volunteered the corrected information.

Mercator had invested this densely detailed, painstakingly corrected map with an intensity which did not recur again in his 'new geography'. Self-referential, denominalized, spatially selective and encoded with its bent Rhenish spire, the cartographic trinity of Brabant, Jülich and Cleves could be read as a personal statement and as an experiment in political mapmaking.

29

Apocalypse

Less than half of the modern maps required for the 'new geography' were in print, and as 1588 approached, Mercator became increasingly preoccupied with the inevitable apocalypse.

One of those who had recently taken an interest in Mercator's knowledge of chronology and astronomy was Heinrich Rantzau, the hugely wealthy viceroy of Holstein, a duchy on the floating margin between the Empire and Denmark. A humanist whose interests spanned astrology, history and cartography, Rantzau was well known as a patron of the arts, supporting among others the Danish mapmaker Mark Jorden. Rantzau was not averse to seeing his name in print, and he had appeared no less than six times on the map of Denmark in Braun and Hogenberg's *Civitates Orbis Terrarum*, while the map's author – the hapless Jorden – had been omitted. At the time, Denmark was enjoying the benefits of being a major European power: she was well governed by Frederick II and had become the focus of a circle of noblemen and amateur scientists who had gathered around the controversial astronomer Tycho Brahe on his island observatory of Uraniborg.

Distinguished by a copper nose which he had fashioned himself after losing his real one in a duel, Brahe had shared Mercator's rejection of Copernican theory, favouring instead a modification of the Ptolemaic system. As one of those under Brahe's spell, Rantzau had included the astronomer in his catalogue of 'Emperors, Kings and Princes who have loved, honoured and practised the Astrological Art'.[1] Gratifyingly for the viceroy, he became one of a select group who received personally dedicated elegies written by Brahe, at least one of which was soon circulating at Frankfurt for all to note.

Mercator would have known of Rantzau through Georg Braun in Cologne, and through Rantzau's *Catalogus imperatorum*, which had been published by Plantin in 1580. Rantzau was a potentially valuable contributor to the 'new geography'; under Frederick II, Denmark had taken control of all of the seas that washed Scandinavian coasts, and Rantzau himself was well placed in the administration to furnish Mercator with maps and descriptions of Denmark's political structures. In return, Rantzau expected recognition in the pages of Mercator's 'new geography'. He also wanted Mercator's views on the imminent end of the world.

'In regard to the Danish maps', Mercator wrote to Rantzau, 'I hope that the glory of your name will be celebrated from help of this kind'.[2] Observing that no one would be 'better able' to describe the 'order of all nobility and the political structure' than the viceroy, Mercator added that if Rantzau could also add 'a fuller description of all its regions, then easily Denmark would hold the principal place and glory among all the records of Europe'.[3]

Through 1585 and 1586, Rantzau and Mercator exchanged letters and information.

From Rantzau, Mercator received a list of the fortresses and royal towns of Fünen, two maps of Sweden, 'various outlines of Dietmarsia',[4] and astronomical news from Uraniborg – including a description of the comet that had been observed in 1585, which, Mercator wrote with evident pleasure, would 'offer much for my speculations'.[5] Mercator was particularly pleased to hear from Rantzau that Brahe was no Copernican: 'I am delighted', enthused Mercator, 'at the exquisite observations of Tycho Brahe concerning the movement of the sun.'[6] (When he came to write the geographical description for Denmark, Mercator praised Uraniborg's 'school of astronomy and meteorology' for its efforts to 'emend that divine art with wonderful and unheard-of observations'.[7])

From Mercator, Rantzau received a long and detailed exposition on 'the motion and orbits of the sun', and the 'substance and the function for which it was established'.[8] Neither the motion of the sun nor that of the planets could be held responsible for the forthcoming apocalypse: 'The planets do not fail in nature and strength, but remain as they were when established, ordained for the procreation and sustenance of the species in this lower world.'[9] Any failure, continued Mercator, would arise 'from a defect of substance and a harmful mixture of the elements, of which the main and almost the only cause is sin'.[10] And sin, reminded Mercator, 'comes not from the planets nor from any inclination of nature created by God, but only from the free will of man'.[11] And that free will had been misdirected: 'There should not have been abuse cast at heaven … for the substance from which we are born and which we live by God, having been abused, brings with it all the ills which accompany the species.'[12]

Since heaven had been 'created for the procreation of the species, their strengthening and conservation', they would be 'kept intact, as long as the world is preserved'.[13] And the world would not be preserved for much longer: 'Its death, about to happen at the end of time, will not happen from any natural weakness or gradually perishing nature but from the will of God alone, cutting off its course and dissolving its composition.'[14] Terrestrial decrepitude was self-inflicted: 'The old age therefore of the world is gathered not from the failing nature of the bodies above, but from the retreat from the first condition of Creation, which happens not from deviation of nature created but from the free will of man, and is strained through the growing abuse of creatures and slander of elements right up to the fall and end of things.'[15]

One year later, Mercator could write that he was 'delighted'[16] by the political information that he had received from Rantzau. By May 1586, the general map of Denmark was complete, and the regional maps of Fünen and Holstein. His energy, he wrote to the viceroy, was concentrated 'now on finishing the entire description of Denmark'.[17] All he lacked was a better description of the island of 'Zeland'.[18] Mercator assured Rantzau that various manuscripts he had lent would soon be returned.

But Mercator's time – and all sublunary time – was nearly run. Exchanging letters in the spring of 1586, the viceroy and the cartographer compared their timings for the 'prophetic year'.[19]

In his *Catalogus imperatorum*, Rantzau had included a treatise on 'climacteric' years, the critical points that recurred every seven years in a human life cycle. Various illustrious men had perished in climacteric years, especially in their forty-ninth, fifty-sixth and sixty-third. From Duisburg, Rantzau received a summary of Mercator's long-held view that the end was due in 1588. Referring the viceroy to Sleidan's history of religion, Mercator explained that by adding the seventy years of Babylonian captivity to 'the date the doctrine of Luther and Zwingli was established', one came up with 'the beginning of the year '88'.[20]

'Farewell,' wrote Mercator. 'May our Lord bring you through the approaching year of climacterics felicitously and with strong health.'[21]

Catastrophe came sooner than Mercator had expected. Just four months after warning Rantzau to expect extraordinary events, Barbara died. It was a summer's day in Duisburg: 24 August 1586.

To Ghim, who knew her well, Barbara had been more than a loyal and 'excellent housewife'.[22] Above all, she had been temperamentally suited to the peculiar exigencies of life as a cosmographer's wife. Wars, plagues and economic disasters were to be expected, but Barbara had been called upon to stand beside a husband who had been shamed and jailed; a husband whose calling had been obsessive and periodically unremunerative. She had borne six 'obedient and talented'[23] children whose character she could claim to have nurtured. Three of those children, the boys, had followed their father's calling with aptitudes and dedication that reflected intelligence and filial respect. Barbara had been married to Mercator for fifty years and three weeks.

Barbara had died as flames licked at the Rhine.

The Spanish military incursion which had begun four years earlier as a response to Cologne's disastrous switch to Calvinism had crept progressively closer to Duisburg. 'On all sides', observed Mercator, 'we are pressed by war'.[24] To the south of Duisburg, Neuss was 'besieged and captured'; to the north, Berck was 'surrounded by an army'.[25] The road to Wesel was cut. The sending of letters was impossible.

Through the summer of 1586, Rantzau waited for Mercator to reply to his latest letter. He wrote a second letter, and then a third. Despite the fighting, Mercator had received all three letters, the first accompanying a map of Sweden, and the second, a 'golden gift'.[26]

When Mercator eventually responded, in September, he omitted to mention Barbara's passing and his letter – which dealt mainly with the maps for the unfinished 'new geography' – was framed in a new vulnerability. 'Embarrassed' by 'such a long silence',[27] Mercator described the mayhem gripping the lower Rhine:

> There is nothing new to tell you, except that everything is sad where the utmost peace and tranquillity used to be. This is a cruel war, in which nothing is spared; friend, neutral, all are treated with the same fate; everywhere there is hunger and a great lack of bread; and unless the Lord cuts this war short it is feared that enormous numbers will perish from famine, especially in Gallia, where they say that this year has seen a very meagre crop. May the Lord think it worthy to put an end to these ills.[28]

Despite bereavement and the prospect of a Spanish attack, Mercator had continued to make progress with the cosmography. The maps covering Italy and Greece were 'for the most part'[29] marked out, and while these were being engraved, he planned to tackle 'the Sarmatian regions and the northern kingdoms'.[30] Denmark was 'ready for engraving'.[31]

Meanwhile, he was waiting for 'ample descriptions' of Poland and Livonia, which had been deposited for him at Cologne 'by a certain noble Pole'.[32] As soon as he received these maps, he would set about 'correctly measuring'[33] them; only when their coordinates had been corroborated would he be able to commit them to copper.

Then Arnold died from pneumonia.

First Barbara; now Arnold. A double tragedy. The price that Mercator was paying for his own longevity was to witness the passing of his own family. As wife and eldest son, Barbara and Arnold had been Mercator's present and his future, the pillars of his mortal being.

Arnold had been the cartographer-in-waiting, the son most able to continue his father's works. He was, wrote Ghim, 'almost unrivalled in constructing accurate and beautiful mathematical instruments', instruments which had been supplied to 'some of the most important dignitaries in Germany'.[34] He had 'hardly an equal in his skill in geography and surveying'.[35] Arnold's surveys in the archbishopric of Trier and in the county of Katzenellenbogen had won him 'an ample honorarium';[36] salaried to the Duke of Cleves, he'd mapped the jurisdiction of Windeck, and worked in the districts of Sittard and Born, and in Wehrmeisterei near Düren; he'd surveyed Cologne and had recently begun a new survey of Hesse for the Landgrave, Wilhelm IV.

Father and son had played a Ptolemaic duet, the geographer and the chorographer mapping Creation at complementary scales. And there had been money in chorography; Arnold's regional maps had been commissioned to resolve and confirm territorial issues, and they had brought with them fees and honoraria. Locally, Arnold was respected for his plans for houses and for dykes, for his triangulated survey of the neighbourhood between Duisburg and Meiderich; for his work on the giant sundial clock for Salvator kirche. Unlike his father, Arnold had become a burgher of Duisburg, and a member of the town council. He died shortly before his fiftieth birthday, leaving thirteen children.

The premature death of another of his children can only have been distressing in the extreme. That it was Arnold, who came the closest of them all to replicating Mercator's multiple aptitudes, had repercussions beyond personal loss. Two of Mercator's most important engravers were now his grandsons, Johann and Gerard. Three weeks after Arnold died, Mercator wrote to Landgrave Wilhelm IV suggesting that Johann should complete his father's work in Hesse.[37]

Arnold had died on 6 July. By the end of that awful summer of 1587, the fighting was closing in. Mercator prepared the family to leave

Duisburg, packing books, maps, papers. When the town Senate asked him to undertake a genealogy of the dukes of Jülich and Cleves, Mercator responded that 'these dangerous times' had driven him to 'confuse my writings and maps and to put things together for flight to a safer place, if it should be necessary'.[38] His affairs, he wrote, were 'scattered'.[39] Mercator was caught in an apocalypse of his own dating.

With a horrible inevitability the war reached the walls of Duisburg. Mercator must have felt that he was in Louvain again, encircled by indiscriminate soldiery. With terrifying efficiency, the Spanish troops besieged and then overran the fortress of Ruhrort on the riverbank just outside town. But, just as Louvain had been spared at the final hour, so too was Duisburg. The Spanish left the town intact.

So apocalypse year was preceded by – and then proceeded with – predictable calamity.

The year 1588 brought plague to Duisburg and for the first time, the council appointed a physician to tend to the town's wealthier citizens.

In other troubled lands, Philip II's long-promised reconquest of Protestant Europe opened with the 'Day of Barricades' in Paris, during which the conciliatory French King Henry III was driven from the city by the extremist Catholic League under the third Duke of Guise, Henri, and his brother Louis – the sons of the dead warrior Duke of Guise, Francis of Lorraine. Simultaneously, a vast Spanish armada of ships sailed against the English.

Infirm and isolated in Duisburg, doubly bereft Mercator knew that the time had come. Notorious for their ease with Calvinism, the towns on the Rhine would be among the first targets of a rampaging Catholic king once he'd snuffed out the heretical populations of England and the Low Countries.

But Philip's reconquest went wrong: his armada was catastrophically scattered by Frobisher, Drake, Hawkins, and a heretical wind. Around 15,000 soldiers and sailors perished and only sixty or so of the 130 ships returned to Spain. At Christmas, Henry III of France rediscovered his nerve and severed the head of the Catholic League: the Guise brothers, Henri and Louis, were assassinated, and their charred remains dumped in the Loire. Moderates across the continent waited for the French monarch to unite his subjects and drive extremists from his land.

Mercator's old university ally and patron Antoine Perronet, did not live to see Spain's catastrophe. He had died in Madrid in September 1586. Some time earlier, Philip II had offered the cardinal the opportunity to return to the Low Countries under Margaret of Parma. Humiliated by the manner of his departure from Brussels, Perronet had refused the invitation and had spent his last years beset with ill-health, diminishing influence and gloom. Spain, he wrote before his death, was 'heading towards the abyss' and 'final ruin'.[40]

30

Atlas

The post-apocalyptic year of 1589 was also Mercator's eleventh climacteric; in March, he turned seventy-seven.

Revived vigour coursed through the house on Oberstrasse: over a few hectic months, Mercator remarried, Rumold married, and another fascicle of modern maps were printed.

Mercator's new wife was Gertrude Vierlings, the widow of one of Duisburg's earlier mayors, Ambrosius Moer. Rumold married Gertrude's daughter. The twenty-two maps in the new batch covered Italy and the Balkans. Despite the progress that he had made with Scandinavia, the maps of the northern latitudes were not ready.

Like the previous set of maps, *Italiae, Sclavoniae, et Graeciae tabulae geographicae* came with their own engraved title page, a portrait of the author, a dedicatory letter, an address to the reader and an index to the places named on the maps. The decision to publish another partial collection of maps may have appeared premature, but nevertheless the customer could buy a handsome, bound volume accompanied by all the usual bibliographic courtesies.

Mercator's short, straightforward address to the 'Gentle reader' was a reminder that the latest maps were part of a systematically programmed world view: as Ptolemy had prescribed, the sheets progressed (broadly) from the west to the east and from the north to the south 'so that nothing in between is omitted'.[1] The first regional map was thus Lombardy and the last, Crete.

For those who had waited a decade or two already, there was a reassurance too, that 'finally we shall produce all the regions of the world, as infrequently as possible repeating what has once been described, lest the purchaser be burdened with a superfluous multiplication of maps'.[2] Echoing St Paul's 'wise master builder'[3] who laid the foundation that cannot be relaid, Mercator stressed that these maps were part of a definitive exercise 'to describe and complete the whole work once and once only, in as pure a form as possible'.[4]

The maps followed the same format as those of Gallia, Germania and Belgii inferioris, framed with scales of latitude and longitude, and decorated with a single title cartouche. Superfluous artistry had been virtually banished from the plates. In the first fascicle of the 'new geography', a modest total of thirteen ships and four monsters could be counted throughout the fifty-one maps. On the twenty-two maps of Italy and the Balkans, there was just one ship and one monster on the general map of Italy, and a solitary lateen-rigged caravel dashing across the regional map that described Italy's 'foot'.

In banishing the carracks and boggling cod, and in reducing 'all things so diligently to the truth',[5] Mercator was progressively removing artificial decoration and replacing it with the geographical patterns of Creation.

There was only one potential wall-map hidden among the latest maps, but it was the most spectacular to date. When cut and pasted, four sheets formed a geographical image of northern Italy. In the steepled Italian towns, cradled to north, west and south by the curving Alps and Apennines, could be seen the map's silent homage to the home of humanism. Fifty years after engraving his first map, Mercator had found a landscape that could be used to display the allegorical world views of Patinir as Ptolemaic truth. Compiled from the latest data Mercator could find, the map contrived a perfect composition through the placing of its margins, three corners framed by mountains, while Italy's limb departed from the fourth. Only in this particular rectangle of Europe, where the highest mountains met the largest floodplain, was geography so well disposed to the full cartographic lexicon. In one sweep, the eye crossed stippled sea and wrinkled coast to fingered vales and marching peaks. Here was nature explained, the seemingly meaningless meanders of shore and hill slope presented from aloft in their intended patterns. And of those patterns none could match the symmetry – or symbolism – of the great river basin which filled the map's heart. The cobbler's son, who had grown up on the continent's other great floodplain, had engraved the Po's multitudinous tributaries so that the great river and its dendritic branches looked like the tree of life.

This second fascicle of the new geography was dedicated by Mercator in elaborate terms to Prince Ferdinando de' Medici, Cardinal, Grand Duke of Tuscany, and Protector of Spain. To the grand duke, Mercator presented 'Italy, the flower of the earth'.[6]

Renowned for his policy of religious toleration in the republic of

Florence, Ferdinando was an avid collector of classical statues and a generous patron of the arts. And of mapmakers, among them, Mercator. 'I pray that this work,' read the dedication, 'will be sacred and commended to your highness, and offer it. I was forcefully driven to such audacity by your favour and protection'.[7]

In Mercator's dedication to Medici could be discerned the suggestion of the cosmography's ultimate identity; the title which would carry it far beyond the *Theatrum* of Ortelius and *Speculum* of de Jode.

Ahead of the brief reference to Medici's 'favour and protection', Mercator had formulated a more dignified reason for dedicating the maps to a distant Italian prince. This had required the creation of a somewhat tortuous genealogy linking a fount of cosmographical wisdom with Ferdinando: 'As I was worrying about this matter,' wrote Mercator, 'my spirit, avid to celebrate Roman antiquity, drew me to the seat of that most ancient and wise king Janus, Etruria. For just as there, under the aegis of Idaean Hercules and of his son Tuscus and his grandson Janus and Janus' tutor Italian Atlas, the whole glory of Italy arose as though from its earliest cradle.'[8]

Renewing the memory of that noble Tuscan king, Mercator hailed Ferdinando as 'the successor of Janus'.[9]

And Atlas, the tutor?

Atlas the tutor became Mercator's cosmography.[10]

'I have set this man Atlas,' explained Mercator, 'so notable for his erudition, humaneness, and wisdom as a model for my imitation.'[11]

In the preface which he prepared for the cosmography, Mercator outlined the genealogy of the Titan condemned to bear the heavens upon his shoulders. The progenitor of Atlas had been the Phoenician king and astronomer Sol, 'the Sun'. Sol's son Terrenus (also known as Caelus, or Heavenly), had produced no less than forty-five sons, seventeen of them (the Titans) through his sister Titea (also known as Terra, or Earth). One of the noblest of the titanic children had been a boy named Atlas, who had become the King of Mauretania. Atlas was not only 'a very skillful astronomer' but was 'the first among men to discourse of the sphere'. It had been the son of Atlas, also named Atlas, who had become King of Iberia and then – on the death of his wise brother Hesperus – the administrator of Etruria and the tutor of Janus.

So Mercator's model was Atlas minor, the great-grandson of Sun, the grandson of Heaven and Earth and tutor to 'that most ancient and wise'[12] Tuscan King Janus, who had taken *his* name from the god of the beginning of all things. That the Etruscan Janus also had a divine role as the god of the vault of heaven made his involvement in Mercator's intricately constructed genealogy all the more appropriate.

But there was more to Mercator's Atlas than mere idolization. Atlas was available in several guises: to Homer he had been 'malevolent Atlas, who knows the sea in all its depths and with his own shoulders supports the great columns that hold earth and sky apart'.[13] Later, Plato made him Neptune's eldest son, and by the time it was Ovid's turn, he had become a rich north African king who had been turned into a mountain by Perseus. By the first century BC, Atlas had matured into a creature of the sublunary world, an astronomer who had 'discovered the spherical nature [or arrangement] of the stars, and for that reason was generally believed to be bearing the entire firmament upon his shoulders'.[14] This was the Atlas of the historian Diodorus Sicilus, an Atlas whose relationship with the firmament had become figurative. Seeking a synthesis as always, Mercator drew from Diodorus and from Eusebius (who supported the Phoenician bloodline for Atlas) to create an Atlas with divine antecedents and contemporary qualities. Just as mythical Atlas could view the sphere through his divine advantage, Mercator's *Atlas* offered a viewpoint for mankind.

The full title of the cosmography would be *Atlas sive cosmographicæ meditationes de fabrica mundi et fabricati figura*: Atlas or Cosmographic Meditations on The Fabric of the World and The Figure of the Fabrick'd.

Mercator was not at Frankfurt's spring fair for the launch of the Italian and Balkan maps.

In Duisburg, he had turned to the next fascicle of the *Atlas* with renewed urgency. These would be the long-awaited maps of the northern lands. Once printed, they would conclude – with the exception of troublesome Iberia – the coverage of Europe.

By now it was apparent that Mercator had – as usual – under-estimated the scale of a project. The total for the modern maps of Europe already exceeded the hundred sheets that he had told Gymnich the new geography would require, and he had not begun on Africa, Asia, America and 'Magellanica'.

But for the time being, all attention was focused on the northern lands. It had always been Mercator's intention that the cosmography's modern maps should be ordered 'in imitation of Ptolemy, the prince of cosmographers, to begin from the pole itself and the regions that lie about it'.[15] These were the maps which would introduce the reader to the rest of the world; the maps which would set a standard for comparison. And the very first map of all would be the Arctic pole.

The Arctic pole was more than a geographical description. Circular, the map stared from the proof like an eye, its inner circle of polar islands the iris. The circular – and spherical – form symbolized Mercator's cosmography; this was the Ptolemaic universe; the sphere supported by Atlas.

The central disc of the map was derived from the small polar inset on the world map of 1569, the quartet of islands transposed intact, with their description of arctic pygmies and of four ceaselessly flowing 'arms of the sea' drawn towards the great rock at the pole and so to the 'bowels of the Earth'.[16]

But Mercator had extended the radius of coverage by ten degrees of latitude, so that the northern tip of Scotland now peeked from the edge of the map. All of Greenland was shown, and the adjacent coast of North America. On the facing hemisphere, the northern regions of Russia and Asia were in view, and the straits of Anian. The first printed map entirely dedicated to the Arctic lands, this was also the cosmographer's glass; the lens through which the reader would be invited to review the process of discovery. In the fretworked straits and inlets between Greenland and America could be read the endeavours of English mariners: Frobisher's dogged enterprise was remembered by the long passage labelled 'Fretum Forbosshers', by 'Locke's Land', 'Beers S.' and 'Contie Warwiks sound'. And the more recent voyages of John Davis could be read in the place-names on the coast beyond the 'Fretum Davis': 'Mont ralegh', 'Cap. Walsingam', 'L. Lumleys Inlet' and the 'furious over fall' where the pinnace *Ellen* had been driven by a gale of wind into waters that were 'whirling and roring, as it were the meeting of tides'.[17]

Unlike all the previous modern maps for the *Atlas*, this one was unashamedly decorated. The circular form of the map had been set out by Mercator within a foliaceous frame with a roundel in each corner. Three of these roundels were occupied by island groups (the Shetlands, Faeroes and Zeno's 'Frislant') and the fourth by the map's title.

Septentrionalium terrarum descriptio was an emblematic map. In its italicized circularity and symmetry it offered all the pleasures of humanist proportion; in its conjectural lands it recorded cartography's mythological roots and its mathematical future. And in its decoration it welcomed and lifted the reader's heart.

Introduced by the Arctic, the *Atlas* maps which followed would be those of the northern lands: Iceland, and then the British Isles and Scandinavia, Sarmatia and Russia. All were regions whose cartography had been reformed by half a century of intensive mapmaking.

The Iceland engraved by Mercator in 1554 was almost unrecognizable. In 1585, a new map of the island had been dedicated to Frederick II of Denmark by the Danish historian Anders Sørensen Vedel.[18] The engraved map had probably been created by Bishop Gudbrandur Thorláksson. Where Mercator had been able to mark around thirty place-names, the bishop had placed about 250. Now, the Iceland Mercator revised for the *Atlas* was ragged with new peninsulas and inlets, and dramatized by exploding Hekla: 'The mountain itself', elaborated Mercator in the supporting text, 'rages, resounds as though with frightful thunderclaps, and sends forth huge rocks, and vomits sulphur.'[19]

By virtue of its northerly location, the British Isles would occupy the first pages of the *Atlas* after the Arctic and Iceland. Britain had an emblematic role too, for Albion possessed the geography of paradise: 'Nature', wrote Mercator, 'endowed Britain with all the goods of heaven and earth. There, neither the rigours of winter are too great ... nor is the summer's heat.'[20] The gifts, he continued, 'of Ceres and of Bacchus abound: the forests are without savage beasts, and the land is free of poisonous serpents. On the contrary, there is a numberless throng of gentle flocks, swollen with milk and laden with fleece'.[21] In Britain

the days are very long, and no night is without a little light, since the extreme flatness of her shores do not draw down shadows, and the face of heaven and of the stars passes the finish-line of night like the sun itself ... Indeed, Britain is the work of joyous nature; nature seems to have created her like another world outside the world, for the pleasure of the human race, and to have limned her singularly like a shape of utmost beauty and a universal ornament, with such gemlike variety and pleasant painting that the eye of whosoever falls on her is refreshed.[22]

Mercator had, of course, never been to Britain, although he had (apparently) met her inhabitants, 'with their perfect physical appearance, most seemly manners, gentlest spirits, and greatest souls, whose virtue, in deeds at peace and in war is most amply attested throughout the entire world'.[23] Led by the 'slighte youthe' who had called from the north forty years earlier, the English had won Mercator's affections.

As Mercator finalized the northern maps, his son Rumold and his grandson Michael[24] were both in London, presumably tasked with gathering last-minute information. While he was there, Michael struck a silver medal to commemorate the circumnavigation of the globe by Sir Francis Drake. A presentation gift whose purpose can only have been to ease the flow of news concerning English discoveries, the medal depicted

Drake's route in a dotted line and played to English ambitions by showing 'Virginea' and describing the western part of North America as 'New Albion' – neither of which had appeared on Mercator's world map of 1569, or on Rumold's recent map of 1587.[25]

The geographical information flowing back from London to Duisburg was of an order unthinkable a couple of decades earlier.[26] Mercator was able to improve his depiction of inland Scotland from information on Lawrence Nowell's excellent manuscript maps of the British Isles, at least one of which had been constructed with a graticule of latitude and longitude. And by now, Alexander Lyndsay's accurate chart of Scotland had been printed in Paris, and this too made its way to Duisburg. From surveys undertaken for Queen Elizabeth's government by Robert Lythe, Mercator was able to update Ireland, while England and Wales could be revised from the splendid new wall-map and set of thirty-four county maps produced by Christopher Saxton. Inconsistent by Mercator's standards, and lacking grids of latitude and longitude, Saxton's maps were immensely useful for their place-names and county boundaries.

England emerged as the most systematically mapped monarchy in the *Atlas*. Sixteen overlapping sheets covered England, Scotland, and Ireland at three principal scales. At the smallest scale was a map of the entire British Isles. Then came three maps describing Scotland, Ireland and England. And at the largest scale were maps of north and south Scotland, north and south Ireland, and then six regional maps of England. Among the British Isles series were a pair of potential wall-maps of Scotland and of Ireland.

The sixteen British maps dominated the coverage of the northern lands. Scandinavia was less coherent, with one general map covering Norway and Sweden but (thanks to Rantzau) four maps covering Denmark. With individual maps for Prussia, Livonia, Russia, Lithuania, Transylvania and Tauric Chersonese (Crimea), also drawn, Mercator had virtually completed a systematic map coverage of greater Europe.

Then, on 5 May 1590, he was struck down. It had to be his heart, of course. Having been carried through apocalypse year by his faith in renewal, his own heart, the symbol of love and of life, had finally faltered.

31

Creation

The thunderbolt extinguished Mercator's power of speech. He couldn't move his left arm or left leg. The left side of his body had died, its muscles immovable, its flesh loose. The nobly symmetrical, bearded countenance engraved by Hogenberg stared from the bed with asymmetrical, contradictory expressions: one resolute, the other ungoverned.

Solenander was called. The duke's personal physician 'did all he could to cure the illness'.[1]

But Mercator was seventy-eight. The paralysis had robbed him of the ability to sleep at night, and to eat normal food. He had become as vulnerable and as needy as an infant.

Solenander's administrations initially appeared to have little effect. It was, wrote Ghim a couple of years later, 'the weakness of advancing age' which 'precluded a successful outcome'.[2]

With time, however, Mercator did grow stronger. Little by little, he found his voice, uttering his only recorded words of speech to be handed down the centuries.

Weeping and striking his breast 'two or three times with his fist', he cried 'Hit, burn, cut your servant, O Lord, and, if you have not hit him hard enough, strike harder and sharper according to your will so that I may be spared in the life to come.'[3]

Taking the 'various remedies' prescribed by Solenander, Mercator began to sleep for longer periods; to eat easily digested food; to 'refresh himself with a draught of wine or beer.'[4]

In an attempt to restore movement to his left leg and arm, Mercator's daughter-in-law massaged the afflicted limbs with 'excellent salves'[5] for an hour each morning and evening. Still immobile, Mercator 'provided himself with a chair'[6] in which he could be carried about the house.

His impatience was palpable. Yet, as a practitioner of chronology, Mercator had to acknowledge that he had doubled his time. In an age when a minority of men survived their forties, Mercator had almost reached his eighties. He had been the beneficiary of two consecutive life-spans. After his rehearsal with fate at thirty-eight, locked in a castle cell waiting for the executioner's call, he had been given a second chance, a second life. And it had been that second life, the Duisburg life, that had produced the monumental works: the map of Europe, the chronology of the universe; the new projection; the definitive Ptolemy; the modern maps for the new geography; the cosmography called *Atlas*.

But his heart had failed too soon; he wasn't ready; there were too many loose ends. The *Atlas* was unfinished: of the hundred or so modern maps that he had anticipated, seventy-three were in print, but over thirty remained in varying states of readiness. Within Europe, Spain and Portugal continued to be a particular problem; the monarchies who had launched the age of exploration were proving the most reluctant to release geographical information. And several sections of text had yet to be written for the *Atlas*.

Mercator also had to finish – and send to the printer – various works in progress on the word of God: a commentary on St Paul's Letter to the Romans, a commentary 'on some chapters of Ezekiel',[7] and on the Apocalypse. There was also an unfinished elaboration of the exercise on Gospel harmonization which Mercator had inserted into the *Chronologia* twenty years earlier.

But none of these works were as important to Mercator as his account of Creation, the subject which had magnetized his imagination as an eager young graduate and led him by degrees towards the *Atlas*.

Hindered by paralysis, Mercator fretted with frustration. At his proven rate of progress when fit and healthy, these works would require several years; perhaps another lifetime.

'The thing he resented most about his illness', recorded Ghim, 'was that its prolongation cost him so much precious time.'[8]

On better days, Mercator was able to pick up his studies 'inasmuch as his small store of strength permitted', reading, writing or 'occupying his mind with meditations on subjects of great moment'.[9]

In a letter to Wolfgang Haller in the summer of 1592, Mercator wrote that he had 'fallen into great debility, so that my weak hands and my eyes, being sometimes struck blind, do not sufficiently perform their own duty'.[10] These were the lines of an anguished artisan, a handworker, a perfectionist who could not delegate: 'I do not dare to use an amanuensis; and so I fear lest you may with difficulty read certain things. For I am compelled, whenever I write, to mark out the lines one by one, and to attend completely to the formation of every single one lest I form anything wrong.'[11] How cruel an affliction for the man who taught the fluent strokes of italic.

Unable to wait for a complete recovery, and 'in order to while away the time when his left arm was paralysed', Mercator turned to his description of 'the creation and making of our world'.[12]

Of the various unfinished parts of the *Atlas*, the account of creation was the one which only Mercator could complete. A decade earlier, he had referred to it as 'the most difficult part of my whole undertaking'.[13] It was the enormity of rewriting Genesis (and the certainty that it would bring him into conflict with certain theologians) that had convinced Mercator that he should produce the first part of the cosmography at the end of his life: 'Although this is the last part of my work, it will nonetheless be the most important, indeed the very base and summit of the whole, which should have been ready before all the rest, if necessity had not demanded firstly resources and then philosophy. This will be the goal of all my labour, this will mark the end of my work.'[14]

The thought that he might die before providing the cosmography with its foundation work was impossible for Mercator to contemplate: 'Often he complained with great sorrow in his heart', wrote Ghim, 'that his illness would prevent him from finishing the works which, as I said before, he had conceived in his mind and, in a sense had at his fingertips'.[15] 'On the Creation and Fabric of the World' was a device for giving the *Atlas* a cosmographical context. It was also a dying man's treatise; the first and last opportunity for a sixteenth-century humanist to unburden his mind of deepest truths.

He began by explaining his life's purpose: 'For this is our goal while we treat of cosmography: that from the marvellous harmony of all things toward's God's sole end, and in the unfathomable providence in their composition, God's wisdom will be seen to be infinite, and his goodness inexhaustible.'[16]

It was God's overflowing fecundity that had led him to will the creation of man 'in order to share with him his glory'.[17] That glory could be seen in the founding and ordering of the world, 'which we have undertaken to contemplate'.[18] In a phrase which recalled his own philosophical fumblings sixty years earlier, readers were reminded that 'whoever wishes to undertake the description of the world' should 'begin from its first beginning, if he wishes to extend it usefully into philosophy'.[19]

The preliminary chapters concluded with Mercator's justification of

a universal geography ... which is still so necessary since merchants do not have access to the noblest and richest regions where they might trade with other nations, and make all lands familiar to Christians, nor are princes able to state anything certainly and solidly about their realms, unless with great labour and the aid of too-faithless servants, since maps are lacking, the ocular witnesses of their realms and domains.

'Some years ago,' added Mercator, 'I made a beginning on this work...'[20]

Returning to Creation as his temporal life faded, Mercator finally completed the investigation which he had begun sixty years earlier when he walked into the sunlight beyond Louvain's walls. Working through the first chapter of Genesis, day by day, he harmonized select authorities with the biblical record. By the time he had finished, he had written 30,000 words. Not only did he manage to complete the 'Creation and Fabric', but he found that he preferred it 'to all the other brain children which he had sired in his life'.[21]

A manuscript copy was sent to Solenander, who had it transcribed and sent on to the learned Jacob Sinstedius. The effect on Sinstedius was dramatic. As he read of the origin of all things arising from 'one and the same mass', he experienced the 'great kinship, affection, and love of the heavens, angels, and all created things among themselves and towards man'.[22] Without having met him, Sinstedius was describing the spiritual mechanics of Mercator's own personality. He confessed to Solenander, 'I indeed will say freely what happened to me as I read: I began to be inflamed by a great desire and ardour toward all things created by God (excepting only devils), as being akin to me and created by the best God for my uses.' Acknowledging that Mercator's views on original sin would 'not satisfy every theologian in all points', Sinstedius finished by exclaiming that 'the science of nature will be incomplete without the addition of this author'.[23]

Solenander was similarly awed: 'Once your arguments have been correctly viewed and understood,' he wrote, 'we shall no longer be seduced by Aristotle's exceedingly vain and monstrous "privation".'[24] Acutely aware of Mercator's age and health, Solenander urged him to 'bring out not only this book, but the other ones, however you may'.[25]

At the end of 1593, Mercator was struck down a second time.

Alarmingly, 'a great discharge' issued 'from the head' and 'obstructed his jaws and throat'.[26] Again, he lost the power of speech, 'and could only swallow with the utmost difficulty any food or drink that was offered to him'.[27]

Again, he fought back, regaining some facility to speak and to eat.

But the ageing could not be stilled and 'the fatal weakness grew upon him'.[28]

Thrice struck by Paul's metaphorical rods, the end was evident. The *'concionatore'*[29] was called, and his close neighbours. Still lucid, Mercator complained of terrible pains in his limbs, invoking several times 'with great confidence'[30] the mercy of God.

He lasted that long December night. Still conscious by morning, he sat 'in his chair before the hearth',[31] the burning wood lending little warmth to his withered limbs. With his remaining strength, he 'earnestly demanded'[32] the *concionatore* to offer public prayers for him at the end of the assembly in church.

'These were the last words of Mercator that those around him could understand,' recorded his neighbour, Walter Ghim, 'and he fell quietly asleep in the Lord shortly after eleven in the morning on 2 December, having lived eighty-two years, thirty-seven weeks and six hours and seen his great-grandchildren. May the Lord', concluded Mercator's friend and future biographer, 'grant him a joyful resurrection on the Day of Judgement.'[33]

Epilogue

Absent in body, Mercator became the editor-in-spirit of his incomplete *Atlas*.

Rumold, and Arnold's three sons, Johann, Gerard and Michael, worked quickly to collate the various parts of the cosmography. Poems and eulogies were sought from friends. Awarded access to Mercator's library and letters, Walter Ghim wrote an affectionate 'Life'.

The *Atlas* would open with a title page and then a dedication to Princes Wilhelm and Johann Wilhelm of Jülich and Cleves. Frans Hogenberg's portrait of Mercator at sixty-two would follow, framed by words of gratitude from Vivianus 'for having shown new stretches of the earth and sea, and the great, all-containing heavens.'[1]

From Bernardus Furmerius of Leeuwarden came a poem eulogizing Mercator's many-faceted qualities: 'He was', wrote Furmerius, 'truly learned, pious, pure, just to all, / as dextrous in his genius as dextrous in his hand.'[2] As a mathematician, 'he described the stars' and 'brought together the broad orb of the earth into maps'.[3] He 'taught history to speak times certain' and 'uncovered the sacred mysteries of the prophets'.[4] In a telling line which concluded this poetic résumé, Furmerius drew attention to the solitary nature of Mercator's quest for perfection: 'And he did these things so as to surpass all past / artists, on his own, and by his own hand.'[5]

Facing the portrait would be a plain epitaph whose carefully chosen words made it impossible to precisely place Mercator on his own map of Lower Belgium. Thus, he was, 'a Fleming of Rupelmonde, / Born in the province of Jülich, / Servant of Charles v / Holy Roman Emperor / to Wilhelm (the father) and Johann Wilhelm (the son), / Dukes of Jülich and Cleve, etc.'[6]

Mercator's posthumous refusal to tag his name with spatial coordinates was symptomatic of his non-aligned nature; it was a small step from locative sentimentality to territorial bigotry. Reconciliation of place (and faith) required the mind of a mediator; a practitioner of harmony. This conscious ambiguity would be picked up in the first line of another poem, typeset below the epitaph. 'You ask who I was?'[7] opened the poem rhetorically. 'With heaven as auspice,' came the response, 'I espied the earth, / reconciling things below with those above. Through me, the stars of heaven shine in maps.'[8]

Less than four months after Mercator died, the *Atlas* was ready.

Weighed in its near-entirety, the cosmography had expanded to a heavy folio volume.[9] The introductory pages of laudation were followed by Ghim's lengthy *Vita*, by the letters from Solenander and Sinstedius praising Mercator's 'study written on the works of the six days',[10] by Mercator's preface outlining the genealogy of *Atlas* and then by the 36,000-word treatise 'On the Creation and Fabric of the World'.

Part II of the *Atlas* was titled 'A New Geography of the Whole World'.

'Finally', began Rumold in his address to the reader, 'we now offer the second part of *Atlas*, the first volume of the new geography, namely, the description of the lands of northern Europe.'[11]

At last, the three fascicles of modern maps were united between covers, the seventy-three maps of 1585 and 1589 joining the twenty-nine that Mercator had been working on when he was struck down. With the exception of Spain and Portugal, the coverage of greater Europe was complete.

And what, the reader might ask, of the rest of the world?

'Once the northern lands are published,' continued Rumold, 'I shall take on the second volume of the new geography, that is the accurate description of Spain, and thence on to Africa, Asia, America, and, if it is discovered, as is our hope, I shall gird myself for the third continent,[12] which is called the Magellanic or southern (Australis) land.'[13]

Rumold believed that he could 'bring to its destined conclusion' the work that his father 'of pious memory left uncompleted'.[14] But he could not do it alone: 'I strenuously beseech again and again the learned students of geography that they may please to aid me with their observations of travels and wanderings over land and sea, and solemnly and reverently pledge that I shall not be forgetful of the favour done to me.'[15]

That the reader should be in no doubt that the Mercator family honoured such favours, the entire 'new geography' had been dedicated to the monarch who had done more than any to further the cause of geographical knowledge. 'Elizabeth,' read the dedication, 'the most serene and most mighty Queen of England, France and Ireland'.[16]

Rumold's dedication was praiseworthy with reason. Through Elizabeth and her subjects 'notice and illumination first dawned to the world of those regions that until now were hidden'. And as the queen commanded further voyages 'through the rough kingdoms of Neptune, shining with ice, to the remote Indies by unheard-of sea voyages' the time would come when these voyages would bear fruit 'not only for your Britons, but for all the rest of the neighbouring inhabitants of Europe'. Through the 'maritime expeditions' of 'noble heroes' such as 'Thomas Cavendish, Francis Drake, Martin Frobisher, and others' Elizabeth would 'uncover and illuminate many unknown and marvellous things throughout the world concerning the various locations of kingdoms and regions, the differing flux and reflux of the sea, and the temperature of the heavens and the air and its variety'.[17]

The 'famously celebrated sea journeys' of the English, continued Rumold, would not only continue to reveal the mysteries of the northern hemisphere, but 'also the Antarctic, which, containing a third of the earth, still lies hidden in shadow, will soon be made plain to the world'.[18] Made plain, moreover, through Rumold's anticipated maps. 'This is greatly to be hoped for, so that at last the face and figure of the whole earth shall be made known in this old age of the world, and that, as the human race is multiplied, so too shall its familiar intercourse and daily use of trade be multiplied, that we may all be kindled with greater wonder, desire, and ardor toward all things created for man's use by God.'[19] (Rumold would have known that the English would pass on the holy motive, but trade . . . trade was a different matter.)

So the English got the dedication and the first slot in 'A New Geography of the Whole World'. In return, Rumold expected the Antarctic information which would allow him to complete his father's life work. In turn, it was understood that the advantage would feed back to the English. Just as Mercator's letter to Dee on the polar regions had furnished Elizabeth with a claim to the northern lands, Rumold's maps of the Antarctic would allow the English to take the southern continent too.

Thus did Rumold ask the 'most serene' Elizabeth to accept 'these maps of the regions belonging to you'.[20]

To the 102 modern maps prepared by Mercator, Rumold and his nephews added another five. From Rumold came his 1587 map of the world, and a map of Europe. Gerard engraved Africa and Asia. The youngest Mercator, Michael, was given the task of engraving the youngest continent, America.

Johann, the eldest of Arnold's boys, wrote a personal epitaph: 'I, who have recently mourned the loss of both my parents, / am forced to go, wretch that I am, to new graves. / Thus, venerable and pious grandfather, who so often / gave me belief, does the black day snatch you from us?'[21]

Atlas sive cosmographicæ meditationes de fabrica mundi et fabricati figura was published in the spring of 1595.

Sales were disappointing.

With Spain and Portugal missing, and only three maps showing regions outside greater Europe, the *Atlas* was hardly the 'Figure of the Fabrick'd' world that the title claimed, a deficiency unwittingly advertised by Rumold's reference to a forthcoming, second volume of modern maps.

Without Mercator, the *Atlas* lost momentum. And Rumold's 'noble heroes' of England passed on: Frobisher died of wounds in 1594, Drake of dysentry in 1596. Geographical titbits ceased to flow from Ortelius; by June 1598, Mercator's old friend was describing himself as 'weak not only in hand but in body'.[22] A month later, Ortelius was buried at St-Michael's in Antwerp.

Discouraged no doubt by the reception of the *Atlas*, Rumold failed to produce any of the promised maps by the time Ortelius died. Only five years after losing his father, Rumold also died – on 31 December 1599 – the last day of the century.

Mercator's heirs attempted to revive the *Atlas* in 1602 with a new edition containing the same 107 maps and a reset text. But it still suffered from comparison with Ortelius' ever-expanding *Theatrum* – now in thirteen Latin editions, as well as in Dutch, French, Italian, German and Spanish. Few who compared the two publications on the stalls of Frankfurt's Buchgasse understood the conceptual basis of Mercator's austere-looking maps.

Not for the first time, a crisis in the Mercator family precipitated a change of fortune. Money was needed to pay for the welfare of Rumold's younger children. In 1604, Mercator's library was sold off at the premises of a Leiden bookseller, and the same year, Gerard Mercator and Tylmann de Neuville (the second husband of Mercator's daughter Dorothea) requested – as guardians of Rumold's children – that the municipal council of Duisburg agree to the sale of the copper plates of Mercator's maps. The plates were bought by Gerard himself, but were soon sold on to an Amsterdam cartographer, Jodocus Hondius, who wasted little time in turning the *Atlas* into a commercial publication. Hondius added text, and maps of Spain, and of Africa, Asia and America, bringing the total to 143. Now resembling the book Mercator had intended, the *Atlas* was republished in 1606 bearing his name and the original title.

The enlarged *Atlas* was an enormous success: between 1609 and 1641, Hondius and his son produced twenty-nine editions in Latin, Dutch, French, German and English, adding additional maps as they did so. Following the example of Ortelius, Hondius also repackaged the *Atlas* in a pocket form as the *Atlas minor*.

At last, Mercator's rigorous cosmography had found its market. And although the maps themselves would eventually be superseded, the concept endured. Volume 1 of 'A New Geography of the Whole World' embodied the principles of future mapmaking. Mercator's italic lettering; his identical map overlaps; his complete coverage of regions at more than one scale; his consistent use of grids of latitude and longitude; his singular editorial control, were all adopted as cartographic standards. *Atlas*, the cosmography, became atlas, the (Oxford English Dictionary) term for 'A collection of maps in a volume'.

And what of Mercator's other works?

After a fifty-year run of success, the terrestrial globe of 1541 was eventually superseded by more up-to-date Amsterdam globes. The ingenious and heretical *Chronologia* enjoyed a mere fourteen years of notoriety before it was overshadowed by Joseph Justus Scaliger's *De emendatione temporum*. Mercator's edition of Ptolemy's *Geography* enjoyed over a century of sales. In 1636, the English essayist Thomas Blundeville was using 'an excellent good Map of Europe, made by Mercator in the yeere of our Lord 1554'[23] as a cartographic exemplar in his *Exercises*. Blundeville praised Mercator's European wall-map as 'a necessary Card ... made with very good Art'.[24]

It was Blundeville who also promoted Mercator's misunderstood world map of 1569. The year that Mercator died, the essayist became the first to publish the tables of secants required to construct the new projection. Blundeville had borrowed the tables from his friend Edward Wright, a bright young mathematician of Caius College, Cambridge. Shortly afterwards, Wright used Mercator's projection to make a map of a voyage he had taken to the Azores, and then in 1599, he published his tables in *Certaine Errors in Navigation*. In the same year, Richard Hakluyt published a world map (also by Wright) on Mercator's projection. By the end of the century, Mercator's projection had become established as a geographical utility. Criss-crossed with its once mysterious rectilinear rhumb lines, his image of the world had at last found its audience: 'He does smile his face', wrote a young English playwright at the time, 'into more lynes than are in the new Mappe with the augmentation of the Indies.'[25]

The new projection won a slower acceptance among seafarers, who eventually came to recognize that the peculiar distortions of space were a necessary price to pay for a chart which could lead them unerringly to a distant port. Nevertheless, it was not until the mid 1600s that Sir Robert Dudley's *Arcano del Mare* became the first sea atlas to use Mercator's projection throughout. Thirty years later, in 1686, Edmond Halley used the projection for the first meteorological chart.

So powerful a tool did the projection become that Mercator became subsumed by his own device. By the twentieth century, the youth who had changed his name from 'Gerhard Kremer' to 'Gerardus Mercator' had become 'Mercator's Projection'. His projection was adopted by state cartographers to map the land that he'd named 'North America'. In 1938, Mercator's Projection was selected by the Ordnance Survey to map Britain anew. And in 1974, the American cartographer Alden P. Colvocoresses used the Space Oblique Mercator Projection for the first satellite map of the USA. When the Jet Propulsion Laboratory sent Mariner 8 and Mariner 9 to map Mars, they undertook their Martian cartography on a standard Mercator Projection and – naturally enough – the first book of maps describing the Red Planet was titled *Atlas of Mars*.

One by one, the mappable orbs of our solar system are appearing on the world wide web, flattened for our screens according to Mercator's principles. Using 'synthetic aperture radar mosaics'[26] sent back to Earth by the Magellan spacecraft, Venus has been mapped on Mercator's Projection. Jupiter can be viewed on Mercator's Projection, and its volcanic moon Io, and Saturn's largest moon, Titan.

Mercator's Projection reconciled the sphere and the plane, while his *Atlas* enveloped the world with an integrated system of maps. In the midst of an era of tumult, he lived, and he engraved, for global harmony. And in his spatial masterworks, the cobbler's boy from the Low Countries inscribed his own, universal epitaph.

Notes

I A LITTLE TOWN CALLED GANGELT

1 Averdunk and Müller-Reinhard (1914, p. 1) suggested that the boy may have been born prematurely, as a result of the arduous journey.

2 One century later, the chronicler Jacob Kritzraedt compared Gangelt to a '*klot-zechtig*' – a roundel of tree-trunk. 'When one approaches from Sittart', added Kritzraedt, the town 'glittered like the Schieferberg' – a reference to a prominent mountain to the east. Gangelt, he claimed, could be seen from as far away as Aachen and Maastricht. (Quoted in Achten, 1995, p. 20.)

3 By 1500, Europe's print centres had issued 40,000 editions, distributing something like 20 million books and booklets among a continental population of around 80 million. Typeface from Cologne was used to print Oxford's first book and it was in Cologne in 1471 that William Caxton – England's first printer – was taught the techniques of movable type.

4 For this use of '*mercator*', see Westfall Thompson, 1911, p. 19, f. 67.

5 *St John Beholding the Seven Golden Candlesticks*, from the *Apocalypse* series, 1498. See Belgrave et al. 1999, pp. 74–5.

6 Quoted in Fry, 1995, p. 37.

7 According to Kritzraedt's *Stadtbuch* of 1644 (see Achten, 1995, pp. 4–5), Gangelt had benefitted from a good Latin school since the thirteenth century, and must therefore have had an elementary school too. At the time, even the smallest Rhenish towns had an elementary school, even if the teaching amounted to little more than a well-intentioned villager with an episcopal licence.

8 Quoted in Jedin, 1980, vol. IV, p. 571.

9 Stabel (Clark, 1995, p. 221) estimated that the average number of children in urban Flemish households between 1510 and 1520 ranged from 3.32 to 4.62.

10 Assuming that Emerentia had given birth every year and a half, the eldest boy, Gijsbrecht, would have been under ten years of age when Gerard was born.

11 The entire family of nine would have required around five kilograms of bread every day. See Blockmans and Prevenier, 1978, pp. 21–2.

12 In May 1521, the Imperial tax schedule required the Duke of Jülich-Berg to contribute 45 horsemen, 270 footsoldiers and 500 gulden for the Roman campaign and for governmental upkeep. The Duke of Cleves-Mark had to contribute the same. The account sheet for the Worms *Reichsmatrikel* show that the two duchies combined had to hand over more money than any other of the 367 contributing electors, archbishops, ruling princes, free and imperial towns, counts and lords. Only the Elector of Bohemia, Archduke of Austria and Duke of Burgundy contributed more horsemen and footsoldiers. See Benecke, 1974, p. 385.

13 In the area of Osnabrück, north-east of Gangelt, the population nearly doubled during the sixteenth century; see *The Cambridge Economic History of Europe*, vol. IV, p. 25.

14 The Erfut chronicle of 1483, quoted in ibid., vol. IV, p. 24.

15 Sebastian Franck, *Germaniae Chronicon* (Augsburg, 1538), quoted in ibid., vol. IV, p. 25.

16 Sixteenth-century *Zimmerische Chronik*, quoted in ibid., vol. IV, p. 25.

17 First printed in Barcelona in April, 1493, the letter existed in seventeen editions, nearly all of them translations, by 1500. The first German translation was printed in 1497, at Strasbourg.

18 Morison (1955) pointed out that the twenty days referred to in the printed letter arose from a misprint and that the actual time taken was thirty-three days; the mass circulation of the lower figure created an impression that the islands Columbus had found were much closer to Europe than was the case.

19 The passages quoted here are all taken from the translation by Morison (1955, pp. 205–13) of the original, Barcelona edition of Columbus' letter of 1493.

20 Vespucci's letter was published as *Mundus Novus* (New World). By the time Mercator was born, the letter had appeared in no less than forty editions, including Flemish, French and Czech. The German edition of 1505 travelled under the title *Von der neüw gefunden Region*. All passages quoted are taken from *Mundus Novus*, trans. in Northrup, 1916, pp. 1–13.

21 Printed in 1505, this was the earliest picture of American Indians.

22 Brant, 1971, p. 169.

23 Letter of 5 June 1516 from Erasmus to John Fisher, Erasmus, 1974–, vol. 3, p. 295.

24 Ibid., vol. 3, p. 295.

25 Letter of 5 March 1518 from Erasmus to William Warham [?], Erasmus, 1974–, vol. 5, p. 319.

26 Letter of 13 March 1518 from Erasmus to Beatus Rhenanus, Erasmus, 1974–, vol. 5, p. 345.

27 Ibid., vol. 5, p. 345.

28 Letter of 24 April 1518 from Erasmus to Cuthbert Tunstall, Erasmus, 1974–, vol. 5, p. 409.

29 Letter of second half of April 1518 from Erasmus to Thomas More, Erasmus, 1974–, vol. 5, p. 401.

30 Letter of 31 May 1518 from Erasmus to Pierre Barbier, Erasmus, 1974–, vol. 6, p. 39.

31 B. de Mandrot (ed.), *Mémoires de Philippe de Commynes* (Paris, 1901), p. 15, quoted in Blockmans and Prevenier, 1999, p. 141.

2 PROMISED LANDS

1 Referring to the rent on Hubert's Rupelmonde 'Hofstatt', unpaid 'for six years since Christmas 1511', Averdunk and Müller-Reinhard (1914, p. 2) surmised that the Kremers returned to Rupelmonde 'in 1517 or 1518'. Of the two dates, the latter is more likely, since an elapsed period of six years would produce a date of Christmas 1517, assuming that the rent was calculated in arrears. Furthermore, the sequence of floods, freezes and high prices that had beset Flanders between 1515 and 1517 (with correspondingly high mortality rates) makes it less likely that the Kremers would have migrated until 1518, the year that conditions (and prices) began to improve in

Flanders. Since winter travel would have been out of the question, the Kremers would have left Gangelt during the drier months of 1518, confronting the same perils of the road that Erasmus faced that same year.

2 Quoted in Fry, 1995, p. 37.

3 Hondschoote's population would explode from 2,500 in 1469 to 15,000 by 1560. See Lis and Soly, 1979, p. 68.

4 The Aachen–Maastricht road lay about 20 kilometres south of Gangelt.

5 Letter of the first half of October 1518 from Erasmus to Beatus Rhenanus, Erasmus, 1974–, vol. 6, p. 120.

6 Four hundred years on, the mill still stands.

7 Stabel, in Clark, 1995, pp. 210–11, estimated that Rupelmonde had a population of 702 in 1469, placing it forty-first equal with Torhout in size among the settlements in the county of Flanders. The largest three settlements were Ghent (60,000), Bruges (45,000) and Ypres (9,900). Before the disastrous battle of 1452, Rupelmonde's population may have been as high as 5,000–6,000.

8 Stabel (Clark, 1995, p. 223) identified 1516 and 1517 as years of 'mortality crisis' in Flanders.

9 See Van der Wee, 1963, vols I–III, for analysis of price fluctuations of these commodities during the early years of the sixteenth century.

10 Martyr had first referred to the 'New World' in a letter of 20 October 1494, while he held a teaching post at the Spanish court.

11 Quoted by Borstin, 1985, p. 257.

12 Ibid.

13 Averdunk and Muller-Reinhard, 1914, p. 1, point out the likelihood that Gisbert attended Louvain, since it was the only university in the non-French part of Belgium, and that the majority of Flemish clerics must have studied there. Furthermore, Louvain was popular among those from the western part of Jülich.

14 See ibid., p. 1.

15 The dialogue (*Dialogus cantoris et regis*) and its wordplays are discussed by Van Zijl, 1963, pp. 75–81.

16 Ibid., p. 76.

17 Ibid., p. 76.

18 Ibid., p. 77.

19 The Schelde had one of the densest concentrations of schools in Europe, with over 150 in Antwerp alone.

20 A woodcut by Dirk Vellert dating from 1526 shows the interior of an Antwerp schoolhouse similar to that described here, with children gathered in small groups being instructed by adults. The impression given is of disciplined learning under the eyes of adequate staff.

21 Elementary education was available to both the merchant's son and the peat-digger's daughter; beyond elementary-level schooling, education for girls and paupers' sons was almost impossible.

22 This is a likelihood rather than recorded fact.

23 Trans. Duff and Duff, 1934, p. 601.

24 Ibid., p. 625.

25 Ibid., p. 625.

26 Ibid., p. 627.

27 Daniel 7:16.

28 Daniel 7:7.

29 Quoted in Roberts, 1995, pp. 578–9.

30 Koenigsberger, Mosse and Bowler, 1989, p. 231.

31 This, and the following quoted passages, are taken from Dürer's description of the procession, trans. in Fry, 1995, pp. 42–4.

32 The mathematician was Johann Stöffler, whose *Almanach* of 1499 predicted a universal flood for the year 1524.

33 Zeydel, 1967, p. 62.

34 Fry, 1995, p. 63. The anguish of Agnes would have been felt by her husband; later, his friend Willibald Pirckheimer would accuse Agnes of driving Albrecht to an early death by her stinginess.

35 Quoted in Van der Stock, 1998, p. 43.

36 Letter of 8 February 1524 from Erasmus to Lorenzo Campeggi, Erasmus, 1974–, vol. 10, p. 170.

3 TO THE WATER MARGIN

1 Post (1968) argued that the only Dutch humanists who flourished between 1480 and 1540 who 'could conceivably be linked with the Modern Devotion as it had then developed' were Erasmus, the cousins Cornelis and William, and Macropedius. Schoeck (1990, p. 48) enlarged the list without contradicting the eminence of Macropedius. Schoeck (p. 267) also included Macropedius on a list of Erasmus' friends who had been schooled by the Brethren.

2 From Rupelmonde, the road distance to 's-Hertogenbosch is approximately 110 kilometres.

3 At Lier, up to 3,500 cattle could change hands in one day. See Van der Wee and Aerts, 1979, pp. 235–40, for an analysis of the Low Countries livestock trade in the early 1500s.

4 In 1526, the Brabant census revealed that the rural poor in the duchy amounted to 27 per cent of the population (see Blockmans and Prevenier, 1978, p. 35). Three years later, in 1529, the prelates of Brabant would write despairingly of their inability to stem the emigration from field to urban slum: 'We find ourselves and the whole land in such great poverty that it is not possible to describe it. We cannot prevent the country dwellers from leaving, as they have already done in many places' (quoted in Lis and Soly, 1979, p. 78).

5 Between 1480 and 1526, the poor hearths in small Brabant towns like Turnhout and Tilburg had risen from 14–19 per cent to 27–29 per cent. See Blockmans and Prevenier, 1978, p. 35.

6 See Keuning, 1952, p. 40.

7 The descriptive material in this paragraph is drawn from Braun and Hogenburg's views in *Civitates Orbis Terrarum*, which were themselves drawn from 1567 woodcut by Lodovico Guicciardini and a drawing – from the late 1550s or early 1560s – by Jacob van Deventer.

8 Dürer, quoted in Fry, 1995, p. 61.

9 Ibid., pp. 61–2.

10 Referred to in later years as *The Garden of Earthly Delights*. In 1517, the painting was seen in Brussels, hanging in the palace of Henry III or Nassau. However, it is not improbable that Mercator was familiar with its content at 's-Hertogenbosch, through copies or alternative versions by Bosch.

11 Letter of August 1516 from Erasmus to Lambertus Grunnius, trans. in Erasmus, 1974–, vol. 4, p. 11.
12 In Macropedius' writing, the influence of Titus Maccius Plautus (*c.* 254–184 BC) is more frequently detected than the younger Publius Terentius Afer (*c.* 195–159 BC). Terence was the better writer, a craftsman of high comedy, whereas Plautus specialized in applying farce to borrowed Greek originals. But in the boom and bust of Europe's new economies, Plautus was a parable of the times; a humble carpenter from Umbria who rose to become ancient Rome's greatest comic dramatist. Having made his money from comedies, Plautus had become a merchant, but lost everything, then returned in poverty to Rome where he took a job turning the handmills for a baker. While grinding corn in the bakehouse, he wrote another three comedies.
13 In the preface to *The Rebels* and *Aluta* (see Best, 1972) – published together in 1535 – Macropedius wrote that he had started to compose his 'trifles' twenty years earlier, i.e. around 1515, twelve or so years before Mercator became his pupil. In another reference, he makes it clear that he had written – though not published – six of his twelve dramas by 1535. Eliminating the later dramas suggests that Mercator was probably familiar with *Asotus*, *The Rebels*, *Aluta*, *Petriscus*, *Andrisca* and *Bassarus*.
14 Epilogue, *The Rebels*, trans. in Lindeman, 1983, p. 105.
15 Ibid., p. 31.
16 Ibid., p. 31.
17 Ibid., p. 31.
18 The tower that was to become a long-distance beacon on the polder had yet to be built; it would be finished around the time – 1561 – that the church became a cathedral.
19 Largely due to its network of Latin schools and the educational achievements of the Brethren of the Common Life, the Low Countries had the highest levels of literacy in Europe.
20 Ghim (trans. in Osley, 1969, p. 185) related that Mercator came to 's-Hertogenbosch 'to complete his study of grammar and to start learning logic', implying that his schooling on the Schelde was insufficient to guarantee a place at university.
21 *De ratione studii ac legendi interpretandique auctores liber* was first published in 1511, in Paris. The following year an enlarged edition was published in Louvain.
22 Erasmus, *De ratione studii ac legendi interpretandique auctores liber*, Erasmus, 1974–, vol. 24, p. 673.
23 Ibid., p. 673.
24 Letter of 3 June 1524 from Erasmus to Willibald Pirckheimer, Erasmus, 1974–, vol. 10, p. 279.
25 Letter of May 1499 from Erasmus to Jacob Batt, Erasmus, 1974–, vol. 1, p. 192.
26 Letter of 20 November 1500 from Erasmus to Fausto Andrelini, Erasmus, 1974–, vol. 1, p. 285.
27 For a list of the books left by the Brethren of 's-Hertogenbosch, see Hyma, 1930, pp. 136–7.
28 Surviving records suggest that 's-Hertogenbosch was the most northerly point in the world that Mercator reached during his lifetime.
29 Pliny the Elder, 1991, p. 44.

4 THE CASTLE

1 Ghim, 1595, trans. in Osley, 1969, p. 185.
2 At the time, an imaginary square with four-mile sides drawn in the centre of the Kempen would have contained sixteen villages; centred on Louvain, the same square would have contained sixty-eight.
3 An ell was equivalent to about 0.7 metre.
4 In 1530, nearly one-third of Louvain's student population were 'foreigners'.
5 J. B. Mullinger, 'Universities', *Encyclopedia Britannica*, 1911, 11th edn, vol. 27, p. 759.
6 Andreas Vesalius, *De Humani Corporis Fabrica*, 1543, quoted by O'Malley, 1964, p. 31.
7 Vesalius had matriculated a few months ahead of Mercator, on 25 February 1530.
8 According to Vanpaemel, 1995, p. 46, fn. 17, 'The curriculum of the Faculty of Arts remained still largely scholastic until the middle of the seventeenth century.'
9 From Dante's reference, in the *Divine Comedy*.
10 *Sententia 1*, quoted in Van Raemdonck, 1869, p. 23, fn. 2.
11 *Acta 1470*, quoted in ibid., p. 24, fn. 2.
12 Juan Luis Vives, 1979, p. 80.
13 Ibid., p. 80.
14 Ibid., p. 80.
15 Andreas Vesalius, *De Humani Corporis Fabrica*, 1543, quoted in O'Malley, 1964, p. 31.
16 Letter of 1500 from Vespucci to Lorenzo di Piero Francesco de' Medici, quoted in Parry, 1968, p. 180.
17 Ibid., p. 180.
18 Ibid., p. 180.
19 Ghim, trans. in Osley, 1969, p. 185.
20 Ibid., p. 185.
21 Mercator, Preface, *Evangelicæ historiæ quadripartita Monas*, quoted in Van Raemdonck, 1869, p. 22, fn. 1.
22 Ibid., p. 22, fn. 2.
23 Hyma (1930, p. 136) lists *Summa Theologiae* and Aquinas' *De Puritate Conscientiae et de Modo Confitendi* among the works in the Brethren house library of 's-Hertogenbosch.
24 Dedicatory letter, *Evangelicæ historiæ*, quoted in Van Raemdonck, 1869, p. 25, fn. 2.
25 Louvain theologians would not have been as lenient as those in Paris two years later, when Petrus Ramus escaped unscathed after writing a thesis which he unambiguously titled 'Everything that Aristotle taught is false'.
26 Mercator, Dedicatory letter, *Evangelicæ historiæ*, quoted in Van Raemdonck, 1869, p. 25, fn. 2.
27 Ibid., p. 21, fn. 2.
28 Ibid., p. 21, fn. 2.
29 Ibid., p. 21, fn. 2.
30 Ibid., p. 21, fn. 2.
31 Such a work surfaced at the end of Mercator's life and was published in the *Atlas* of 1595. The foundations for his 36,000-word treatise 'On the Creation and Fabric of the World' must have been laid during this two-year period of contemplation in his twenties. Van Raemdonck, 1869, p. 25, appears convinced that Mercator returned to Louvain with 'his new doctrine', which his detractors 'hoped would be published immediately'.
32 Mercator, Dedicatory letter, *Evangelicæ historiæ*, quoted in Van Raemdonck, 1869, p. 25, fn. 2.

33 Ibid., p. 25, fn. 2.

5 TRIANGULATION

1 Antoon Verdickt, a Calvinist prisoner in 1558, quoted in Marneff, 1966, p. 3 (from Adriaen Van Haemstede, *Historie der Martelaren* (Dordrecht, 1659, fol. 320r.).

2 See Van Isacker and Van Uytven, *c.* 1986, p. 85: Antwerp's population rose from an estimated 33,000 in 1480 to 40,000 by 1496. By 1526 – the year before Mercator left his home near Antwerp for 's-Hertogenbosch – the city's estimated population was 55,000, and by 1568, it had soared to 114,000.

3 See Febvre and Martin, 1976, p. 187.

4 Christophe Plantin, quoted in ibid., p. 125.

5 Mercator, Dedicatory letter, *Evangelicæ historiæ*, quoted in Van Raemdonck, 1869, p. 25, fn. 2.

6 Praefatio, *Tabulae geographicae Cl: Ptolemei . . .*, 1578, quoted in Babicz, 1994, p. 62.

7 Keuning, 1952, p. 43, regarded Mechelen as the oldest cartographic centre in the Netherlands.

8 Denucé (1941, quoted in Van der Krogt, 1993, p. 44) referred to Monachus as 'one of the greatest geographers of his time'.

9 The terrestrial globe has not survived, but it is referred to in an undated booklet, published in the form of a letter from Monachus to Jean Carondelet (Monachus' patron, the Archbishop of Palermo, the chairman of the Privy Council of Mechelen and a friend of Erasmus'). After summarizing the available evidence, Van der Krogt (1993, pp. 43–4) narrowed the period of publication for the booklet and the globe to 1526–7.

10 This translation of *Geōgraphikē hyphēgēsis* is taken from Berggren and Jones (2000, p. 4). Dilke (1991, p. 295) rendered Ptolemy's title as 'Manual of Geography'.

11 Berggren and Jones, 2000, p. 57.

12 Ibid., p. 57.

13 Ibid., p. 93.

14 Johannes Cochlaeus, *Brevis Germaniae Descriptio*, 1512, quoted in Englisch, 1996, p. 116, fn. 16.

15 Ibid., p. 116, fn. 16.

16 Quoted by Wilford, 1981, p. 70.

17 Letter from the historian Johannes Trithemius 12 August 1507, quoted in Karrow, 1993, p. 571.

18 Quoted in ibid., p. 579.

19 From the text of the Dürer-Stabius map, 1515, quoted in Whitfield, 1994, p. 52.

20 Schöner, *Luculentissima quaedam terrae totius descriptio . . .*, 1515, trans. Van der Krogt, 1993, p. 31.

21 Letter accompanying Schöner's globe of 1523, trans. Van der Krogt, 1993, p. 32.

22 Monachus, *De Orbis Situ . . .*, 1526/7, trans. in Van der Krogt, 1993, p. 42.

23 This land was not part of a continent as Monachus imagined, but the island of South Georgia.

24 Trans. in Karrow, 1993, p. 408.

25 Quoted in Kish, 1967, p. 6.

26 This reference to Gemma's first globe comes from the dedication (dated 18 January 1553) he wrote in *De principiis astronomiae*. Quoted in Van der Krogt, 1993, pp. 48–9.

27 Gemma Phrysius, *De principiis astronomiae*, 1530, quoted in Van der Krogt, 1993, p. 51.

28 Gemma Phrysius, *De principiis astronomiae*, 1530, trans. in Andrewes, 1993, p. 390.

29 Ibid., p. 390.

30 Ibid., p. 391.

31 Ibid., p. 391.

32 Richard de Benese, *The Boke of Measuring of Lande*, London, 1537, quoted in Kiely, 1947, p. 104.

33 Anthony Fitzherbert, *The Boke of Surveyinge*, London, 1523, quoted by Kiely, 1947, p. 104.

34 Quoted in Karrow, 1993, p. 413.

35 *Libellus de locorum* ..., 1533, quoted in Haasbroek, 1968, p. 13.

36 When the English eventually experimented with 'triangulation' twenty-six years later, they did so in the flatlands of Norfolk.

37 Noted in Van der Krogt, 1993, p. 38.

38 *Legatio ... Presbyteri Joannis ad Emanuelem* (Antwerp, 1532).

39 If the 'Jacobus lantmetere de Mechlina' who matriculated at Louvain in 1523, is the Van Deventer who went on to map the Low Countries, Gemma may have met the future surveyor in the city following his own matriculation in 1526.

40 Berggren and Jones, 2000, p. 63.

41 Psalm 104:24.

42 *Atlas sive cosmographicæ meditationes de fabrica mundi et fabricati figura* (hereafter *Atlas*), 1595, p. 35; all subsequent references are to the Lessing J. Rosenwald Collection CD-ROM edition, translated by D. Sullivan, 2000.

43 Ghim, trans. in Osley, 1969, p. 185.

44 It is equally possible that Gisbert had been unable to afford Mercator's further education through the Faculty of Theology, since the higher faculties of the university required board and lodging, tuition fees, ceremonial costs and bookshop bills.

45 The only surviving chronological record of this mysterious period of Mercator's life is the 'Life' penned sixty years later by Walter Ghim, who described the philosophical studies as lasting 'for some years'. In order for Mercator to have subsequently mastered mathematics 'within a very few years' (Ghim) prior to his work in 1535–6 on Gemma's globe, he cannot have been engaged in his philosophical adventure for much more than the twenty-four months between October 1532 and October 1534. Although he does not explain his reasoning, Van Raemdonck (1869, p. 25, fn. 1) writes that Mercator's return to Louvain 'very probably took place in 1534'.

46 Ghim, *Atlas*, p. 5.

6 THE MATHEMATICAL JEWEL AND OTHER SUITABLE TOOLS

1 During their researches in Antwerp, Averdunk and Müller-Reinhard (1914, p. 4, fn. 17) were unable to confirm Van Raemdonck's assertion that Mercator had lived in the city for a single, extended period rather than one or more shorter periods.

2 Letter of 3 March 1581 from Mercator to Haller, quoted in Van Durme, 1959, p. 166.

3 Ibid., p. 166.

4 Juan Luis Vives, 1979, p. 80.

5 In Sanderus, *Flandria Illustrata*, vol. II, p. 548, quoted in Osley, 1969, p. 20.

6 Letter of 3 March 1581 from Mercator to Haller, quoted in Van Durme, 1959, p. 166.

7 Ibid., p. 165.

8 Ibid., p. 165.

9 Ibid., p. 165.

10 Ibid., p. 165.

11 Ibid., p. 165.

12 Ibid., p. 166.

13 Ibid., p. 166.

14 Ibid., p. 166.

15 Ibid., p. 166.

16 Ibid., p. 166.

17 Ibid., p. 166.

18 Ibid., p. 166.

19 As Karrow (1993, p. 168) has pointed out, there has been some debate concerning the correct vernacular spelling of this name. Consistent with Karrow (who preferred 'Fine' because it accorded with contemporary poetry and with local usage in the area of Fine's birth), I have omitted the acute accent above the final letter.

20 Letter of 3 March 1581 from Mercator to Haller, quoted in Van Durme, p. 166.

21 Ibid., p. 166.

22 Erasmus, *De recta Latini Graecique sermonis pronuntiatione*, quoted in Erasmus, 1974–, vol. 26, pp. 390–1.

23 Ibid., vol. 26, p. 391.

24 See Osley, 1969, p. 34 for an example of Grapheus' hand.

25 The 1534 edition of this book is catalogued in Mercator's library, Watelet, 1994, p. 412.

26 From Bordone's application of 1508 to the Venetian Senate, to publish a world map. Quoted in Karrow, 1993, p. 89.

27 Quoted in Gaur, 1994, p. 169.

28 Ghim, trans. in Osley, 1969, p. 185. The material used would have been brass rather than bronze.

29 There is no hard evidence that Mercator learned his instrument-making skills with Van der Heyden, but a working relationship is strongly suggested by the mutual interests they shared with Monachus and Gemma Frisius, and by the fact that Mercator's earliest (surviving) credited work was a collaborative effort whose engraving role he shared with Van der Heyden. Disagreeing with Van Ortroy (1920), Turner and Decker (1993, pp. 423–4) make the point that Gemma Frisius probably played a 'less than active' role in the actual fabrication of instruments, relying rather on the craftsmanship available in Van der Heyden's workshop. Ghim does not mention Mercator's training. Given the skills that Mercator was displaying in 1536, he must have begun working with metal no later than 1534.

30 The term 'mathematical jewel' originally applied to the ingenious portable astrolabe, and it was used in 1585 by John Blagrave for the title of one of the first printed English books on mathematics: *The Mathematical Jewel, shewing the making, and most excellent use of a singular instrument so called, etc.* (London, 1585).

31 The paper, *Usus annuli astronomici*, was published in 1534. The earliest surviving edition is in the 1539 edition of Apian's *Cosmographicus Liber*. W. R. Martin (*Encyclopedia Britannica*, 1911, 11th edn, vol. 19, p. 286) provides an illustration of the device.

32 Gemma Frisius, *Usus annuli astronomici*, 1534, quoted in Konrad Gesner, *Bibliotheca universalis*, Zürich, 1545 (Karrow, 1993, p. 208).

33 Ibid., p. 209.

34 Ibid., p. 209.

35 Ibid., p. 208–9.

36 Roeland Bollaert, Advice to the reader, *Solidi ac spherici corporis sive Globi Astronimici...*, 1527, quoted by Van der Krogt, 1993, p. 41.

37 Quoted in Van der Krogt, 1993, p. 41.

38 Van der Krogt, 1993, p. 575.

39 Ibid., p. 575.

40 Ibid., p. 575.

41 Ibid., p. 575.

42 Ibid., p. 575.

43 Ibid., p. 575.

7 NEITHER KNOWN NOR EXPLORED

1 Maximilianus Transylvanus, 1969, p. 112.

2 Ibid., p. 119.

3 Japan.

4 Gemma's globe of 1536 measured 370 mm in diameter.

5 Van der Krogt, 1993, p. 55.

6 Ibid., p. 54.

7 Mercator located 'Matonchel' on the western coast of 'Hispania nova' (now Mexico), at the mouth of the only river he marked on the entire Pacific coast of South and Central America. Matonchel later became 'matanchel' or 'Chacala' (see Lanzas, 1900, Book II, pp. 216–8), a busy Pacific port. The river is now the Rio Grande de Santiago.

8 Van der Krogt, 1993, p. 55.

9 Trans. in ibid., p. 54.

10 Ibid., p. 54.

8 CELESTIAL MAIDENS

1 Johanna was to outlive her husband Jan by forty years.

2 Ghim, trans. in Osley, 1969, p. 90.

3 Ghim later wrote that Barbara had been 'an excellent housewife', trans. in Osley, 1969, p. 190.

4 A view based on a letter from Molanus to Mercator of 24 March 1566, in which Molanus refers to Johanna's 'false opinions', Van Durme, 1959, p. 64.

5 Andreas Vesalius, *Epistola, rationem modumque propinandi radicis Chynae decocti...*, Basel, 1546, quoted in O'Malley, 1964, p. 63.

6 Andreas Vesalius, *De humani corporis fabrica libri septem*, Basel, 1543, quoted in O'Malley, 1964, p. 64.

7 Ibid., p. 64.

8 Ibid., p. 64.

9 Ibid., p. 64.

10 An edition of *Hyginus fabularum liber* was printed in Basel in 1535. See Dekker, 1999, p. 133, fn. 7.

11 Van der Krogt, 1993, p. 55.

12 Ghim, *Atlas*, p. 6.

13 'Geographical maps': letter of June 1534 from Viglius of Aytta to Hadrianus Marius, quoted in Waterbolk, 1977, p. 45.

14 Letter of 18 September 1534 from Viglius of Aytta to Hadrianus Marius, quoted in Waterbolk, 1977, p. 46.

15 Ibid., p. 46.

16 Viglius van Aytta to Joachim Hopper, quoted in Motley, 1900, p. 285.

17 Letter of 18 September 1534 from Viglius of Aytta to Hadrianus Marius, quoted by Waterbolk, 1977, p. 46.

18 Cranevelt was one of only eight recipients in the Low Countries to receive a fine limited edition of Erasmus' collected works.

19 Adrien Amerot de Quenville, also known as Adrianus Amerocius (in Mercator's letter of 4 August 1540 to Antoine Perronet), Amerotius and Amoury (see Bietenholz, 1985, vol. I, p. 48).

20 De Vocht, 1951–5, p. 137.

21 Letter of 6 May 1545 from Algoet to Dantiscus, de Vocht, 1961, p. 372. The maritime chart was compiled in Augsburg, in August 1530.

22 Letter from Laurin to Dantiscus, quoted by Van der Krogt, 1993, p. 39. Laurin's huge globe had been constructed in 1531.

9 TERRAE SANCTAE

1 Quoted in Karrow, 1993, p. 605.

2 Quoted in Nissen, 1956, p. 47.

3 Letter of 22 May 1567 from Mercator to Andreas Masius, Van Durme, 1959, p. 75.

4 Quoted in Nissen, 1956, p. 52.

5 *Man with a Fish*, Dirk Vellert, Antwerp, 16 August 1522, reproduced in Van der Stock, 1998, p. 103.

6 Matthew 17.

7 Micah 6:3.

8 Deuteronomy 8:7–10.

9 As separate sections, the map's booklet measured 275 x 180 mm; assembled with its decorative borders, the wall-map measured 1216 x 666 mm.

10 Many years later, Mercator complained in a letter to Masius that the map had been compromised by haste and by the inconsistency of his primary source.

11 Ghim, trans. in Osley, 1969, p. 186.

12 Matthew 17.

13 Ziegler was not the first German to illustrate this magnetic quirk. Twenty years earlier, the Nuremberg compass-maker Erhard Etzlaub had attached a map to the lid of a hinged sundial/compass which showed a magnetic deviation of 10 degrees.

10 NAMING AMERICA

1 Note to the reader, on Mercator's untitled world map of 1538, usually referred to as '*Orbis imago*'.

2 Berggren and Jones, 2000, p. 57.

3 Ibid., p. 57.

4 Mercator, Praefatio, *Tabulae geographicae Cl: Ptolemei...*, 1578, quoted in Babicz, 1994, p. 62.

5 Ibid.

6 Berggren and Jones, 2000, p. 93.

7 Ibid., p. 92.

8 Sylvanus' heart-shaped world map appeared in his edition of Ptolemy's *Geography*, published in Venice in 1511.

9 Johannes Werner's *Libellus de quatuor terrarum orbis in plano figurationibus* was published in Nuremberg in 1514, and contained three suggested heart-shaped projections.

10 Fine's map reappeared only four years later, in Glareanus' *De Geographia*, published in Freiburg in 1536.

11 Oronce Fine, Address to the reader, *Nova, et integra universi orbis descriptio*, 1531.

12 Ibid.

13 Ibid.

14 Proverbs 4:23.

15 Philip Melanchthon, *Apology to the Augsburg Confession*, quoted in Aune, 1994, p. 1.

16 Karrow, 1993, p. 174.

17 Ibid., p. 174.

18 Ibid., p. 175.

19 This conjecture is based on Van Raemdonck's identification of the 'J. Drosius' on Mercator's map with the 'Joannes Droeshaut de Bruxellis' of Louvain's *Liber immatriculationum* and the Jan Drusius/Jean Drusius/Droeshout (Campan, 1862, vol. 1, pp. 304–5) who was implicated in the Louvain heresy trials of 1543–4. Campan (1862, vol. 1, p. 304, fn. 1) took the view that 'Droeshout' was probably an ecclesiast.

20 Only two copies of Mercator's 'Orbis imago' of 1538 are known. One was discovered in a 1578 edition of Ptolemy, and the other turned up in a copy of Simon Grynaeus' *Novus Orbis Regionum*, lying loose beside Fine's hearts.

21 According to Nordenskiöld (1889, p. 107), this was one of the earliest indications of the river Plate and its principal tributaries.

22 Tibet, Cathay and the Land of the Lequii.

23 Berggren and Jones, 2000, p. 95. The 'Tanais' is now known as the river Don.

24 'Sipangi' is modern Japan.

25 Given the birth dates of her immediate siblings, Emerentia could have been born between the summer of 1538 and the summer of 1539.

26 Letter of 4 August 1540 to Perronet, in Van Durme, 1959, p. 15.

11 THE FALL OF GHENT

1 Quoted in Arnade, 1996, p. 197.

2 Van der Beke's map is the oldest surviving printed map of Flanders (Van der Gucht, 1994, p. 289).

3 Trans. in Delano Smith, 1985, p. 25.

4 Ibid., p. 25.

5 Ibid., p. 25.

6 Letter of 4 August 1561 from Richard Clough to Thomas Gresham, quoted in Marneff, 1996, p. 30.

7 Van Dis, *Reformatorische rederijkersspelen*, 33, quoted in Arnade, 1996, p. 200.

8 Richard Clough, quoted in Arnade, 1996, p. 200.

9 Letter of 6 May 1545 from Algoet to Dantiscus (de Vocht, 1961, p. 372), trans. in Karrow, 1993, p. 36.

10 Ghim, *Atlas*, p. 6.

11 The only remaining copy of Mercator's Flanders map is incomplete and bears no date. Ghim relates that Mercator produced the map after his Holy Land map of 1537 and before his manual on italic. The moment when a group of Ghent merchants would have been most motivated to commission an Imperial image of Flanders was during the autumn of 1539, once it had become clear that the rebellion was going to lead to Imperial retribution. De Smet (1962, p. 57), suggested that in the weeks preceding the emperor's arrival in Ghent, the merchants struck a secret deal with the councillors in Brussels, which led to Mercator's commission to produce a map of Flanders dedicated to the emperor.

12 Ghim, trans. in Osley, 1969, p. 186.

13 The panel which might have definitively identified the merchants is missing from the only surviving example of the map. The panel would have contained the address to the reader, and the map's date. When Mercator's map was printed, the association of three booksellers was however encoded at the foot of the map. There were no names, but the insignia and town names identified the three as Guillaume Van den Berg of Antwerp, Pierre de Keyzere, the publisher of Van der Beke's image of Imperial fragmentation and Barthelémi de Grave/Bartholomeus Gravius of Louvain. The printer of the map is not recorded on the surviving copy, but it may have been Gravius, for whom Mercator later designed a printer's device.

14 Ghim, trans. in Osley, 1969, p. 186.

15 Van Deventer left no map of Flanders. See de Smet, 1962, pp. 55–6, for a summary of the evidence suggesting that Flanders had been triangulated by Van Deventer prior to 1540. Kirmse (1957, pp. 24–5) concluded that Van Deventer would have had time to survey Flanders between 1536 and 1539, and that Van Deventer possibly drew his Flanders map ahead of his map of Holland.

16 Letter of 6 May 1545 from Algoet to Dantiscus (de Vocht, 1961, p. 372), trans. in Karrow, 1993, p. 36.

17 Osley, 1969, p. 65, took the view that the lettering on Mercator's map of Flanders bore 'some signs of haste'.

18 The assumption that Mercator was copying rather than originating is based on three factors: 1) Mercator's Flanders map was engraved at a similar scale (1:172,000) to Van Deventer's regional maps (1:180,000); 2) Mercator's Flanders map is the only one of his maps not to have latitude and longitude scales – which are not required by a cartographer using data collected by triangulated survey, or by a cartographer making a tracing; 3) Many of Mercator's vignettes – for towns, villages, abbeys and woods – are those used by Van Deventer, vignettes which Mercator would not employ on later maps.

19 De Smet (1962, p. 54) suggested that Mercator's source for this information may have been Jacques Meyer (or Meyerus), whose book (*Jacobi Meyeri Baliolani Flandricarum rerum tomi X*) was published in Bruges and Flanders in 1531, a suggestion supported by the presence in the list of Mercator's library titles of 'Compendium Chronicorum Flandriae Meyeri, 1538' (Watelet, 1994, p. 406).

20 Examining 100 measurements of distance between towns, Kirmse (1957) demonstrated that the average deviation from actual distance was only 34 metres per kilometre, or 3.4 per cent, while a sample of 100 angular measurements showed an average error of only 2 degrees 20 minutes. De Smet (1962, p. 55) pointed out that such accuracy was far in advance of anything achieved by other cartographers at the time.

21 Ghim, trans. in Osley, 1969, p. 186.
22 Mercator's map of Flanders (including the decorative borders) measured 1,230 x 950 mm; Van der Beke's map measured 990 x 750 mm.
23 Marneff, 1996, p. 30.
24 Romans, 11:20.
25 Among Mercator's many maps, this was the only one to have such a panel.
26 Letter of 4 August 1540 to Antoine Perronet, Van Durme, 1959, p. 15.
27 See Van de Krogt, 1993, p. 62 for a summary of the long debate concerning the year that this (undated) letter was published.
28 Letter of 4 August 1540 written from Mercator to Antoine Perronet, Van Durme, 1959, p. 15.
29 Letter of 4 August 1540 written from Mercator to Antoine Perronet trans. in Van der Krogt, 1993, p. 60.
30 Ibid., p. 60.
31 Although Van Durme (1959, p. 17, fn. 9) proposed that the 'Secretary Morillon' referred to by Mercator was 'without doubt Antoine Morillon', de Smet (1962, p. 47, fn. 65) suggested that the Morillon supporting Mercator's new globe was Antoine's father, Gui. Gui Morillon's relationship with Rescius and the geographer Damião de Gois supports de Smet's view. And Mercator would have been more likely to advertise the support of the influential Imperial secretary Gui Morillon than his son.
32 Letter of 4 August 1540 to Antoine Perronet, trans. in van der Krogt, 1993, p. 60.
33 Ibid., p. 60.

12 LATIN LETTERS

1 Doubt surrounds the publication date of *Literarum latinarum*, and also therefore the period during which the book was written and engraved. Although the publication date in the book is given as March 1540, de Smet (1962, pp. 59–60, fns. 92, 93, 94) believed that the book's printers may have been using the 'old style' Easter calendar; if that were the case, the 'new style' date of publication would have been March 1541. The fact that Mercator's printer Rescius had fallen out with the university (who had already switched to the new style year) adds to the likelihood that the book was indeed dated according to the old, Easter year, which was still followed by the town of Louvain.
2 Osley (1969, p. 27, f. 1) noted that Mercator's description 'Italic' was the earliest recorded use of the expression that he knew in the Low Countries.
3 Sigismondo dei Fanti, *Theorica et practica ... de modo scribendi*, 1514, noted in Osley, 1969, p. 27, fn. 1.
4 Ghim, trans. in Osley, 1969, p. 186.
5 Bietenholz, 1985, vol. III, p. 143.
6 Erasmus, *De recta latini graecique sermonis pronuntiatione*, 1528, Erasmus, 1974–, vol. 26, p. 392.
7 Osley, 1969, p. 177, fn. 1.
8 See Osley, 1969, p. 45, fn. 2.
9 Mercator, *Literarum latinarum, quas italicas, cursoriasque vocat, scribendarū ratio*, 1540/41, trans. in Osley, 1969, p. 123.
10 Ibid., p. 123.
11 Ibid., p. 123.

12 Erasmus, *De recta latini graecique sermonis pronuntiatione*, 1528, Erasmus, 1974–, vol. 26, p. 390.

13 Ibid., vol. 26, p. 391.

14 Mercator, *Literarum latinarum*, trans. in Osley, 1969, p. 123.

15 Ibid., p. 123.

16 Ibid., p. 123.

17 Ibid., p. 123.

18 Ibid., p. 124.

19 Ibid., p. 125.

20 Ibid., p. 128.

21 Ibid., p. 134.

22 Ibid., p. 163.

23 Ibid., p. 163.

24 Ibid., p. 171.

13 A MORE COMPLETE GLOBE

1 Ghim, trans. in Osley, 1969, p. 186.

2 No copies of this map have survived. See Karrow, 1993, pp. 210–11 for a discussion of Gemma's 'influential' world map.

3 When Ghim came to write his friend's 'Life' fifty years later, the only major work to be excluded was Mercator's double-cordiform map of 1538. Given Ghim's otherwise thorough catalogue of Mercator's works, the omission of the 1538 world map suggests that Mercator wanted to suppress its memory.

4 Letter of 4 August 1540 from Mercator to Perronet, trans. in Van der Krogt, 1993, p. 60.

5 Berggren and Jones, 2000, p. 57.

6 Gemma's globe of 1536/7 had a diameter of 370 mm; Mercator's globe of 1541 had a diameter of 420 mm. The 50 mm increase in diameter yielded Mercator a 28.85 per cent increase in surface area. (I thank Hol Crane for this figure.)

7 Letter of 4 August 1540 from Mercator to Perronet, trans. in Van der Krogt, 1993, p. 60.

8 Ibid., 1993, p. 60.

9 Ibid., 1993, p. 60.

10 Ibid., 1993, p. 60.

11 Mercator's 'Mangi' was Polo's 'Manzi', or China.

12 Letter of 4 August 1540 from Mercator to Perronet, trans. in Van der Krogt, 1993, p. 60. 'Scythia' was roughly equivalent to modern Ukraine and southern Russia; 'the two Sarmatias' were the 'Sarmatia Asie' and 'Sarmatia Europe' of his 1538 map, and corresponded roughly with modern Muscovy and Lithuania. 'Scondia' was modern Scandinavia, and the 'Deucaledonius Sea' was the Norwegian Sea.

13 Letter of 4 August 1540 from Mercator to Perronet, trans. in Van der Krogt, 1993, p. 60.

14 Mercator's 'Beach' is called 'Locach' or 'Lochac' in some copies of Marco Polo. (See Taylor, 1955, p. 104 and Skelton, 1958, p. 14.) Latham (1958) uses 'Lokak'. Beach/Locach corresponds with what became Siam, and is now Thailand. The shape Mercator gave to Beach and Maletur later gave rise to speculation that the north coast of Australia had been visited in the early sixteenth century.

15 Marco Polo's 'idolaters' are taken to be Buddhists.

16 Polo, 1958, p. 251.

17 Mercator's 'Maletur' is Marco Polo's 'Malaiur' or 'Malayur', now known as Malaya. Confusing Antarctica with Malaya and Thailand was one of Mercator's more consequential misconceptions.

18 Japan.

19 The Isle of Wight.

20 The Orkney islands.

21 Mercator's Britain was pretty much the shape we know today. See Barber, 1998, pp. 49–51, for a discussion of the sources Mercator may have used.

22 Also known as a loxodrome.

23 Following the chronological hypothesis outlined by Van de Krogt (1993, p. 62), the earliest date that the globe could have been completed was the first day of 1541, Easter style, i.e. 17 April in that particular year.

24 Ghim, *Atlas*, p. 6.

25 Karrow, 1993, p. 383.

14 ENEMY AT THE RAMPARTS

 1 Quoted in Brandi, 1939, p. 476.

 2 Quoted in Henne, 1859, vol. III, p. 179.

 3 See Edwards, 1994, p. 28. The figures quoted relate to the years 1518–44, during which period there were 'at least' 2,551 printings and reprintings of Luther's publications, compared to 514 Catholic printings.

 4 See ibid., p. 39.

 5 Matthew 22:21.

 6 Quoted in Verberckmoes, 1999, p. 85.

 7 Francisco Enzinas adopted the pen-name Dryander.

 8 *Fides, religio moresque Aethiopium* (Louvain, 1540).

 9 Details of this edition in Osley, 1969, pp. 54 and 177.

10 The reference to this meeting occurs in the letter of 23 February 1544 from Pierre de Corte to Maria of Hungary, in which de Corte associates Mercator's meeting with the two clerics with the most recent visit of the emperor to Brussels. Prior to the November 1543 visit, the emperor had last visited the city in the autumn of 1540, well before de Corte wrote the letter.

15 THE MOST UNJUST PERSECUTION

 1 Campan, 1862, vol. I, part 2, p. 298.

 2 Ibid., vol. I, part 2, pp. 298–9.

 3 Henne (1859, vol. IX, p. 58) suggested that 'while discoursing on the harmony of the works of God', Mercator 'allowed himself to be diverted ... into digressions on the subject of the influence that the word of God would henceforth exert on the destiny of the world', and that these digressions 'were enough of a sign for the inquisitors'. Henne's hypothesis has the ring of truth, especially since Monachus would have been a prime candidate with whom to share such digressions.

 4 Ghim, trans. in Osley, 1969, p. 193.

 5 Quoted in de Leyn, 1863, p. 30.

 6 Quoted in ibid., p. 30.

 7 Philippians 1: 21–4.

8 Campan, 1862, vol. 1, part 2, p. 445.

9 Ibid., vol. 1, part 2, p. 447.

10 Ibid., vol. 1, part 2, p. 445.

11 Only one letter (to Antoine Perronet) pre-dating Mercator's arrest has survived, and no autographed texts. It is not beyond possibility that Barbara – or Mercator himself – destroyed his own correspondence and disposed of any written texts which might be used as evidence to support the charge of heresy.

12 Campan, 1862, vol. 1, part 2, p. 321.

13 Ibid., vol. 1, part 2, p. 320.

14 Quoted in ibid., vol. 1, part, 2, p. 302, fn. 3.

16 THE SLIGHT YOUTH FROM THE NORTH

1 Letter of 9 October 1544 from Mercator to Perronet, quoted in Van Durme, 1959, p. 29.

2 In his letter to Perronet of 9 October 1544, Mercator gives as a reason for the late delivery of an astronomical ring: 'a lack of materials'.

3 Melanchthon to Camerarius, 17 March 1545, quoted in Boehmer, 1874, vol. 1, p. 144.

4 Quoted in Aubert, 1976, p. 135.

5 This was one of the contemporary sayings illustrated by Bruegel in his oil painting of 1559, *Flemish Proverbs*. See Vöhringer, 1999, pp. 54–7.

6 Letter of 9 October 1544 from Mercator to Perronet, Van Durme, 1959, p. 26.

7 Ibid., p. 29.

8 Ibid., p. 29.

9 Ibid., p. 26.

10 Ibid., p. 26.

11 Ibid., p. 29.

12 Ibid., p. 29.

13 Letter of 18 March 1545 to Antoine Perronet, Van Durme, 1959, p. 31.

14 Ghim, *Atlas*, p. 6.

15 The faulty astrolabe turned up in Brno four hundred years later, and is described in Turner, 1995, pp. 131–45.

16 Sir William Pickering, quoted in Deacon, 1968, p. 23.

17 Letter of 20 July 1558 from John Dee to Mercator, trans. in Shumaker, 1978, p. 112.

18 'The Compendious Rehearsall of John Dee ...', 9 November 1592, quoted in Taylor, 1930, p. 256.

19 Ibid., p. 256.

20 Letter of 20 July 1558 from John Dee to Mercator, trans. in Shumaker, 1978, p. 112.

21 'The Compendious Rehearsall of John Dee ...', 9 November 1592, quoted in Taylor, 1930, p. 256.

22 Ibid., p. 256.

23 Letter of 20 July 1558 from John Dee to Mercator, trans. in Shumaker, 1978, p. 112.

24 Petrus Peregrinus, or Pierre de Maricourt, lived in the thirteenth century and was greatly admired by Roger Bacon for being the only writer of his time to understand perspective. Probably from Picardy, Petrus Peregrinus wrote his treatise on the laws of magnetism in 1269. It was an immensely popular work, and Mercator would have known it well. In 1572, Jean Taisner published a plagiarized version of the treatise, from the Cologne press of Johann Birkmann, a printer known to Mercator.

25 Of Mercator's letter to Perronet dated February 1546, Harradon (1943, p. 200) wrote

that 'we find for the first time the view expressed and substantiated that the Earth has a magnetic pole'.

26 Letter of 23 February 1546 from Mercator to Antoine Perronet, trans. in Harradon, 1943, p. 201.

27 Ibid., p. 201.

28 Trans. in Shumaker, 1978, p. 117.

29 John Dee's diary entry for 7 December 1549, quoted in Fenton, 1998, p. 305.

30 Ghim, trans. in Osley, 1969, p. 186.

31 Ibid., p. 186.

32 John Dee's diary entry for 4 July 1549, quoted in Fenton, 1998, p. 305.

33 Inventory of Michiel Van der Heyden (1552), trans. by Marga Emlyn Jones, in Englander et al., 1990, pp. 138–40.

34 'The Compendious Rehearsall of John Dee ...', 9 November 1592, quoted in Woolley, 2001, p. 23.

35 Mercator's celestial globe of 1551 proved to be the only astronomical work he undertook.

36 Dekker (1999, p. 415) pointed out that Mercator was 'the first to design his celestial gores in equatorial coordinates', adding that his reason for doing so 'still eludes explanation'. But as Van der Krogt (1993, p. 67) noted, the reoriented gores contributed to the globe's aesthetic qualities. Given Mercator's attitude to perfection, and his supreme globe-making skills by this time, such an aesthetic effect would have been justified in spite of the engraving difficulties – difficulties that he was uniquely positioned to overcome.

37 Coma Berenices.

38 Osley, 1969, p. 65, wrote that Mercator's lettering on the 1551 globe possessed a 'serene confidence'.

39 Dekker (1999, p. 91) referred to Mercator's terrestrial globe of 1541 and celestial globe of 1551 as 'the most important pair of globes made in the sixteenth century' and (p. 415) 'highlights in the history of globe making'. Mercator's globe stand later became known as the 'Dutch style' and was used through to the eighteenth century (Dekker, 1999, pp. 415–16).

40 Trans. in Stevenson, 1921, p. 131.

41 Ibid., p. 131.

42 Quoted in Milz, 1994, p. 103.

43 See Milz (1994, p. 103) for a summary of these discussions.

44 Ghim, trans. in Osley, 1969, p. 187.

45 Quoted in Fry, 1995, p. 60.

17 SOMEWHERE WORTHY OF THE MUSES

1 Dedicatory letter of August 1585 to Johann Wilhelm the Younger, Duke of Jülich, Cleve, etc., *Atlas*, p. 185.

2 Ibid., p. 185.

3 Ghim, trans. in Osley, 1969, p. 190.

4 Ibid., p. 190.

5 Letter of 2 May 1559 from Molanus to Mercator, Van Durme, 1959, p. 40.

6 Ibid.

7 Henri Estienne, 1574, quoted in Westfall Thompson, 1911, p. 143.

8 Peter Beausart, the Louvain professor, writing in 1553, de Smet, 1962, p. 73, n. 133, trans. in Karrow, 1993, p. 383.

9 Ghim, trans. in Osley, 1969, p. 190.

10 Ghim, trans. in Osley, 1969, p. 193.

11 Ibid., p. 186.

12 Ibid., p. 187.

13 Trans. in Stevenson, 1921, p. 129.

14 Marino Cavalli, 1551, *Relazioni degli ambasciatori veniti al senato*, ed. Eugenio Albèri, series I, vol. II (Florence, 1840), quoted in Ross and McLaughlin, 1968, p. 300.

15 The measurements were 1,650 x 1,340 mm.

16 A cordiform projection devised by Johannes Stabius.

17 See Barber, 1998, pp. 51–4, for his detailed description of the source material probably used by Mercator.

18 *Nova totius Galliae descriptio*, first printed in 1525, in Paris.

19 Tschudi writing about his teenage journeys, quoted in Karrow, 1993, p. 547.

20 Ghim, trans. in Osley, 1969, p. 187.

21 Ibid., p. 187.

22 There is no evidence that Mercator completed and printed his guide to transcription.

18 FRANKFURT FAIR

1 This could have been the first time Mercator visited the fair; in 1552 Frankfurt had been under siege to Maurice of Saxony, and the fair had been postponed until Martinmas.

2 Two annual fairs had been held at Frankfurt since at least the thirteenth century. Book fairs were held here from at least 1485. See Westfall Thompson, 1911.

3 Josias Maler, quoted in ibid., p. 24.

4 Josias Maler, quoted in ibid., pp. 24–5.

5 Listed in Osley, 1969, p. 85.

6 Wied's map was published in 1555 and its association with Mercator is conjecture. Wied had been born on the Rhine upstream of Duisburg. The use of Dutch and Latin on the map suggested that the engraver had worked from the Low Countries, whilst the lettering followed Mercator's manual. Karrow (1993, p. 585) took the view that it was 'highly likely' that Mercator was the engraver; Osley (1969, p. 101) restricted himself to pointing out that the 'magnificent italic hand' bore a 'marked similarity to Mercator's early hand ... eg map of Flanders [1540]' and that the engraver was 'thoroughly imbued with Mercator's style'.

7 Francis Sweert, The Life of Abraham Ortell, *Ortelius*, 1606, unpaginated. The English spelling has been modified in this quote.

8 Gemma Frisius died on 25 May 1555.

9 Quoted in Febvre and Martin, 1976, p. 125.

10 The date on which the order was received by Plantin was recorded in the ledger as 19 March 1558. See Voet, 1962, p. 189.

19 SPIES AND CARDINALS

1 Dedicatory letter to Johann Wilhelm the Younger, Duke of Jülich, Cleve etc., *Atlas* p. 183.

2 Ibid., p. 183.

3 Ghim's description of Mercator's library, *Atlas*, p. 14.
4 Ghim, trans. in Osley, 1969, p. 192.
5 Ibid., p. 192.
6 Ibid., p. 193.
7 Ghim, 1595, quoted in ibid., p. 193. Johannes Molanus (1533–85) was also known as Jan Vermeulen.
8 Letter of 2 May 1559 from Molanus to Mercator, Van Durme, 1959, p. 40.
9 Ibid., p. 40.
10 Ibid., p. 40.
11 Ibid., p. 40.
12 Ibid., p. 40.
13 Ghim, 1595, trans. in Osley, 1969, p. 193.
14 Ibid., p. 191.
15 The purpose of Arnold's *Thule insula* remains a mystery. The view (repeated in Watelet, 1998, p. 13) that Arnold was correcting an error made by Mercator in his map of Europe ('on which he had represented the island the wrong way round') does not stand up to scrutiny, since Mercator maintained his version of Iceland on both the 1569 world map and on the second edition of his wall-map of Europe.
16 Letter of 20 July 1558 from Dee to Mercator, trans. in Shumaker, 1978, p. 113.
17 Quoted in Woolley, 2001, p. 23.
18 Ibid., p. 39.
19 The letters have not survived, but Dee refers in his letter to Mercator of 20 July 1558, to 'all the letters' he had 'at hand', which had been sent by Mercator (Shumaker, 1978, p. 113).
20 Letter of 20 July 1558 from John Dee to Mercator, trans. in Shumaker, 1978, p. 113; Mercator's original request for a list of Dee's publications may have been made as early as 1557, given Dee's illness.
21 The epidemics were probably influenza, and the total death toll may have been the heaviest for the century. See Williams, 1998, p. 110.
22 Letter of 20 July 1558 from John Dee to Mercator, trans. in Shumaker, 1978, p. 112.
23 Ibid., pp. 113, 115.
24 Ibid., p. 115.
25 Quoted in Woolley, 2001, p. 40.
26 Quoted in letter of 20 July 1558 from Dee to Mercator, trans. in Shumaker, 1978, p. 117.
27 Ibid., p. 115.
28 There seems little doubt that Dee did call at Duisburg between 1562 and 1564. Not only did the town lie close to his route, but Mercator was precisely the kind of bibliophile whom Dee was seeking. Taylor (1930, p. 85) took the view that Dee 'could hardly have failed to visit Mercator', although Taylor's hypothesis that Dee had been the 'English friend' who had supplied the manuscript of the map of Britain to Mercator has been disputed by Barber (Watelet, 1998, pp. 43–77). When Dee returned to London and published his improved edition of *Propaedeumata aphoristica*, he dedicated it to Mercator, a gesture which implied that the Englishman owed more to his Duisburg friend than to any other of his contemporaries.
29 Mercator's *Atlas* map of 1585, *Bolonia & Guines Comitatus*, was a copy of Nicolay's map of 1558.
30 See Moir, 1973, vol. 1, pp. 20–21; and Karrow, 1993, p. 439.

31 Quoted in Moir, 1973, vol. 1, pp. 12–13.

32 Ibid., p. 12.

33 John Elder's *Letter*, dated New Year's Day 1555, quoted in Starkey, 2000, p. 175.

34 Nicholas Throckmorton to William Cecil, 19 September 1559, *Calendar of State Papers Foreign, 1559*, London, 1864, p. 562, no. 1335; quoted in Barber, 1998, p. 70.

35 Letter of 24 December 1561 from Nicholas Throckmorton to Sir William Cecil, *Calendar of State Papers Foreign, 1561–62*, London, 1886, p. 455, no. 743; quoted in Barber, 1998, p. 70.

36 Quoted in Barber, 1998, p. 70.

37 Letter of 9 February 1561 or 1562 from Thomas Byschop to Sir William Cecil, *Calendar of the State Papers Relating to Scotland*, 1898, vol. 1, p. 602; quoted in Moir, 1973, vol. 1, p. 13.

38 Nobody has illuminated the provenance of Mercator's British Isles map of 1564 more thoroughly than Peter Barber (1998, pp. 43–77). Elder's authorship of the map which Mercator copied is not absolutely certain, but Barber argues that – of the several candidates who might have been responsible – it is Elder who 'would therefore seem to be the most likely person (possibly in association with John Rudd) to have created the prototype'.

39 Barber (1998, p. 70) suggests that the 'most likely hypothesis' is that the map was sent to Mercator 'shortly before ... December 1561'.

40 See Barber, 1998, p. 62.

41 See Barber (1998, pp. 55–63) for a fascinating examination of the probable sources used by the map's creator. Barber suggests (p. 66) that the prototype of Mercator's map had been drawn between 1548 and 1556.

42 Mercator, *Angliae & Scotiae & Hibernie nova descriptio*, 1564.

43 With due deference I join Peter Barber's 'parlour game' (Watelet, 1998, p. 55) by contributing yet another name to the list of candidates who might have been the mysterious 'friend'. Barber has very satisfactorily provided the likely identity of the map's original creator, yet there is no evidence that Elder and Mercator ever communicated, let alone regarded each other as 'friends' (unlikely, given Elder's openly extreme Catholic associations). Barber's suggestion that a third party might have been involved in transmitting the map between Elder and Mercator encourages me to introduce Cardinal de Granvelle as a significant player in the parlour game. Intriguingly, another of Mercator's friends – Christophe Plantin – was afoot in France between 1562 and 1563, having fled Antwerp for publishing heretical works. It is conceivable that Mercator was kept unaware of the map's original source, knowing only the 'friend' who was responsible for dispatching it to Duisburg. Ghim's statement that the map had been sent 'from England' (Osley, 1969, p. 187) can be taken as little more than an understandable guess, contributed thirty years after the event.

44 Quoted in O'Malley, 1964, p. 285.

45 In the four years since Plantin had been selling Mercator's works, orders had amounted to a modest twenty-five maps and three globes, but the sudden interruption in business caused by Plantin's flight undoubtedly affected Mercator's income.

46 See Verberckmoes, 1999, p. 97.

47 Henne, A. (ed.), *Memoires de Pontus Payen*, Brussels, 1860–61, vol. 1, p. 64, quoted in Verberckmoes, 1999, p. 97.

48 *Letters and Papers*, xiv. i. 404, 530–1, and xiv. ii. 151–3, quoted in Mackie, 1952, p. 399.

49 This was one of the sayings which Bruegel had included five years earlier on his oil painting *Flemish Proverbs* (1559), a painting familiar to the Bruegel collector Granvelle. See Vöhringer, 1999, pp. 54–7.

50 The largest single order which Plantin received for Mercator's British Isles map came from Paris in January 1566 and amounted to forty (uncoloured) copies. This single order represented more than half of the seventy-seven copies Plantin sold in the first five years of the map's existence (see de Voet, 1962, pp. 189 and 201–5).

51 Mercator, *Angliae & Scotiae & Hibernie nova descriptio*, 1564.

20 RENÉ'S DOMAIN

1 Quoted in Skelton, 1966, p. vii.

2 From the address to the reader in the 1513 Ptolemy, quoted in ibid., p. xvi.

3 St Dié.

4 Waldseemüller's 'address to the reader' in the 1513 Ptolemy, quoted in Skelton, 1966, p. xv.

5 Published in 1552, and containing descriptions of roads, with distances between places.

6 Quoted in Hellwig, 1994, p. 306.

7 There is no confirmed link between Granvelle and Mercator's map of Lorraine, but I find it impossible to believe that the cardinal's presence in Nancy that spring can have been unrelated to the commission which followed.

8 In a typical year, the Vosges could expect a rainfall four times the total of the lower Moselle. See Departments of State and Official Bodies ..., 1919.

9 Mercator's decision to accept the surveying commission of Lorraine had never been adequately explained. Given Mercator's age, standing and inclinations, the most plausible motive has to be the promise of access to the map archives and libraries of Nancy and St Dié.

10 Quoted in Hellwig, 1994, p. 297.

11 Alix to Duke Charles IIII, quoted in ibid., p. 311.

12 Lorraine appeared on two sheets in the first, 1585, edition of the *Atlas* maps.

13 These villages appear on modern French maps as Herbaupaire, Lusse, La Pariée, haute Merlusse and Les Trois Maisons.

14 Now St Pierremont.

15 Errors and absences in the mapping of the southern Vosges suggest, for example, that neither Mercator nor his son pushed all the way up the valley of the 'Lestraye'. Lorraine's borders are also missing from the Vosges.

21 HUNTERS IN THE SNOW

1 Ghim, trans. in Osley, 1969, p. 187.

2 Ibid., p. 187.

3 By the duke's doctor, Antoine le Pois. See Hellwig, 1994, p. 299.

4 Quoted in Buisseret, 1992, p. 107.

5 Thirty years later, Alix revealed his enduring grudge in a letter to Charles III in which he reminded the duke that Mercator's map of Lorraine, still to be found in the duke's cabinet, was flawed and incomplete. On Mercator's map of Lorraine in his *Atlas* of 1595, the disputed town of Sarrebourg was labelled as a 'town of the Empire', a snub (perhaps inadvertent) to Lotharingians which must have added a dose of Sar salt to

Alix's wounds, had that label been transferred from Mercator's original map of 1564.

6 For those holding Mercator's views on religion, it was not unusual for illness to be seen as an outcome of God's anger, an interpretation which frequently led to depression. The Emperor Maximilian II was such a sufferer, beseeching his Maker in 1563 to rid him of his 'weakness' (see Fichtner, 2001, p. 103).

7 Secretary Bave to Cardinal Granvelle, 4 December 1565, E. Poullet, *Correspondence ... de Granvelle*, I, 27, quoted in Parker, 1990, p. 57 and p. 285, fn. 33.

8 Letter of 24 March 1566 from Molanus to Mercator, Van Durme, 1959, p. 63.

9 Ibid., p. 63.

10 Ibid., p. 64.

11 Ibid., p. 64.

12 In summarizing the dominant characteristic of exiled humanists on the lower Rhine, Forster (1971, p. 68) quoted Van Dorsten's phrase 'non-disputative religio-political engagement'. Exiles in Utenhove's circle were, wrote Forster (p. 68), 'tolerant towards Catholics but nonetheless [exercised] a firmly protestant engagement'.

13 Quoted in Vermij, 1995, p. 190.

14 Matthew 24:22.

15 See letters of 31 May 1585, 14 April 1586, 18 May 1586 from Mercator to Heinrich Rantzau, Van Durme, 1959, pp. 191–4, 201–3.

16 Quoted in Parker, 1990, p. 76.

17 Hosea 8: 4–7.

18 Hosea 8: 14 and 10:2.

19 This was one of the popular contemporary sayings illustrated seven years earlier by Bruegel in his oil painting *Flemish Proverbs* (1559). See Vöhringer, 1999, pp. 54–7.

20 Abraham Ortelius to Emanuel Demetrius, 27 August 1566, Hessels, 1887, p. 37.

21 Ibid., p. 37.

22 A STUDY OF THE WHOLE UNIVERSE

1 Given Mercator's commitments in Lorraine in 1564, and his subsequent 'derangement', it would seem unlikely that he began working on the cosmography in earnest until 1565 at the earliest – although he may have conceived its structure (as Keuning, 1947, suggests) whilst teaching at the Duisburg *Gymnasium*.

2 See dedicatory note to Johann Wilhelm the Younger, Duke of Jülich, Cleve, etc., *Atlas*, p. 183.

3 Praefatio ad lectorem, *Chronologia*, 1569, p. 3.

4 Ibid., p. 3.

5 Ibid., p. 3.

6 Ibid., p. 3.

7 Ibid., p. 3.

8 Ibid., p. 3.

9 Ibid., p. 3.

10 Ibid., p. 3.

11 Ibid., p. 3.

12 Ibid., p. 3.

13 Ibid., p. 3.

14 Ibid., p. 3.

15 Ibid., p. 3.

16 Mercator, Prolegomenon to The fabric of the world, *Atlas*, p. 27.

17 Letter of 19 May 1567 from Molanus to Mercator, Van Durme, 1959, p. 70.
18 Ibid., p. 70.
19 Ibid., p. 70.
20 Ibid., p. 70.
21 Ibid., p. 71.
22 Ibid., p. 71.
23 Ibid., p. 71.
24 Ibid., p. 72.
25 Ibid., p. 73.
26 Ibid., p. 73.
27 Ghim, trans. in Osley, 1969, p. 192.

23 TIME ...

1 The authors are listed in the fourth part of the *Chronologia*, 1569.
2 Mercator, 'Praefatio ad lectorem', *Chronologia*, 1569, p. 4.
3 Erasmus, Ep. 784, 1518, quoted in Bietenholz, 1985, vol. 1, p. 61.
4 Quoted in Hale, 1993, p. 591.
5 Mercator, *Chronologia*, 1569, p. 335.
6 Hosea 10:12.
7 Hosea 14:4.
8 Hosea 14:7.
9 Henricus Oliverius, or Heinrich Bars (*c.* 1500–75) was also known as Olisleger.
10 Ghim, trans. in Osley, 1969, p. 189.
11 Ibid., p. 187.
12 Metellus, or Jean Matal, (1520–97) had encouraged Panvini to study antiquity. Later, he would be one of those to contribute to the *album amicorum* of Ortelius. His eulogy to Mercator was printed in the *Atlas* of 1595. Metellus also joined Guillaume Postel on the pages of an *album amicorum* compiled by the German humanist Arnold Andre Sittard (Newberry Center for Renaissance Studies: Newberry MS 5055).
13 Johannes Metellus, *Atlas*, p. 3.
14 Quoted by Ghim, trans. in Osley, 1969, p. 189.
15 Mercator's *Chronologia* was listed in the *Index librorum prohibitorum* of Parma (1580) and of Rome (1596).

24 ... AND PLACE

1 Letter of 23 February 1546 from Mercator to Antoine Perronet, trans. in Harradon, 1943, p. 201.
2 In recent years, connections have been sought between the projection employed and described by Mercator on his world map and a similar projection which was used sixty years earlier by the Nuremburg compass-maker Erhard Etzlaub, for a map of Europe and North Africa (see, for example, Englisch, 1996, pp. 103–18). Etzlaub's map was attached to the hinged lid of a compass/sundial and – like Mercator's – its meridians were parallel and the latitudinal markings were progressively spaced. Etzlaub's map was small (the entire instrument measured 84 x 116 mm) but remarkably accurate: the deviation of the meridian segments from modern values averages no more than 0.17 mm. While Etzlaub was first to illustrate the projection's use, there is no evidence that Mercator had ever seen or owned one of these compass/sundials.

The two men approached the problem from different angles: Etzlaub's purpose was to design a way-finding device for land-travel; Mercator's was to devise a projection which would harmonize spherical and planar depictions of the earth's surface. Mercator's projection shares a far closer genealogy with Pedro Nuñez, and the light he cast on the inherent errors of sea-charts which failed to account for the convergence of the meridians (see Taylor, 1930, pp. 83–4).

3 Mercator's method was almost certainly empirical: it was not until 1599 that the Englishman Edward Wright demonstrated how to plot 'Mercator's Projection' with his table of meridional parts, in *Certaine Errors in Navigation*.

4 Ghim, trans. in Osley, 1969, p. 187.

5 Mercator, *Nova et aucta orbis terrae descriptio ad usum navigantium emendatè accommodata* (hereafter *Nova et aucta orbis*), Duisburg, 1569.

6 Ghim, trans. in Osley, 1969, p. 187.

7 The world map measured 2,024 x 1,236 mm, and was published in Duisburg, in August 1569.

8 Mercator, *Nova et aucta orbis*, trans. in *The Hydrographic Review*, vol. IX, no. 2, November 1932, p. 11.

9 Ibid., p. 11.

10 Hakluyt's Reference in the *Discourse of Western Planting*, 1584, p. 102, quoted in Taylor, 1930, p. 272.

11 They appear on the sale catalogue of Mercator's library (Watelet, 1994, p. 410) as 'Volume primo delle Navigatione & Viaggi fol. Volume secundo delle navigatione. fol. Volume terzo delle navigatione. fol.'

12 Quoted in Lucas, 1898, p. 8.

13 Quoted in Hobbs, 1949, p. 16, fn. 3. Hobbs referred to Ruscelli's Italian edition of 1561, and it was this edition which Moletius revised and published the following year. Mercator used the Moletius edition as a sources while researching his own edition of Ptolemy's *Geography*.

14 A long and entertaining controversy surrounds the Zeno map. Modern views range from that of Lucas (1898), who regarded Zeno as a fraud, to Hobbs (1949), who felt that his own argument, 'conclusively proves them [the Zeno brothers] to have been honest and reliable explorers who were far in advance of their age' (Hobbs, 1949, p. 19). Karrow (1993, p. 602) concluded that Zeno's map and book describing a voyage along the shores of Greenland 'can in no wise be considered a record of such an event'.

15 Since perhaps the mid 1550s. If Mercator had read Cnoyen's travels before 1551, he would surely have shared its astonishing contents with John Dee – who knew so little of Cnoyen's account by 1577 that he had to write to Mercator for further information.

16 Letter of 28 July 1580 from Mercator to Hakluyt, Hakluyt, 1907, vol. 2, p. 226. Taylor (1956, p. 61) took the view that this friend was 'probably Ortelius'. Karrow (1993, p. 409) asked whether it was possible that the friend was Monachus.

17 Letter of 28 July 1580 from Mercator to Hakluyt, Hakluyt, 1907, vol. 2, p. 216.

18 Mercator, *Nova et aucta orbis*, trans. in *The Hydrographic Review*, 1932, p. 27. Watelet (1994, p. 410) recorded that Mercator's library contained 'The history of the moste noble and worthey King Arthur. fol. 1557'.

19 Mercator, *Nova et aucta orbis*, trans. in *The Hydrographic Review*, 1932, p. 27.

20 Ibid., p. 27.

21 Ibid., p. 27.

22 Ibid., p. 27.

23 2 Corinthians 12:2–4.

25 'LIKETH, LOVETH, GETTETH AND USETH'

1 Mercator, Praefatio ad lectorem, *Chronologia*, 1569, p. 3.

2 In his letter of 9 May 1572 to Ortelius, Mercator mentioned that he had 'been occupied for some years in correcting Ptolemy and the recent maps' (trans. in Hessels, 1887, p. 88). 'Some years' might suggest that Mercator began working on the cartographic elements of the cosmography no later than 1570.

3 See *Chronologia*, 1568, Nij.

4 Quoted by Babicz, 1994, p. 62.

5 Mercator, Praefatio, *Tabulae geographicae Cl:Ptolemei* ..., 1578, quoted by Babicz, 1994, p. 62.

6 Ibid., p. 63.

7 Ibid., pp. 62–3.

8 Ibid., p. 63.

9 Ibid., p. 63.

10 Ibid., p. 63.

11 The friend in Cologne was the jurist Johannes Helmann (Averdunk and Müller-Reinhard, 1914, p. 77).

12 Servetus (*c.* 1511–53) changed his name to Villanovanus in the 1530s.

13 Servetus burned despite the fact that the statement concerning the infertility of the Holy Land had been on the verso of the map since its appearance in the Lorenz Fries edition of 1522.

14 See de Voet, 1962, pp. 189–221 for figures describing Mercator's sales to Plantin during this period.

15 Hessels, 1887, p. 30.

16 See Bagrow, 1948, pp. 18–20, for an inventory of Viglius' map collection.

17 John Dee, *Preface to the English Euclid*, 1590, quoted in Taylor, 1930, p. 283.

18 Abraham Ortelius ... to the courteous Reader, Ortelius, 1606, unpaginated. The English has been modified in this and the following quotes.

19 Twenty-five years later, Mercator's first biographer Walter Ghim countered the view that Ortelius had conceived his book of maps before Mercator: 'Moreover,' claimed Ghim, 'long before Abraham Ortelius, Mercator had formed the idea of publishing further particular and general maps of the world in a small format. He had drawn a considerable number of models with his pen and had measured out the distance between the places in their proper proportion, so that it only remained to engrave them. Since however, this Ortelius was an intimate friend of his, he purposely held up the enterprise he had begun, until Ortelius had sold a large quantity of his *Theatrum Orbis Terrarum* ... before he published his own small-scale maps' (trans. in Osley, 1969, p. 188). Mercator may well have been thinking about a volume of maps long before he announced his intention to proceed with the cosmography, but by 1570, he was in no position to publish such a volume. Ghim, who was fastidiously accurate, and who wrote his biography in the difficult months after Mercator's death, may have been fed a loyal embellishment of the truth by Mercator's son Rumold, or grandsons, all then engaged in pulling together the disparate elements of the *Atlas* as Ortelius' *Theatrum* continued its extraordinary run of success.

20 Abraham Ortelius ... to the courteous Reader, Ortelius, 1606, unpaginated.

21 Adolphi Mekerchi, Frontispicii Explicatio, Ortelius, 1570, unpaginated, trans. in Van den Broecke, Van der Krogt and Meurer, 1998, p. 169.

22 Abraham Ortelius ... to the courteous Reader, Ortelius, 1606, unpaginated.

23 Abraham Ortelius, A description of the whole world, Ortelius, 1606, unpaginated.

24 Letter of *c.* 1570 from Petrus Bizarus to Ortelius, trans. in Hessels, 1887, p. 75.

25 Letter of 30 October 1570 from Johannes Crato von Krafftheim to Abraham Ortelius, trans. in Hessels, 1887, p. 70.

26 Ibid., p. 70.

27 Letter of 22 November 1570 from Mercator to Ortelius, trans. in Hessels, 1887, p. 73.

28 Abraham Ortelius ... to the courteous Reader; Ortelius, 1606, unpaginated.

29 Watelet (1998, p. 7) suggested that this client might have been Werner von Gymnich.

30 Watelet (1998, p. 10) pointed out that – according to Gymnich's biography – the marshal was travelling in Italy between 1572 and 1575.

31 Trans. in Hakluyt, 1907, vol. 2, p. 368 (the English in this and the following quotes has been modified). The new legend was placed in a redesigned cartouche off the north-west coast of Scandinavia.

32 Trans. in ibid., vol. 2, p. 368.

33 Trans. in ibid., vol. 2, p. 368.

34 John Dee, *Preface to the English Euclid*, 1570, quoted in Taylor, 1930, p. 105.

35 Ibid., p. 105.

36 Ibid., p. 105.

37 Ibid., p. 105.

38 Letter of 9 May 1572 from Mercator to Ortelius, trans. in Hessels, 1887, pp. 87–8.

39 Letter of 9 May 1572 from Mercator to Ortelius, trans. in Ortelius, 1955, p. 22.

40 Letter of 9 May 1572, from Mercator to Ortelius, trans. in Hessels, 1887, p. 88.

41 Letter of 20 October 1572 from Daniel Rogers to Abraham Ortelius, trans. in Hessels, 1887, p. 100.

42 Ibid., p. 100.

43 Letter of 30 September 1571 from Hieronymus de Rhoda to Abraham Ortelius, trans. in Hessels, 1887, p. 81.

44 The 1598 edition of *Civitates Orbis Terrarum* carried an engraving of a famous stone slab outside Poitiers. Celebrated by Rabelais in *Pantagruel*, the slab had been a symbolic rendezvous for Poitiers students, who used to scratch their names on its surface. In the version which appeared in Braun and Hogenberg's book, the names on the slab were those of geographers and publishers, among them Mercator, Ortelius, Hogenberg and Hoefnagel, accompanied by the dates 1560 and 1561. The engraving and the dates have given rise to speculation that Mercator travelled to Poitiers with his humanist friends, and scratched his name on the Rabelaisian Pierre Levée. But there is no other evidence that Mercator made the trip, and neither was it remotely likely to have occurred. Mercator was not the kind of man to make a long, hazardous journey across France during the wars of religion, to Poitiers (of all places).

45 Simon van den Neuvel adopted the name Novellanus.

46 Letter of 5 September 1564, from Andreas Masius to Georg Cassander, trans. in Van Durme, 1959, p. 60.

47 Abraham Ortelius, A description of the whole world, Ortelius, 1606, unpaginated.

48 Abraham Ortelius, Catalogus Auctorum, Ortelius, 1606, unpaginated.

49 Quoted in Taylor, 1955, p. 104.

50 Francis Sweertius, Life of Abraham Ortell, Ortelius, 1606, unpaginated.

51 Letter of 13 August 1573, from Mercator to Johannes Vivianus, Van Durme, 1959, p. 107.

52 Ibid., p. 108.

53 Ibid., p. 108.

54 Ibid., p. 108.

55 The unspecified illness is mentioned by Mercator in a letter of 23 October 1578 to Johannes Crato von Krafftheim. In it, Mercator reports that he had fallen seriously ill 'a little after' he had first received a letter from Crato, a letter which had been written in 'the sixth year' – that is *c.* 1572 – before Mercator eventually replied in October 1578. The illness therefore can be dated approximately to 1752–73.

56 The date of the journey – 1575 – was recorded by Ortelius' first biographer, his friend Francis Sweertius (see Life of Abraham Ortell, Ortelius, 1606, unpaginated).

57 Ortelius and Vivianus, 1584, trans. Schmidt-Ott, in Van den Broecke, Van der Krogt and Meurer, 1998, p. 375.

58 Mercator, *Nova et aucta orbis*, trans. in *The Hydrographic Review*, vol. ix, no. 2, November 1932, p. 45.

59 Listed as 'Le livre du Chevallier Monsieur Ian Mandeville' by Cherton and Watelet, 1994, p. 409.

60 Ortelius and Vivianus, 1584, trans. Schmidt-Ott, in Van den Broecke, Van der Krogt and Meurer, 1998, p. 375.

61 Giraffes, Mandeville, 1983, p. 177.

62 Four of the miles Ortelius was using in his *Itinerarium* was roughly equivalent to 22 kilometres.

63 For understandable reasons, Antoine Le Pois – 'Anthony Pea' – used the pen-name Antonius Piso.

64 Letter of 25 March 1575 from Antonius Piso to Abrahamus Ortelius, trans. in Hessels, 1887, p. 125.

65 Ibid., p. 125. The sketch map was presumably Waldseemüller's woodcut in the Strasbourg Ptolemy of 1513.

66 Ortelius and Vivianus, 1584, p. 41.

67 Schmidt-Ott, in Van den Broecke, Van der Krogt and Meurer, 1998, p. 370, estimated that the journey lasted for approximately two months.

68 Letter of 9 May 1572 from Mercator to Ortelius, trans. in Hessels, 1887, p. 88.

69 Hogenberg's portrait of 1574 reappeared to great effect in the *Atlas* of 1595.

70 Mercator's description of Crato's letter; in a letter of 23 October 1578 from Mercator to Crato von Krafftheim, Van Durme, 1959, pp. 152–3.

71 Letter of 23 October 1578 from Mercator to Crato von Krafftheim, Van Durme, 1959, p. 152.

72 Fichtner, 2001, p. 84.

26 PTOLEMY CORRECTED

1 Letter of 24 March 1574 from Mercator to Camerarius, Van Durme, p. 109.

2 Letter of 30 August 1574 from Mercator to Camerarius, Van Durme, 1959, p. 110.

3 Ibid., p. 110.

4 Ibid., p. 110.

5 The nature of the gift was not disclosed in Mercator's reply. Van Durme (1959, p. 114, fn. 1) suggested that the gift may have been a manuscript account of Ortelius'

and Vivianus' recent journey, but such a journey could not have been undertaken so early in the year. Perhaps the 'little gift' was simply one of Ortelius' coins or medals, or a book.

6 Letter of 26 March 1575 from Mercator to Ortelius, Van Durme, 1959, p. 113.

7 Ibid., p. 113.

8 Jean Bodin, *Methodus ad facilem historiarum cognitionem*, 1566, quoted in Van Raemdonck, 1869, p. 283, fn. 1.

9 Ibid., p. 283.

10 Ibid., p. 283.

11 Letter from Mercator to Ortelius 26 March 1575, Van Durme, 1959, pp. 113–14.

12 Ibid., p. 114.

13 Ibid., p. 114.

14 Letter of 20 August 1575 from Mercator to Camerarius, Van Durme, 1959, p. 124.

15 Ibid., p. 124.

16 Letter of 2 April 1576 from Mercator to Camerarius, Van Durme, 1959, p. 126.

17 Letter of 14 November 1576 from Geeraert Janssen to Jacop Cool (Senior), trans. in Hessels, 1887, p. 146.

18 Letter of 22 June 1577 from Mercator to Langer, trans. in Boyce, 1948, p. 132.

19 Ibid., p. 132.

20 Ibid., p. 132.

21 Ibid., p. 132.

22 Ibid., p. 132.

23 Letter of 31 August 1577 from Mercator to Zwinger, trans. in Boyce, 1948. p. 134.

24 Ibid., p. 134.

25 Ibid., p. 134.

26 Ibid., p. 134.

27 Letter of 4 September 1577 from Mercator to Camerarius, trans. in Boyce, 1948, p. 135.

28 It was not until 10 March 1578 that Mercator wrote to Camerarius, informing him that the 'vexations' of the Ptolemy were past.

29 Letter of 23 October 1578 from Mercator to Crato von Krafftheim, trans. in Boyce, 1948, p. 135.

30 Ibid., p. 135.

31 Letter of 10 March 1578 from Mercator to Camerarius, Van Durme, 1959, p. 150.

32 Ibid., p. 150.

33 Ibid., p. 150.

34 Ibid., p. 150.

35 In 1580; de Voet, 1962, p. 195.

36 The licence was dated 3 February 1578, Brussels.

37 Letter of 23 October 1578 from Mercator to Crato von Krafftheim, Van Durme, 1959, p. 152, trans. in Boyce, 1948, p. 135.

38 Letter of 23 October 1578 from Mercator to Crato von Krafftheim, Van Durme 1959, p. 152.

39 Ibid., p. 152.

40 Ibid., p. 152.

41 Ptolemy's *Geography*, trans. in Stevens, 1908, p. 157.

42 Among the elements of Mercator's Flanders map which de Jode copied were the four bears rampant, the defensive dyke and cannon down by St Omer, and the designations

for male and female abbeys and priories. There is no evidence that Mercator and de Jode ever corresponded.

43 Letter of 1579 from Guillaume Postel to Abrahamus Ortelius, trans. in Hessels, 1887, p. 186.

27 ADORN YOUR BRITTANIA!

1 Skelton (1962, p. 162), considered that the letter 'should probably be dated soon after 1576'. Given Rumold's connections with Dee and Hakluyt, it would seem likely that Mercator's son heard about Frobisher's voyage very shortly after Frobisher's return to England in October 1576, and knowing the significance of the news, passed it on to his father without delay. On this basis, the letter may have been written as early as November 1576.

2 Letter of *c.* 1576–7 from Mercator to Rumold, trans. in Hakluyt, 1582, p. 1. The letter from Rumold has not survived, and only a single passage from Mercator's response is known. The passage was transcribed and translated into the dedicatory epistle ('To the right worshipfull and most vertuous Gentleman master Phillip Sydney Esquire') of *Divers foyages touching the discoverie of America etc.*, 1582. The English spellings have been modified.

3 In January 1577, Dee had instructed Ortelius that an urgent letter could reach him 'through the servants of the Birckmann's' (John Dee to Abraham Ortelius, 16 January 1577, Hessels, 1887, p. 158).

4 Letter of *c.* 1576–7 from Mercator to Rumold Mercator, trans. in Hakluyt, 1582, p. 2.

5 Ibid., p. 2.

6 Quoted by Skelton, 1962, p. 160.

7 Quoted in *Encyclopedia Britannica*, 1910, 11th edition, vol. 11, p. 237.

8 *Michael Lok's Notes on the Frobisher Voyage*, 26th Jan, 1578/9, quoted in Taylor, 1930, p. 271.

9 Quoted in *Encyclopedia Britannica*, 1910, 11th edition, vol. 11, p. 237.

10 *Michael Lok's Notes on the Frobisher Voyage*, 26th Jan, 1578/9, quoted in Taylor, 1930, p. 271.

11 Quoted in Kelsey, 2000, p. 76.

12 Letter of 16 January 1577 from John Dee to Abraham Ortelius, trans. in Hessels, 1887, p. 158.

13 Quoted in Sherman, 1995, p. 149.

14 John Dee, *General and Rare Memorials ...*, 1577, quoted in Sherman, 1995, p. 150.

15 Quoted in Taylor, 1956, p. 68. Dee summarized the evidence of Queen Elizabeth's title to her Arctic lands on the back of a map which he compiled, dated 1580.

16 Letter of 16 January 1577 from John Dee to Abraham Ortelius, trans in Hessels, 1887, p. 158.

17 Ibid., p. 158.

18 Ortelius visited London in spring 1577, and was with Dee at Mortlake on 12 March (see Taylor, 1956, p. 61).

19 Hakluyt's reference in the *Discourse of Western Planting*, 1584, p. 102, quoted in Taylor, 1930, p. 272.

20 Letter of 1577 from Mercator to John Dee, quoted in Taylor, 1956, p. 56.

21 See Taylor, 1956, pp. 56–68.

22 Dee, 1577, quoted in Taylor, 1956, p. 56.

23 Ibid., p. 56.

24 Ibid., p. 56. One of the many curious circumstances surrounding this episode is Dee's claim that Mercator responded to the request for information 'spedily'. For a man whose epistolary response time once extended to six years, this promptness was out of character.

25 Letter of 20 April 1577 from Mercator to Dee, transcribed by Dee and quoted in Taylor, 1956, p. 57.

26 Ibid., p. 58.

27 Ibid., p. 59.

28 Ibid., p. 57.

29 Letter of 13 May 1577 from Dee to Mercator. See Van Durme, 1959, pp. 140–1.

30 Quoted in *Encyclopedia Britannica*, 1910, 11th edn, vol. 11, p. 238.

31 Fenton, 1998, p. 2.

32 Quoted in Taylor, 1956, p. 68.

33 The letter has not survived, but Mercator referred to it in the opening of his response to Camden, dated 31 January 1579. See Watelet, 1994, p. 127.

34 Letter of 4 August 1577 from Camden to Ortelius, trans. in Hessels, 1887, p. 167.

35 See ibid., pp. 167–8.

36 Letter of 24 October 1578 from Camden to Ortelius, trans. in Hessels, 1887, p. 181.

37 Letter of 31 January 1579 from Mercator to Camden, Van Durme, 1959, p. 154.

38 Ibid., p. 154.

39 Ibid., p. 154.

40 Ibid., p. 154.

41 Ibid., p. 154.

42 Ibid., p. 154.

43 Letter of 12 December 1580 from Mercator to Ortelius, trans. in Kraus, 1970, p. 86.

44 Arthur Pitt: Arthur Pet, or Pett. Ibid., p. 86.

45 Not for the first time, questions arise regarding the identity of Mercator's English friends. Richard Hakluyt (the younger) did write to Mercator concerning Pet's voyage, and his letter is that noted by Mercator as arriving in Duisburg on 19 June 1580 (see Letter of 28 July 1580 from Mercator to Hakluyt, trans. in Hakluyt, 1907, vol. 2, p. 224). But in a letter to Ortelius dated 12 December 1580, Mercator stated that 'in April this year' he had been informed of Pet's proposed voyage. Van Durme (1959, p. 157) has suggested that this earlier informant might have been John Dee. But it seems more likely that Mercator was referring to Hakluyt's original inquiry, received by Mercator on 19 June, and answered on 28 July (letter of 28 July 1580, Hakluyt, 1907, pp. 224–6). Since Pet sailed on 30 May, it would have been reasonable for Hakluyt to have written to Mercator in April, if he was expecting a reply which could be acted upon prior to Pet's departure. Taylor (1930, pp. 127–8) suggested that Mercator may have deliberately delayed his response to Hakluyt because he 'was not ready to promote a foreign discovery adventure'. But nothing in Mercator's character suggests that he was competent of guile; it seems more likely that the combination of disrupted communications on the continent, and Mercator's habitual slowness in responding to letters, were responsible for the delayed dispatch of a response to Hakluyt – a response which could not be said to have withheld geographical information.

46 Quoted by Coote and Beazley, *Encyclopedia Britannica*, 1910, 11th edn, vol. 12, p. 828.

47 Letter of 28 July 1580 from Mercator to Hakluyt, trans. in Hakluyt, 1907, vol. 2, p. 224 (the English in this and the following quotes has been modified).

48 Ibid., p. 224.

49 Ibid., p. 224.
50 Ibid., p. 224.
51 Ibid., pp. 224–5.
52 Ibid., p. 226.
53 Mercator seems unlikely to have been deliberately concealing his source from Hakluyt (and Dee) and neither did he have an apparent motive for doing so, beyond perhaps a concern that sharing the manuscript would reveal that the configuration of the Arctic owed as much to Mercator's interpretation as it did to the specifics of Cnoyen's text. The Cnoyen manuscript had indeed probably been lost. Sixteenth-century libraries could be confusing: Dee notoriously ordered his collection of over 3,000 books by size, and many were unbound, stacked face up and lacking titles. Mercator's own library may have been similarly difficult to navigate. One week after Mercator wrote to Hakluyt explaining that the Cnoyen had gone missing, Gregor Braun wrote to Ortelius complaining that books he had lent to Cusantus had disappeared: 'when I asked for them back, [he] calmly replied he had lost them' (trans. in Ortelius, 1955).
54 Letter of 28 July 1580 from Mercator to Hakluyt, trans. in Hakluyt, 1907, vol. 2, p. 226.
55 Ibid., p. 226.
56 Letter of 12 December 1580 from Mercator to Ortelius, trans. in Kraus, 1970, p. 86.
57 Ibid., p. 86.
58 Ibid., p. 86.
59 Ibid., p. 86.
60 Letter of 20 February 1581 from John Balak to Mercator, trans. in Hakluyt, 1907, vol. 2, pp. 364–7. The identity of John Balak is something of a mystery; Van Durme (1969, p. 164) suggested that Mercator and Balak must have lived together at the university of Louvain, and it is true that the geographical texts Balak recalls them enjoying were those of authors whom Mercator would regard as outdated soon after leaving university. Nevertheless, it is strange (but entirely characteristic of Mercator's dealings with the English) that 'John Balak' should suddenly reappear fifty years later with revelatory, highly topical geographical information conveyed by the mysterious Low Country traveller 'Alferius'. In common with the mysterious Cnoyen, Alferius brought news guaranteed to encourage the exploratory efforts of the English in the northern lands and oceans. It is worth noting that 'Balak' is an odd name for a sixteenth-century humanist; as the idolatrous, fornicating king of the Moabites who 'cast a stumbling-block before the children of Israel' (Revelations 2: 14), the name 'Balak' carried such extreme connotations that the authorship of the letter has to remain in doubt.
61 The curious Balak wrote from a curious location: Van Durme (1959, p. 164) suggested that 'Arusburg upon the river of Osella' was 'Arnsburg on the Baltic island of Osel'. In his *Atlas* of 1595, Mercator showed the island of 'Osel' on the map of 'Livonia', with the coastal town of 'Arensburgk' beside a river on the island's east coast. In the accompanying text, Mercator calls the island 'Osilia'. At the time, Arensburgk was the island's main trading centre, protected by a castle founded in the thirteenth century. The island (which is in the Gulf of Riga) has exchanged its Swedish name 'Osel' for the Estonian Saaremaa, while 'Arensburgk' is now Kuressaare. At the time the letter from Osel was dated, the Livonian war was reaching its climax; what Balak was doing in the war-zone, and why he failed to mention the troubles in his letter, adds to the mystery.

62 Letter of 20 February 1581 from John Balak to Mercator, trans. in Hakluyt, 1907, vol. 2, p. 364. The English has been modified in this and the following quotes.
63 Ibid., p. 364.
64 Ibid., p. 364.
65 Ibid., p. 364.
66 Ibid., p. 364.
67 Ibid., p. 365.
68 Ibid., p. 366.
69 Ibid., p. 366.
70 Ibid., p. 366.
71 Ibid., p. 367.
72 Ibid., p. 367.

28 THE NEW GEOGRAPHY

1 Letter of 14 July 1578 from Mercator to Werner von Gymnich, Van Durme, 1969, p. 151.
2 Mercator, to the Studious and Benevolent Reader, *Atlas*, p. 162.
3 Mercator, Useful Prefatory Instruction to the Maps of Germany, *Atlas*, p. 233.
4 Quoted in Parker, 1990, p. 35.
5 Mercator, To the Studious and Benevolent Reader, *Atlas*, pp. 161–2.
6 Ibid., p. 162.
7 Mercator, Dedicatory letter to Johann Wilhelm the Younger, Duke of Jülich, Cleve etc., August, 1585, *Atlas*, p. 184.
8 Ibid., p. 184.
9 Mercator's intention to extend the scope of the 'new geography' to Antarctica was inferred by his son Rumold; see To the amiable reader, *Atlas*, p. 103.
10 See de Vries, 1992, pp. 3–10 for the revelation that Mercator had included such 'dual function' maps in his *Atlas*.
11 Letter of 14 July 1578 from Mercator to Gymnich, Van Durme, 1959, p. 151.
12 Ibid., p. 151.
13 Letter of 24 March 1583 from Mercator to Ludgerus Heresbachius, Van Durme, 1959, p. 184.
14 Ibid., p. 184.
15 Ibid., p. 184.
16 Ibid., p. 184.
17 Ibid., p. 184.
18 Ibid., p. 183.
19 Letter of 31 July 1587 from Mercator to William IV, Landgrave of Hesse Kassel, Van Durme, 1959, p. 205.
20 Letter of 24 March 1583 from Mercator to Ludgerus Heresbachius, Van Durme, 1959, p. 183.
21 Ibid., p. 183.
22 Ibid., p. 183.
23 Ibid., p. 183.
24 Ibid., p. 183.
25 Ibid., p. 183.
26 Ibid., p. 183. A thaler (later, the dollar) was the silver equivalent of the gold 'gulden'.
27 Ibid., p. 184.

28 Ibid., p. 184.
29 See Parker, 1990, p. 215.
30 Ortelius and Vivianus, 1584, p. 3.
31 Ibid., p. 3.
32 Ibid., p. 3.
33 Ibid., p. 3.
34 Ibid., p. 69.
35 Ibid., p. 69.
36 Mercator, dedicatory letter to Johann Wilhelm the Younger, Duke of Jülich, Cleve etc., *Atlas*, p. 183.
37 Ibid., p. 183.
38 Ibid., p. 183.
39 Ibid., pp. 183–4.
40 Ibid., p. 184.
41 Ibid., p. 184.
42 Ibid., p. 184.
43 Ibid., p. 184.
44 Ibid., p. 184.
45 Mercator, Advice on Using the Maps, *Atlas*, p. 180.
46 Ibid., p. 180.
47 Ibid., p. 180.
48 Rumold Mercator, To the amiable reader, *Atlas*, p. 103.
49 Mercator, On the Political Status of the Kingdom of France, *Atlas*, p. 167.
50 Ibid., p. 167.
51 Meurer, in Van den Broecke, Van der Krogt and Meurer, 1998, p. 269, referred to Mercator's separation of Belgii Inferioris and Switzerland from the Empire as 'a subdued but extremely daring political statement'.
52 Mercator, The Polity of Belgium under the Burgundians, *Atlas*, p. 208.
53 Mercator, To the studious reader, p. 208.
54 Mercator, The Polity of Belgium under the Burgundians, *Atlas*, p. 208.
55 Mercator, To the studious reader, p. 209.
56 Ibid., p. 209.
57 Ibid., p. 209.
58 Ibid., p. 209.
59 Letter of 12 December 1585 from Mercator to Heinrich Rantzau, Van Durme, 1959, p. 197.
60 Ibid., p. 197.

29 APOCALYPSE

1 *Catalogus imperatorum, regum ac principum qui astrologicam artem amarunt, ornarunt et exercuerunt . . .*, Antwerp, 1580.
2 Letter of 12 December 1585 from Mercator to Rantzau, Van Durme, 1959, p. 197.
3 Ibid., p. 197.
4 Letter of 18 May 1586 from Mercator to Rantzau, Van Durme, 1959, p. 203. Diet marsia, or Dithmarschen, was part of the duchy of Holstein.
5 Ibid., p. 203.
6 Ibid., p. 203.
7 Mercator, The Kingdom of Denmark, *Atlas*, p. 146.

8 Letter of 31 May 1585 from Mercator to Rantzau, Van Durme, 1959, p. 192.

9 Ibid., p. 193.

10 Ibid., p. 193.

11 Ibid., p. 193.

12 Ibid., p. 193.

13 Ibid., p. 193.

14 Ibid., p. 193.

15 Ibid., pp. 193–4.

16 Letter of 14 April 1586 from Mercator to Rantzau, Van Durme, 1959, p. 201.

17 Letter of 18 May 1586 from Mercator to Rantzau, Van Durme, 1959, p. 203.

18 Zeland, now Zealand.

19 Letter of 14 April 1586 from Mercator to Rantzau, Van Durme, 1959, p. 201.

20 Ibid., p. 201.

21 Ibid., p. 202.

22 Ghim, trans. in Osley, 1969, p. 190.

23 Ibid., p. 191.

24 Letter of 7 September 1586 from Mercator to Rantzau, Van Durme, 1959, p. 204.

25 Ibid., p. 204.

26 Ibid., p. 204.

27 Ibid., p. 204.

28 Ibid., p. 204.

29 Ibid., p. 204.

30 Ibid., p. 204.

31 Ibid., p. 204.

32 Ibid., p. 204.

33 Ibid., p. 204.

34 Ghim, trans. Osley, 1969, p. 191.

35 Ibid., p. 191

36 Ibid., p. 191

37 See letter of 31 July 1587 from Mercator to Wilhelm IV of Hesse, Van Durme, 1959, pp. 205–6.

38 Letter of 7 November 15987 from Mercator to Senate of Duisburg, Van Durme, 1959, p. 207.

39 Ibid., p. 207.

40 Quoted in Kamen, 1998, p. 265.

30 ATLAS

1 Address to the 'Gentle reader', March 13 1589, *Atlas*, p. 271.

2 Ibid., p. 271.

3 1 Corinthians 3:18.

4 Mercator, address to the 'Gentle reader', 13 March 1589, *Atlas*, p. 271.

5 Ibid., p. 271.

6 Mercator, dedicatory letter, 13 March 1589, To the most serene Prince Ferdinando de' Medici, *Atlas*, p. 270.

7 Ibid., p. 271.

8 Ibid., pp. 270–1.

9 Ibid., p. 271.

10 Although it would be another six years before the word 'Atlas' would be associated in

print with a book of maps, Mercator's reference in his dedicatory letter to Prince Ferdinando de' Medici suggests that the idea could have occurred by 1589.

11 Mercator, Preface to *Atlas*, p. 25.

12 Mercator, dedicatory letter, 13 March 1589, To the most serene Prince Ferdinando de' Medici, *Atlas*, p. 271.

13 Homer, 1946, Book i, p. 26.

14 Diodorus Siculus, in Page, Capps, Rouse, 1933–67, book iii, ch. 27, vol. 2, p. 431.

15 Mercator, The Arctic Pole, *Atlas*, p. 112.

16 Mercator, *Nova et aucta orbis*, 1569, trans. in *The Hydrographic Review*, vol. 9, no. 2, 1932, p. 27.

17 Quoted in Savours, 1999, p. 17.

18 In 1590, Ortelius added a copy of Vedel's Iceland map to a supplement of the *Theatrum orbis terrarum*, but – given Mercator's Danish connections through Rantzau – it seems reasonable to suppose that the Vedel map might have reached Ortelius by way of Mercator.

19 Mercator, Iceland, *Atlas*, p. 113.

20 Mercator, The British Isles, *Atlas*, pp. 114–15.

21 Ibid., p. 115.

22 ibid., p. 115.

23 Ibid., p. 115.

24 Michael Mercator was Arnold's third and youngest son.

25 P. Barber, in Buisseret, 1992, p. 78, suggested that Michael Mercator's medal was 'probably meant only for limited presentation in court circles'.

26 See Barber, 1998, pp. 73–6, for a detailed analysis of Mercator's sources for his British maps of 1595.

31 CREATION

1 Ghim, trans. in Osley, 1969, p. 193.

2 Ibid., p. 193.

3 Ibid., p. 194. The words Mercator may have had in mind were those of St Paul ('O wretched man that I am! who shall deliver me from the body of this death?'), words which Mercator would reproduce verbatim at the end of his treatise on Creation.

4 Ibid., p. 194.

5 Ibid., p. 194.

6 Ibid., p. 194. Ghim does not describe this chair, but – given Mercator's practical aptitudes – it would seem likely that Mercator had devised a contraption with fore and aft carrying handles.

7 Ibid., p. 189.

8 Ibid., p. 194.

9 Ibid., p. 194.

10 Letter of 31 August 1592, from Mercator to Wolfgang Haller, Van Durme, 1969, p. 223.

11 Ibid., p. 223.

12 Ghim, trans. in Osley, 1969, p. 190.

13 Letter of 24 March 1583 to Heresbach, Van Durme, 1959, p. 183.

14 Ibid., pp. 183–4.

15 Ghim, trans. in Osley, 1969, p. 194.

16 Mercator, Prolegomenon to The fabric of the world, *Atlas*, p. 26.

17 Ibid., p. 27.

18 Ibid., p. 27.

19 Ibid., p. 27.

20 Ibid., p. 36.

21 Ghim, trans. in Osley, 1969, p. 190.

22 Undated letter from Jacob Sinstedius to Reinhardus Solenander, *Atlas*, p. 22.

23 Ibid., p. 22.

24 Letter of 1 July 1594 from Reinhardus Solenander to Mercator, *Atlas*, p. 17.

25 Letter of 1 July 1594 from Reinhardus Solenander to Mercator, *Atlas*, p. 18.

26 Ghim, *Atlas*, p. 16. The symptoms suggest a cerebral haemorrage.

27 Ghim, trans. in Osley, 1969, p. 194.

28 Ibid., p. 194.

29 Ibid., p. 194.

30 Letter of 26 December 1594 from Arnold Mylius to Abraham Ortelius, Van Durme, 1959, p. 229, trans. in Osley, 1969, p. 194.

31 Ghim, trans. in Osley, 1969, p. 194.

32 Literally, the 'convenor of the assembly' (see *Atlas*, p. 16), and therefore not a Catholic priest.

33 Ghim, trans. in Osley 1969, p. 194.

EPILOGUE

1 J. Vivian, *Atlas*, p. 1.

2 Bernardus Furmerius, *Atlas*, p. 2.

3 Ibid., p. 2.

4 Ibid., p. 2.

5 Ibid., p. 2.

6 Epitaph of Gerardus Mercator, *Atlas*, p. 3.

7 Johann Metellus, *Atlas*, p. 3.

8 Ibid., p. 3.

9 The actual dimensions of the Lessing J. Rosenwald copy of the *Atlas* are 437 x 293 mm.

10 Doctor Reinhard Solenander, *Atlas*, p. 17.

11 Rumold Mercator, To the amiable reader, *Atlas*, p. 102.

12 Mercator's concept of continents allowed for a total of three: the first comprised the landmasses recorded in the Bible (Europe, Asia and North Africa), the second was America and the third was counter-balancing Antarctica.

13 Rumold Mercator, To the amiable reader, *Atlas*, p. 103.

14 Ibid., p. 103.

15 Ibid., p. 103.

16 Rumold Mercator, dedicatory letter of 1 April 1595, p. 103.

17 Ibid., p. 104.

18 Ibid., p. 104.

19 Ibid., pp. 104–5.

20 Ibid., p. 105.

21 Epitaph on the death of Gerardus Mercator, *Atlas*, p. 17.

22 Quoted in Karrow, 1993, p. 26.

23 Thomas Blundeville, 1636, quoted in Karrow, 1993, p. 386.

24 Ibid., p. 386.

25 William Shakespeare, *Twelfth Night*, act III, sc. 2.
26 http://spacelink.nasa.gov/
Instructional.Materials/Curriculum.Support/
Space.Science/Our.Solar.System/Venus/Venus.Projection.txt.

Chronology of Mercator's Principal Works

1512	**b. Rupelmonde, county of Flanders**
1536/37	Collaborated on a terrestrial globe: 370 mm diameter
1537	Collaborated on a celestial globe: 370 mm diameter
1537	Wall-map of the Holy Land: 6 sheets; 434 x 984 mm
1538	World map: 1 sheet; 355 x 545 mm
1539/40	Wall-map of Flanders: 9 sheets; 872 x 1,166 mm
1540/41	Manual of italic lettering (*Literarum latinarum*...)
1541	Terrestrial globe: 420 mm diameter
1551	Celestial globe: 420 mm diameter
1554	Wall-map of Europe: 15 sheets; 1,200 x 1,469 mm
1564	Wall-map of the British Isles: 8 sheets; 876 x 1,271 mm
1569	Chronicle of world history (*Chronologia*)
1569	Wall-map of the world on a new projection: 18 sheets; 1,236 x 2,024 mm
1572	Revised edition of wall-map of Europe
1578	Ptolemy's *Geography*: 28 maps
1584	Ptolemy's *Geography*, complete with text
1585	51 modern maps covering France, Low Countries and Germany
1589	22 modern maps covering Italy and the Balkans
1592	Harmonization of the Gospels (*Evangelicæ historiæ quadriparta Monas*...)
1594	**d. Duisburg, duchy of Cleves**
1595	29 modern maps covering the Arctic, Iceland, the British Isles, Scandinavia, the Baltic region; Russia, Transylvania and Crimea. To Mercator's 102 modern maps, his son and grandsons add 5 additional maps describing the world, Europe, Africa, Asia and America. All 107 modern maps are published – together with Mercator's treatise on Creation – as the cosmography called *Atlas*.

Select Bibliography

This is not a comprehensive list of sources consulted during the research for this book. It does, however, include all those works cited in the book's footnotes. It also includes the examples of Mercator's surviving maps and globes which I was able to see, and it includes a small selection of the more important titles which illuminate Mercator's life and works.

Achten, M., *Gerhard Mercator: Sein Leben und Wirken* (Gangelt, *c.* 1995)

Akerman, J. R., 'Atlas, Birth of a Title' (in Watelet, 1998)

Andrewes, W. J. H. (ed.), *The Quest for Longitude: Proceedings of the Longitude Symposium, Harvard University* (Cambridge, Mass., 1993)

Arnade, P., *Realms of Ritual: Burgundian Ceremony and Civic Life in Late Medieval Ghent* (Ithaca, 1996)

Aubert, R. et al., *The University of Louvain 1425–1975* (Leuven, 1976)

Aune, M. B., *To Move the Heart: Philip Melanchthon's Rhetorical View of Rite and its Implications for Contemporary Ritual Theory* (San Francisco, 1994)

Averdunk, H. and J. Müller-Reinhard, *Gerhard Mercator und die Geographen unter seinen Nachkommen* (Gotha, 1914)

Bibicz, J., 'La Résurgence de Ptolémée' (in Watelet, 1994)

Bagrow, L., 'A Page from the History of the Distribution of Maps', *Imago mundi*, 5, 1948

—, 'Old Inventories of Maps', *Imago mundi*, 5, 1948

Barber, P., 'The British Isles' (in Watelet, 1998)

Belgrave, R., V. Castro, C. Donnellan and U. Kuhlemann, *Prints as Propoganda: The German Reformation* (London, 1999)

Benecke, G., *Society and Politics in Germany 1500–1750* (London, 1974)

Berggren, J. L. and A. Jones, *Ptolemy's Geography: An Annotated Translation of the Theoretical Chapters* (Princeton, 2000)

Best, T. W., *Macropedius* (New York, 1972)

Bietenholz, P. G. (ed.), *Contemporaries of Erasmus: A Biographical Register of the Renaissance and Reformation* (Toronto, 1985)

Binder, C. and I. Kretschmer, 'La Projection mercatorienne' (in Watelet, 1994)

Blockmans, W. and W. Prevenier, *Poverty in Flanders and Brabant from the Fourteenth to the Mid-Sixteenth Century: Sources and Problems* (The Hague, 1978)

—, *The Promised Lands: The Low Countries Under Burgundian Rule, 1369–1530* (Philadelphia, 1999)

Blotevogel, H. H. and R. Vermij (eds), *Gerhard Mercator und die geistigen Strömungen des 16. und 17. Jahrhunderts* (Bochum, 1995)

Boehmer, E., *Spanish Reformers of Two Centuries from 1520: Their Lives and Writings,*

According to the Late Benjamin B. Wiffen's Plan with the Use of his Materials, vol. 1 (Strasbourg, 1874)

Boorstin, D. J., *The Discoverers* (New York, 1985)

Boyce, G. K., 'A Letter of Mercator Concerning his *Ptolemy*', *Papers of the Bibliographical Society of America*, vol. 42, 1948

Brandi, K., *The Emperor Charles V: The Growth and Destiny of a Man and of a World-Empire* (London, 1939)

Brant, S., *The Ship of Fools*, trans. W. Gillis, (London, 1971)

Buisseret, D. (ed.), *Monarchs Ministers and Maps: The Emergence of Cartography as a Tool of Government in Early Modern Europe* (Chicago, 1992)

Büttner, M., 'The Significance of the Reformation for the Reorientation of Geography in Lutheran Germany', *History of Science*, 17, 1979

Büttner, M. and R. Dirven (eds), *Mercator und Wandlungen der Wissenschaften im 16. und 17. Jahrundert*, series Duisburger Mercator-Studien, 1 (Bochum, 1993)

Cam, G. A., 'Gerhard Mercator: His "Orbis Imago" of 1538', *Bulletin of The New York Public Library*, 41, no. 5, May 1537

Campan, C.-A., *Mémoires de Francisco de Enzinas: texte Latin inédit avec la traduction Française du XVIe siècle en regard 1543–1545*, vol. 1 (Brussels, 1862)

Campbell, T., *Early Maps* (New York, 1981)

—, *The Earliest Printed Maps, 1442–1500* (London, 1987)

Cherton, A. and M. Watelet, '*Catalogus*' (in Watelet, 1994)

Clark, P. (ed.), *Small Towns in Early Modern Europe: Themes in International Urban History* (Cambridge, 1995)

Deacon, R., *John Dee, Scientist, Geographer, Astrologer and Secret Agent to Elizabeth I* (London, 1968)

De Clercq, C. 'Le commentaire de Gérard Mercator sur L'Epître aux Romains de saint Paul', *Duisburger Forschungen*, 6 (Duisburg, 1962)

Dekker, E., *Globes at Greenwich: A Catalogue of the Globes and Armillary Spheres in the National Maritime Museum, Greenwich* (Oxford, 1999)

Dekker, E. and P. van der Krogt, *Globes from the Western World* (London, 1993)

De Lang, M. H., 'De godsdienstige ideeën van Gerardus Mercator', (in Nave, Imhof and Otte, 1994)

Delano Smith, C., 'Cartographic Signs on European Maps and their Explanation before 1700', *Imago mundi*, 37, 1985

Delano Smith, C. and E. M. Ingram, 'La carte de la Palestine' (in Watelet, 1994)

De Nave, F., D. Imhof and E. Otte, *Gerard Mercator en de Geografie in de Zuidelijke Nederlanden / Gerard Mercator et la géographie dans les Pays-Bas Meridionaux* (Antwerp, 1994)

Denucé, J. (ed.), *The Treatise of Gerard Mercator, Literarum Latinarum, quas Italicas, cursoriasque vocant, scribendarum ratio*... (Antwerp, 1930)

—, *De geschiedenis van de Vlaamsche kaartsnijkunst* (Antwerp, 1941)

Departments of State and Official Bodies, Admiralty ... Naval Intelligence Division, *A Manual of Alsace-Lorraine* (London, 1919)

De Smet, A., 'Mercator à Louvain (1530–1552)', *Duisburger Forschungen*, 6, (Duisburg, 1962)

De Vocht, H., *History of the Foundation and Rise of the Collegium Trilingue Lovaniense, 1517–1550* (Louvain, 1951–55)

De Vries, D., 'Die Helvetia-Wandkarte von Gerhard Mercator', *Cartographia helvetica*, 5, 1992

Dilke, O. A. W., 'Latin Interpretations of Ptolemy's "Geographia"', in A. Dalzell, C. Fantazzi and R. J. Schoeck (eds), *Acta Conventus Neo-Latini Torontorensis*, Medieval and Renaissance texts and studies, 86 (Binghampton, 1991)

Duff, J. W. and A. M., *Minor Latin Poets* (London, 1934)

Dürst, A., 'The Map of Europe' (in Watelet, 1998)

Edwards, M. A., *Printing, Propoganda and Martin Luther* (Berkeley, 1994)

Englander, D., D. Norman, R. O'Day and W. R. Owens, *Culture and Belief in Europe 1450–1600: An Anthology of Sources* (Oxford, 1990)

Englisch, B., 'Erhard Etzlaub's Projection and Methods of Mapping', *Imago mundi*, 48, 1996

Erasmus, Desiderius, *Collected Works of Erasmus* (Toronto, 1974–)

Febvre, L. and H.-J. Martin, *The Coming of the Book: The Impact of Printing 1450–1800* (London, 1976)

Fenton, E. (ed.), *The Diaries of John Dee* (Charlbury, 1998)

Fichtner, P. A., *Emperor Maximilian II* (New Haven, 2001)

Forster, L., 'Charles Utenhove and Germany', in *European Context. Studies in the History and Literature of the Netherlands Presented to Theodoor Weevers* (Cambridge, 1971)

Fry, R., *Dürer's Record of Journeys to Venice and the Low Countries* (New York, 1995)

Gaur, A., *A History of Calligraphy* (London, 1994)

Haardt, R., 'The Globe of Gemma Frisius', *Imago mundi*, 9, 1952

Haasbroek, N. D., *Gemma Frisius, Tycho Brahe and Snellius and Their Triangulations* (Delft, 1968)

Hakluyt, R., *Voyages* (London, 1907)

—, *Divers voyages touching the discoverie of America and the islands adjacent unto the same, made first of all by our Englishmen and afterwards by the Frenchmen and Britons: with two mappes annexed hereunto* (London, 1582)

Hale, J., *The Civilization of Europe in the Renaissance* (London, 1993)

Hantsche, I. (ed.), *Mercator – ein Wegbereiter neuzeitlichen Denkens*, series Duisburger Mercator-Studien, 2 (Bochum, 1994)

Harradon, H. D., 'Some Early Contributions to the History of Geomagnetism – VI. Gerhard Mercator of Rupelmonde to Antonius Perrenotus, Most Venerable Bishop of Arras, A.D. 1546', *Terrestrial Magnetism and Atmospheric Electrity*, 48, 1943

Hellwig, F., 'La carte de Lorraine' (in Watelet, 1994)

Henne, A., *Histoire du règne de Charles-Quint en Belgique* (Brussels, 1859)

Hessels, J. H. (ed.), *Abrahami Ortelii (geographi antverpiensis) et virorum eruditorum ad eundem et ad Jacobum Colium Ortelianum ... epistulae* (Cambridge, 1887)

Hobbs, W. H., 'Zeno and the Cartography of Greenland', *Imago mundi*, 6, 1949

Homer, *The Odyssey*, trans. E. V. Rieu (London, 1946)

Hyma, A., *The Youth of Erasmus* (Ann Arbor, 1930)

Imhof, D., ' "De Officina Plantiniana" als verdeelcentrum van de globes, kaarten en atlassen van Gerard Mercator' (in de Nave, Imhof and Otte, 1994)

Jedin, H. (ed.), *History of the Church* (London, 1980)

Kamen, H., *Philip of Spain* (New Haven, 1998 edn)

Karrow, R. W., *Mapmakers of the Sixteenth Century and Their Maps* (Chicago, 1993)

Kelsey, H., *Sir Francis Drake: The Queen's Pirate* (New Haven, 2000 edn)

Keuning, J., 'The History of an Atlas: Mercator-Hondius', *Imago mundi*, 4, 1947

—, 'XVIth Century Cartography in the Netherlands', *Imago mundi*, 9, 1952

Kiely, E. R., *Surveying Instruments: Their History and Classroom Use* (New York, 1947)

Kirmse, R., 'Die grosse Flandernkarte Gerhard Mercators (1540) – ein Politicum?', *Duisburger Forschungen*, 1 (Duisburg, 1957)

—, 'Zu Mercators Tätigkeit als Landmesser in seiner Duisburger Zeit', *Duisburger Forschungen*, 6 (Duisburg, 1962)

Kish, G., *Medicina, mensura, mathematica. The Life and Works of Gemma Frisius, 1508–1555* (Minneapolis, 1967)

Koeman, C., *The History of Abraham Ortelius and his Theatrum Orbis Terrarum* (Lausanne, 1964)

Koenigsberger, H. G., G. L. Mosse and G. Q. Bowler, *Europe in the Sixteenth Century* (2nd edn, Harlow, 1989)

Kraus, H. P., *Sir Francis Drake: A Pictorial Biography* (Amsterdam, 1970)

—, 'The History of the Gospel Synopsis and Gerardus Mercator's Evangelica Historia' (in Blotevogel and Vermij, 1995)

Lanzas, P. T., *Relación Descriptiva de los Mapas, Planos, & de México y Floridas existentes en el Archivo General de Indias*, Book II (Seville, 1900)

Leyn, A. de, *Biographe Nationale. Esquisse biographique de P. de Corte (Curtius), premier Evêque de Bruges, ancien professeur de l'Université de Louvain* (Louvain, 1863)

Lindeman, Y., *Macropedius* (Nieuwkoop, 1983)

Lis, C. and H. Soly, *Poverty and Capitalism in Pre-Industrial Europe* (Hassocks, 1979)

Löffler, R. and G. Tromnau (eds), *Gerhard Mercator: Europa und die Welt* (Duisburg, 1994)

Lucas, F. W., *The Voyages of the Brothers Zeni: The Annals of the Voyages of the Brothers Nicolo and Antonio Zeno in the North Atlantic about the End of the Fourteenth Century and the Claim Founded Thereon to a Venetian Discovery of America* (London, 1898)

Lynam, E., *The First Engraved Atlas of the World* (Jenkinstown, 1941)

Mackie, J. D., *The Earlier Tudors 1485–1558* (Oxford, 1952)

Mandeville, Sir J., *The Travels of Sir John Mandeville*, trans. C. W. R. D. Moseley (London, 1983)

Marneff, G., *Antwerp in the Age of the Reformation: Underground Protestantism in a Commercial Metropolis 1550–1577* (Baltimore, 1996)

Mercator, G., terrestrial globe, (Louvain, 1536/7; no extant example; Photographs of the terrestrial globe constructed *c.* 1537 by Gemma Frisius, with the assistance of Gerard Mercator and Gaspar à Myrica, British library, Maps 8.bb.10(4))

—, celestial globe (Louvain, 1537; National Maritime Museum, Greenwich, GLB0135)

—, *Amplissima Terrae Sanctae descriptio ad utriusque testamenti intelligentiam* (Louvain, 1537; reproduced in Watelet, 1994)

—, world map known as *'Orbis imago'* (Louvain, 1538; reproduced in Nordenskiöld, 1889)

—, *Vlaenderen. Exactissima [Flandriae descriptio]* (Louvain, 1539/40; Museum Plantin-Moretus, Antwerp)

—, *Literarum latinarum, quas italicas, cursoriasque vocat, scribendarū ratio* (Louvain, 1540/41; 1549 edn, British Library, 648, 1–3; facsimile of 1540/41 edn in Denucé, 1930)

—, terrestrial globe (Louvain, 1541; National Maritime Museum, Greenwich, GLB0096)

—, celestial globe (Louvain, 1551; National Maritime Museum, Greenwich GLB0097)

—, wall-map of Europe (Duisburg, 1554; no extant example; reproduction of 1st and 2nd edn in Watelet, 1998)

—, *Angliae Scotiae & Hibernie nova descriptio* (Duisburg, 1564; reproduced Watelet, 1998)

—, *Chronologia. Hoc est, Temporum demonstratio exactissima ab initio mundi, usque ad annum*

Domini M.D.LXVIII et eclipsibus et observationibus astronomicis omnium temporum, sacris quoq[ue] Biblijs, & optimis quibusq[ue] Scriptoribus summa fide concinnata (Cologne, 1569; British Library C.74.g.11)

—, *Nova et aucta orbis terrae descriptio ad usum navigantium emendatè accommodata* (Duisburg, 1569; reproduced in Watelet, 1998). See also, *Map of the World (1569): In Atlas-form in the Maritiem Museum 'Prins Hendrik'* (Rotterdam, 1961) and *The Hydrographic Review*, 9, no. 2 (Monaco, 1932)

—, *Tabulae geographicae Cl: Ptolemei ad mentem autoris restitutae & emendat[a]e per Gerardum Mercatorem Illustriss: Ducis Clivi[a]e &c: Cosmographu* (Cologne, 1578; British Library, Maps C.1.d.14 and C.1.d.15; Royal Geographical Society, London, 265 G 12)

—, *Cl. Ptolemaei Alexandrini, Geographiae libri octo, recogniti iam et diligenter emendati cum tabulis geographicis ad mentem auctoris restitutis ac emendatis, per Gerardum Mercatorem, Illustriss. Ducis Clivensis etc. Cosmographum* ... (Cologne, 1584; British Library, Maps C.1.d.16)

—, *Galliae tabulae geographicae* (Duisburg, 1585; British Library, Maps C.3.c.1)

—, *Italiae, Sclavoniae, et Graeciae tabulae geographicae* (Duisburg, 1589; British Library, Maps C.3.c.2)

—, *Evangelicae historiae quadripartita Monas, sive Harmonia Quatuor Evangelistarum, in qua singuli integri, inconfusi, impermixti & soli legi possunt, & rursum ex omnibus una universalis & continua historia ex tempore formari* (Duisburg, 1592)

—, *Atlas sive cosmographicæ meditationes de fabrica mundi et fabricati figura* (Düsseldorf, 1595; British Library, Maps C.3.c.3, Maps C.3.c.4, Maps c.3.c.5; Royal Geographical Society, London, 264 H4, 263 G9. Reproduced, with translation to English by D. Sullivan of complete *Atlas* text: the Lessing J. Rosenwald Collection, Library of Congress, on CD-ROM (Oakland, California, 2000)

Meurer, P. H., 'Le territoire allemand' (in Watelet, 1994)

—, 'Les fils et petits-fils de Mercator' (in Watelet, 1994)

Milz, J., *Duisburg* (in Watelet, 1994)

Moir, D. G. et al., *The Early Maps of Scotland to 1850*, vol. 1 (3rd edn: Edinburgh, 1973)

Morison, S. E., *Christopher Columbus, Mariner* (Boston, 1955)

Motley, J. L., *The Rise of the Dutch Republic* (New York, 1900)

Münster, S., *Cosmographia* ... (Basel, 1550)

Nissen, K., 'Jacob Ziegler's Palestine Schondia Manuscript', *Imago mundi*, 13, 1956

Nordenskiöld, A. E., *Facsimile-Atlas to the Early History of Cartography with Reproductions of the Most Important Maps Printed in the XV and XVI Centuries* (Stockholm, 1889; reprinted New York, 1973)

O'Malley, C. D., *Andreas Vesalius of Brussels 1514–1564* (Berkeley, 1964)

Ortelius, A., *Theatrum Orbis Terrarum* (Antwerp, 1570)

—, *Theatrum Orbis Terrarum: The Theatre of the Whole World* (London, 1606)

—, *Catalogue of the Highly Important Correspondence of Abraham Ortelius (1528–98)* (London, 1955)

—, *Album amicorum*, facsimile, annotated and trans. J. Puraye, (Antwerp, 1969)

Ortelius, A. and I. Vivianus, *Itinerarium per nonnullas Galliæ Belgicæ partes* (Antwerp, 1584)

Osley, A. S., *Mercator: A Monograph on the Lettering of Maps, etc. in the 16th Century Netherlands with a Facsimile and Translation of his Treatise on the Italic Hand and a Translation of Ghim's Vita Mercatoris* (London, 1969)

Page, T. E., E. Capps and W. H. D. Rouse, *Diodorus of Sicily*, trans. C. H. Oldfather (London, 1933–67)

Parker, G., *The Dutch Revolt* (revised edn, London, 1990)

Parry, J. H. (ed.), *The European Reconnaissance* (London, 1968)

Pastoureau, M., 'Entre Gaule et France: la "Gallia"' (in Watelet, 1994)

—, 'The 1569 World Map' (in Watelet, 1998)

Penneman, T., *Astrologie in de eeuw van Mercator: Eeen reeks schetsen* (Sint-Niklaas, 1994)

—, (ed.), *Mercator en zijn boeken* (Sint-Niklaas, 1994)

—, 'La bibliothèque de Mercator' (in Watelet, 1994)

Pliny the Elder, *Natural History*, trans. J. F. Healy (London, 1991)

Polo, M., *The Travels of Marco Polo*, trans. R. Latham (London, 1958)

Post, R. R., *The Modern Devotion* (Leiden, 1968)

Rich, E. E. and C. H. Wilson (eds), *The Cambridge Economic History of Europe*, vol. IV (Cambridge, 1967)

Roberts, J. M., *History of the World* (first pub. 1976, London 1995)

Ross, J. B., and M. M. McLaughlin, *The Portable Renaissance Reader* (first pub. 1953, revised edn London, 1968)

Santing, C., 'Gerardus Mercator (1512–1594): The Creation of an Image' (in Blotevogel and Vermij, 1995)

Savours, A., *The Search for the North West Passage* (New York, 1999)

Scafi, A., 'Mapping Eden: Cartographies of the Earthly Paradise', in D. Cosgrove (ed.), *Mappings* (London, 1999)

Schoeck, R. J., *Erasmus of Europe: The Making of a Humanist* (Edinburgh, 1990)

Sherman, W. H., *John Dee: the Politics of Reading and Writing in the English Renaissance* (Amherst, 1995)

Shirley, R. W., *The Mapping of the World: Early Printed World Maps 1472–1700* (London, 1984)

Shumaker, W. (ed. and trans.), *John Dee on Astronomy: 'Propaedeumata aphoristica' 1558 and 1568, Latin and English* (Berkeley, 1978)

Skelton, R. A., *Explorer's Maps: Chapters in the Cartographic Record of Geographical Discovery* (London, 1958)

—, 'Mercator and English Geography in the 16th Century', *Duisburger Forschungen*, 6 (Duisburg, 1962)

—, Bibliographical Note, *Claudius Ptolemaeus Geographia, Strasburg, 1513* (facsimile edition, Amsterdam, 1966)

Sotheby's, *The Mercator Atlas of Europe: To be Sold as a Single Lot in the sale of Valuable Autograph Letters, Literary Manuscripts and Historical Documents on 13th March, 1979* (London, 1979)

Starkey, D., *Elizabeth: Apprenticeship* (London, 2000)

Stevens, H., *Ptolemy's Geography: A Brief Account of All the Printed Editions Down to 1730* (London, 1908)

Stevenson, E. L., *Terrestrial and Celestial Globes: Their History and Construction Including a Consideration of Their Value as Aids in the Study of Geography and Astronomy* (New Haven, 1921)

Taylor, E. G. R., *Tudor Geography 1485–1583* (London, 1930)

—, 'The Earliest Account of Triangulation', *The Scottish Geographical Magazine*, 43, 1947.

—, 'John Dee and the Map of North-East Asia', *Imago mundi*, 12, 1955

—, 'A Letter Dated 1577 from Mercator to John Dee', *Imago mundi*, 13, 1956

Transylvanus, Maximilianus, *De Moluccis Insulis*, ed. C. Quirino, trans. H. Stevens (Manila, 1969)

Select Bibliography

Turner, G. L'E., 'Gerard Mercator as Instrument Maker' (in Blotevogel and Vermij, 1995)

Turner, G. L'E. and E. Decker, 'An Astrolabe attributed to Gerard Mercator, *c.* 1570', *Annals of Science*, 50, 1993

Tyacke, S., 'The Atlas of Europe Attributed to Gerard Mercator', *Imago mundi*, 31, 1979

Van den Broecke, M., P. Van der Krogt and P. Meurer, *Abraham Ortelius and the First Atlas: Essays Commemorating the Quadricentennial of his Death 1598–1998* (Utrecht, 1998)

Van der Gucht, A., 'La Carte de la Flandre' (in Watelet, 1994)

Van der Krogt, P., *Globi Neerlandici: The Production of Globes in the Low Countries* (Utrecht, 1993)

Van der Stock, J., *Printing Images in Antwerp: The Introduction of Printmaking in a City: Fifteenth Century to 1585* (Rotterdam, 1998)

Van der Wee, H., *The Growth of the Antwerp Market and the European Economy* (The Hague, 1963)

Van der Wee, H. and E. Aerts, 'The Lier Livestock Market and the Livestock Trade in the Low Countries from the 14th to the 18th Century', in *Internationaler Ochsenhandel 1350–1750*, Beiträger zur Wirtschaftsgeschichter, 9 (Stuttgart, 1979)

Van Durme, M. (ed.), *Correspondence Mercatorienne* (Antwerp, 1959)

Van Isacker, K. and R. Van Uytven, *Antwerp: Twelve Centuries of History and Culture* (Antwerp, *c.* 1986)

Van Ortroy, F. G., 'L'Oeuvre géographique de Mercator', *Revue des questions scientifiques*, series 2, 2 (Brussels, 1892–3)

—, *Bio-Bibliographie de Gemma Frisius* (Brussels, 1920)

Vanpaemel, G. H. W., *Mercator and the Scientific Renaissance at the University of Leuven* (in Blotevogel and Vermij, 1995)

Van Raemdonck, J., *Gérard Mercator: sa vie et ses oeuvres* (Saint Nicolas, 1869)

Van Zijl, T. P., 'Gerard Groote, Ascetic and Reformer (1340–1384)', *Studies in Medieval History*, 18 (Washington, 1963)

Verberckmoes, J., *Laughter, Jestbooks and Society in the Spanish Netherlands* (London, 1999)

Vermij, R., 'Mercator and the Reformation' (in Büttner and Dirven, 1993)

—, 'Mercator's Stoic Picture of the World' (in Hantsche, 1994)

—, 'Typus Univeritatis' (in Watelet, 1994)

—, 'Gerard Mercator and the Science of Chronology' (in Blotevogel and Vermij, 1995)

—, (ed.), *Gerhard Mercator und seine Welt* (Duisburg, 1997)

Vespucci, A., *Mundus Novus: Letter to Lorenzo Pietro di Medici*, trans. G. T. Northrup (Princeton, 1916)

Vives, J. L., *In Pseudodialecticos*, trans. C. Fantazzi (Leiden, 1979)

—, *John Dantiscus and His Netherlandish Friends as Revealed by Their Correspondence 1522–1546* (Louvain, 1961)

Voet, L., 'Les Relations commerciales entre Gérard Mercator et la maison Plantinienne à Anvers', *Duisburger Forschungen*, 6 (Duisburg, 1962)

—, 'Le Monde de l'édition et du livre' (in Watelet, 1994)

Vöhringer, C., *Pieter Bruegel 1525/1530–1569* (Cologne, 1999)

Watelet, M. (ed.), *The Mercator Atlas of Europe* (includes: *Facsimile of the Maps by Gerardus Mercator Contained in the Atlas of Europe, circa 1570–1572*) (Pleasant Hill, 1998)

Watelet, M. (ed.), *Gérard Mercator cosmographe: le temps et l'espace* (Antwerp, 1994)

Waterbolk, E. H., 'Viglius of Aytta, Sixteenth Century Map Collector', *Imago mundi*, 29, 1977

Westfall Thompson, J. (ed. and trans.), *The Frankfort Book Fair: The Francofordiense Emporium of Henri Estienne* (Chicago, 1911)

Whitfield, P., *The Image of the World: 20 Centuries of World Maps* (London, 1994)

Wilford, J. N., *The Mapmakers: The Story of the Great Pioneers in Cartography – from Antiquity to the Space Age* (New York, 1981)

Williams, P., *The Later Tudors: England 1547–1603* (Oxford, 1998)

Woolley, B., *The Queen's Conjuror: The Science and Magic of Dr Dee* (London, 2001)

Zeydel, E. H., *Sebastian Brant* (New York, 1967)

Index

nomus dictus
Onuphitis & Plinio
emphus

batur & Hircus.

Menbatur & Hircus

de sie no: mus

Pelusiacum os tium

Heracleum

Aegy

Xois

Neut no mus

Pelusium quod & Damiatam vocant

Xoytes nomus

E G Y P T

Magdolum

Pathoscentum

Mendethis

Setroi tes

Sirbo nis lacus

Gerenorium

Busirite nomus

Ficus

Bubasti

Syle nomus

Busiris

Buba sen

Vicus Iudeorum
Hebreis Pibeseth

Serapium

Bubastus

tes

Herculis

Arabiæ

nomus

sis Orij. In huius agro loco patriarcha habitauit.
Onias Iudeorum sacerdos templum emulum illi Hierosolimitano
sub Ptolemeo construxit, & ij parentes Iesu a Iudea exules
plus minus quinquenio morati sunt
num

Thou

Taubasium

P nomus

T tum

uicta hec linea
raicè Sohan,
culis memorabile

Pharbetites

Pahudes

Pharbetus

re nomus

gio siue Gosen

de ser

Raimesses pagus quem filij Israbel sub Aegyptia
captiuitate ædificauerunt duarum & quadra
ginta mansionum prima

A R

Trauanus amnis

Sur

Soroth

z

Phihahiroth

Ethan desertum
3

Posidium siuus loco nunc Quez siue
Ziem emporium

Mara

Elim

Heroe:
polites sinus

quem Aeant Arabes vocant

A P

His castris filij Israbel sub vesperum coturnice
& postera luce manna pas ti sunt.

Sin

8 mansio sin appellatur

de

Babalse:
phon

Daphchah

tum

Daneus portus
Cleopatris

ser

Arsinoe

Charandra
sinus

PARS SINVS

Atus

Hores montes

R. aphidim

de

Amalech in deserto Rap
riuit tas tra feth
quo ibi tr vit to Moses
aram Domino quam Dominu
exaltatio mea vocauit.

ARABICI &Ru
bri maris quod Suph
Hebrei dicunt